Inorganic Electrode Materials in High-Performance Energy Storage Devices

Inorganic Electrode Materials in High-Performance Energy Storage Devices

Guest Editor

Ting Deng

MDPI Basel • Beijing • Wuhan • Barcelona • Belgrade • Novi Sad • Cluj • Manchester

Guest Editor
Ting Deng
School of Materials Science
and Engineering
Jilin University
Changchun
China

Editorial Office
MDPI AG
Grosspeteranlage 5
4052 Basel, Switzerland

This is a reprint of the Special Issue, published open access by the journal *Inorganics* (ISSN 2304-6740), freely accessible at: https://www.mdpi.com/journal/inorganics/special_issues/LJNS78I477.

For citation purposes, cite each article independently as indicated on the article page online and as indicated below:

Lastname, A.A.; Lastname, B.B. Article Title. *Journal Name* **Year**, *Volume Number*, Page Range.

ISBN 978-3-7258-6051-7 (Hbk)
ISBN 978-3-7258-6052-4 (PDF)
https://doi.org/10.3390/books978-3-7258-6052-4

Contents

About the Editor

Ting Deng

Ting Deng is currently an associate professor at the School of Materials Science and Engineering, Jilin University. He obtained his bachelor's degree in materials science and engineering in 2012 and a PhD degree in material physics and chemistry in 2017 from Jilin University. He then joined the national postdoctoral program for innovative talents from 2018 to 2020 at Jilin University. His present research interests focus on high-performance electrode materials and electrolytes for hybrid and novel electrochemical energy storage devices.

Preface

As the global energy system shifts toward sustainability, high-performance energy storage technologies have become pivotal for integrating renewable energy and powering advanced electronics, electric vehicles, and grid storage. Inorganic electrode materials stand out for their high theoretical capacity, structural stability, abundant reserves, and tunable properties, making them key to pushing the limits of energy storage devices. However, critical bottlenecks—including low conductivity, significant volume expansion, sluggish ion diffusion, and the shuttle effect in lithium–sulfur batteries—persist, driving urgent needs for innovative material design and optimization. This Reprint aims to showcase cutting-edge research progress in this field, providing a platform for knowledge exchange and inspiring future advancements.

The collected studies cover a diverse range of energy storage systems, including lithium–sulfur batteries, lithium-ion batteries, aqueous zinc-ion batteries, sodium-ion batteries, hybrid supercapacitors, and methanol-mediated water splitting. Featured works address core challenges through innovative strategies: structural engineering (hollow architectures, carbon nanofiber encapsulation, porous nanosheet design), composition optimization (bimetallic synergistic effects, heteroatom doping, high-entropy alloying), and simplified scalable synthesis (EDTA-based synchronous carbonization–doping, two-step thermal annealing). These efforts deliver remarkable improvements in cycling stability, rate capability, and energy density—from long-cycle sodium-ion battery anodes to low-overpotential water-splitting catalysts—while shedding light on structure–performance relationships and practical industrialization pathways.

We sincerely thank all authors for their high-quality contributions and reviewers for their rigorous feedback, which have elevated the Reprint's quality. Gratitude also goes to the *Inorganics* editorial team for their professional support. Looking ahead, inorganic electrode materials will remain central to advancing sustainable energy storage. We hope this Reprint fosters interdisciplinary collaboration, accelerates the translation of fundamental research into practical technologies, and unlocks new possibilities for high-efficiency, low-cost energy storage solutions to address global energy challenges.

Ting Deng
Guest Editor

inorganics

MDPI

Editorial

Inorganic Electrode Materials in High-Performance Energy Storage Devices

Ting Deng

Key Laboratory of Automobile Materials of MOE, School of Materials Science and Engineering, and Jilin Provincial International Cooperation Key Laboratory of High-Efficiency Clean Energy Materials, Jilin University, Changchun 130012, China; tdeng@jlu.edu.cn

1. Introduction

This special issue focuses on the design, synthesis, optimization, and application of inorganic electrode materials in high-performance energy storage devices, covering lithium–sulfur batteries [1], lithium-ion batteries [2,3], aqueous zinc-ion batteries [4], sodium-ion batteries [5], hybrid supercapacitors [6,7], and methanol-mediated water-splitting systems [8]. It systematically presents the latest research progress in addressing key challenges such as the low conductivity, poor cycling stability, volume expansion, and sluggish reaction kinetics of inorganic electrode materials, providing comprehensive insights for the development of advanced energy storage technologies [9,10].

2. Core Research on Battery Electrode Materials

2.1. Lithium–Sulfur (Li-S) Batteries

A critical challenge regarding Li-S batteries lies in the shuttle effect of lithium polysulfides (LiPSs) and slow redox kinetics [11–14]. Against this background, Tan et al. designed a two-step thermal annealing strategy to derive wormlike N-doped porous carbon-nanotube-supported low-crystalline Co nanoparticles (a-Co-NC@C) from binary Zn-Co ZIF [1]. Ammonia-induced thermal annealing not only reduces the crystallinity of Co nanoparticles but also promotes the growth of highly graphitized N-doped carbon nanotubes. The composite structure provides sufficient sulfur accommodation space and enhanced LiPSs adsorption/catalytic activity, enabling a specific capacity of 559 mAh g^{-1} after 500 cycles at 1 C and 572 mAh g^{-1} at 3 C, with excellent stability even at high sulfur loadings (e.g., 579 mAh g^{-1} after 400 cycles at 1 C with a 2.55 mg cm^{-2} sulfur loading).

2.2. Lithium-Ion Batteries (LIBs)

For LIB anode materials, addressing volume expansion and improving conductivity are key [15–22]. Jin et al. synthesized a novel spinel high-entropy oxide $(Cr_{0.2}Mn_{0.2}Co_{0.2}Ni_{0.2}Zn_{0.2})_3O_4$ via a simple solution combustion method [3]. Benefiting from the entropy stabilization effect, the material exhibits outstanding cycling stability (132 mAh g^{-1} after 100 cycles at 100 mA g^{-1} and 107 mAh g^{-1} after 1000 cycles at 1 A g^{-1}) and rate performance (96 mAh g^{-1} at 2 A g^{-1}).

Xie et al. encapsulated ultrafine In_2O_3 particles in a carbon-nanofiber framework through electrospinning and thermal annealing [2]. The 1D carbon network inhibited In_2O_3 agglomeration, buffered volume changes, and accelerated ion/electron transport, achieving 571 mAh g^{-1} after 200 cycles at 0.3 A g^{-1} and 264 mAh g^{-1} at 3.2 A g^{-1}, significantly outperforming pure In_2O_3 electrodes.

2.3. *Aqueous Zinc-Ion Batteries (AZIBs)*

AZIBs are gaining attention for their safety and inexpensiveness, but their cathode stability and ion storage capacity need improvement [23–27]. Li et al. developed a $Co(OH)_2/CoOOH$ mixed-phase cathode via two-step electrochemical preparation [4]. The mixed phase reduces local structural disorder and enhances thermal stability and mechanical strength, delivering a maximum capacity of 164 mAh g^{-1} at 0.05 A g^{-1}, a high energy density of 275 Wh kg^{-1}, and 78% capacity retention after 200 cycles.

2.4. *Sodium-Ion Batteries (SIBs)*

Sodium-ion batteries are considered a potential alternative to lithium-ion batteries because of the abundance and inexpensiveness of sodium resources [28–33]. However, the larger ionic radius of Na^{+} leads to significant volume changes in electrode materials and slow ion diffusion kinetics. Wang et al. designed and prepared carbon-shell-wrapped hollow-structured cobalt–nickel bimetallic transition metal selenides (HS-$Co_xNi_\gamma Se_2$@C) via a three-step approach of PAN coating-ZIF-67 ion exchange selenization [5]. The hollow structure provides sufficient space for buffering volume expansion during charge–discharge processes, and the synergism between Co and Ni enhances material conductivity and electron mobility. The performance is closely related to the Co-Ni ratio: an excessive amount of Ni^{2+} leads to hollow structure collapse and a reduced electrochemical surface area. The optimized material (Co:Ni $\approx$ 1.05:0.95) exhibits excellent electrochemical performance—a reversible capacity of 334 mAh g^{-1} after 1000 cycles at a high current density of 5.0 A g^{-1}, and a discharge capacity of 428 mAh g^{-1} at the same rate—providing new ideas for the structural design and compositional optimization of SIB electrode materials.

3. Advances in Supercapacitor and Electrocatalyst Materials

3.1. *Hybrid Supercapacitors (HSCs)*

Gao et al. provide a comprehensive review on HSCs based on transition metal oxides (TMOs) and carbon materials, summarizing optimization strategies, including heteroatom doping, heterostructure construction, nanocomposite formation, and MOF derivation [7]. The representative approaches include Nd-doped α-Mn_2O_3 microspheres with 862.14 F g^{-1} at 0.5 A g^{-1}, NiCoP-$NiCoO_2$/NiCo-PO_x crystalline/amorphous heterostructures with 78.7% capacity retention after 10,000 cycles, and Ti_3C_2-ZrO_2 nanocomposites achieving 75.6 Wh kg^{-1} energy density. These approaches address the trade-off between energy density and power density in HSCs, expanding their application in flexible electronics and large-scale energy storage.

Xie et al. report on a $LiClO_4$-doped polypyrrole/titanium ($LiClO_4$-PPy/Ti) electrode for use in supercapacitors [2]. The strong electrostatic interaction between pyrrole N and Ti (2.450 Å) enhances electronic conductivity (DOS: 57.321 electrons/eV) and interfacial interaction, resulting in a specific capacitance of 0.123–0.0122 mF cm^{-2} at 0.01–0.10 mA cm^{-2}, outperforming pure $LiClO_4$/Ti electrodes in this respect.

3.2. *Methanol-Mediated Water Splitting*

Teli et al. synthesized a Pd-MoS_2 catalyst via solvothermal methods for methanol-mediated overall water splitting [8]. The catalyst exhibits low overpotentials (133 mV for MM-OER at 10 mA cm^{-2} and 224.6 mV for MM-HER at -10 mA cm^{-2}) and excellent stability (18 h for MM-OER and 15.5 h for MM-HER). The Pd-MoS_2 | | Pd-MoS_2 cell achieves a small potential of 1.581 V and durable operation over 18 h, providing a cost-effective alternative to noble metal catalysts for hydrogen production.

4. Conclusions and Outlook

This issue showcases the diversity of and innovations in inorganic electrode materials in high-performance energy storage devices, addressing critical challenges in different systems through material design and process optimization. Future research directions include determining how to precisely control material microstructure, developing low-cost scalable synthesis methods, increasing low/high-temperature performance, and integrating multi-functional properties. The studies collected in this Special Issue provide valuable references for academic research on and industrial application of advanced energy storage materials, promoting the transition towards sustainable and high-efficiency energy storage technologies.

We would like to express our sincere gratitude to all the authors who contributed to this Special Issue for sharing their high-quality, cutting-edge research results. We also thank all the reviewers for their rigorous efforts, ensuring the papers were of high quality. Additionally, we sincerely appreciate the professional support and assistance provided by the editorial team of *Inorganics* during the organization and publication of this Special Issue.

Conflicts of Interest: The authors declare no conflict of interest.

References

1. Tan, Q.; Liu, H.; Liang, G.; Jiang, K.; Xie, H.; Si, W.; Lin, J.; Kang, X. Two Birds with One Stone: Ammonium-Induced Carbon Nanotube Structure and Low-Crystalline Cobalt Nanoparticles Enabling High Performance of Lithium-Sulfur Batteries. *Inorganics* **2023**, *11*, 305. [CrossRef]
2. Xie, Y.; Xu, J.; Lu, L.; Xia, C. Electrochemical Investigation of Lithium Perchlorate-Doped Polypyrrole Growing on Titanium Substrate. *Inorganics* **2024**, *12*, 125. [CrossRef]
3. Jin, C.; Wang, Y.; Dong, H.; Wei, Y.; Nan, R.; Jian, Z.; Yang, Z.; Ding, Q. A Novel Spinel High-Entropy Oxide (Cr0.2Mn0.2Co0.2Ni0.2Zn0.2)3O4 as Anode Material for Lithium-Ion Batteries. *Inorganics* **2024**, *12*, 198. [CrossRef]
4. Li, F.; Zhu, Y.; Ueno, H.; Deng, T. An Electrochemically Prepared Mixed Phase of Cobalt Hydroxide/Oxyhydroxide as a Cathode for Aqueous Zinc Ion Batteries. *Inorganics* **2023**, *11*, 400. [CrossRef]
5. Wang, C.; Si, W.; Kang, X. Hollow-Structured Carbon-Coated CoxNiySe2 Assembled with Ultrasmall Nanoparticles for Enhanced Sodium-Ion Battery Performance. *Inorganics* **2025**, *13*, 96. [CrossRef]
6. Xie, W.; An, Z.; Li, X.; Wang, Q.; Hu, C.; Ma, Y.; Liu, S.; Sun, H.; Sun, X. Encapsulating Ultrafine In2O3 Particles in Carbon Nanofiber Framework as Superior Electrode for Lithium-Ion Batteries. *Inorganics* **2024**, *12*, 336. [CrossRef]
7. Gao, L.; Liu, F.; Qi, J.; Gao, W.; Xu, G. Recent Advances and Challenges in Hybrid Supercapacitors Based on Metal Oxides and Carbons. *Inorganics* **2025**, *13*, 49. [CrossRef]
8. Teli, A.M.; Mane, S.M.; Mishra, R.K.; Jeon, W.; Shin, J.C. Unveiling the Electrocatalytic Performances of the Pd-MoS2 Catalyst for Methanol-Mediated Overall Water Splitting. *Inorganics* **2025**, *13*, 21. [CrossRef]
9. Khan, M.H.; Lamberti, P.; Tucci, V. Multi-Dimensional Inorganic Electrode Materials for High-Performance Lithium-Ion Batteries. *Inorganics* **2025**, *13*, 62. [CrossRef]
10. Wei, D.; Sun, Z.; Xu, L. Simplified Preparation of N-Doped Carbon Nanosheets Using EDTA Route. *Inorganics* **2025**, *13*, 148. [CrossRef]
11. Zhang, S.S. Liquid electrolyte lithium/sulfur battery: Fundamental chemistry, problems, and solutions. *J. Power Sources* **2013**, *231*, 153–162. [CrossRef]
12. Chung, S.H.; Manthiram, A. Current Status and Future Prospects of Metal–Sulfur Batteries. *Adv. Mater.* **2019**, *31*, e1901125. [CrossRef]
13. Chen, Y.; Wang, T.; Tian, H.; Su, D.; Zhang, Q.; Wang, G. Advances in Lithium–Sulfur Batteries: From Academic Research to Commercial Viability. *Adv. Mater.* **2021**, *33*, 2003666. [CrossRef]
14. Ali, T.; Yan, C. 2 D Materials for Inhibiting the Shuttle Effect in Advanced Lithium–Sulfur Batteries. *ChemSusChem* **2019**, *13*, 1447–1479. [CrossRef]
15. Yoo, S.; Lee, J.I.; Shin, M.; Park, S. Large-Scale Synthesis of Interconnected Si/SiOx Nanowire Anodes for Rechargeable Lithium-Ion Batteries. *ChemSusChem* **2013**, *6*, 1153–1157. [CrossRef]
16. Yu, S.H.; Lee, S.H.; Lee, D.J.; Sung, Y.E.; Hyeon, T. Conversion Reaction-Based Oxide Nanomaterials for Lithium Ion Battery Anodes. *Small* **2015**, *12*, 2146–2172. [CrossRef]

17. Gu, M.; Rao, A.M.; Zhou, J.; Lu, B. In situ formed uniform and elastic SEI for high-performance batteries. *Energy Environ. Sci.* **2023**, *16*, 1166–1175. [CrossRef]
18. Eshetu, G.G.; Elia, G.A.; Armand, M.; Forsyth, M.; Komaba, S.; Rojo, T.; Passerini, S. Electrolytes and Interphases in Sodium-Based Rechargeable Batteries: Recent Advances and Perspectives. *Adv. Energy Mater.* **2020**, *10*, 2000093. [CrossRef]
19. De Silva, K.K.H.; Huang, H.-H.; Joshi, R.; Yoshimura, M. Restoration of the graphitic structure by defect repair during the thermal reduction of graphene oxide. *Carbon* **2020**, *166*, 74–90. [CrossRef]
20. Li, Y.; Leung, K.; Qi, Y. Computational Exploration of the Li-Electrode | Electrolyte Interface in the Presence of a Nanometer Thick Solid-Electrolyte Interphase Layer. *Acc. Chem. Res.* **2016**, *49*, 2363–2370. [CrossRef] [PubMed]
21. Lin, D.; Liu, Y.; Cui, Y. Reviving the lithium metal anode for high-energy batteries. *Nat. Nanotechnol.* **2017**, *12*, 194–206. [CrossRef]
22. Nayak, P.K.; Erickson, E.M.; Schipper, F.; Penki, T.R.; Munichandraiah, N.; Adelhelm, P.; Sclar, H.; Amalraj, F.; Markovsky, B.; Aurbach, D. Review on Challenges and Recent Advances in the Electrochemical Performance of High Capacity Li- and Mn-Rich Cathode Materials for Li-Ion Batteries. *Adv. Energy Mater.* **2017**, *8*, 1702397. [CrossRef]
23. Liu, H.; Zhou, Q.; Xia, Q.; Lei, Y.; Long Huang, X.; Tebyetekerwa, M.; Song Zhao, X. Interface challenges and optimization strategies for aqueous zinc-ion batteries. *J. Energy Chem.* **2023**, *77*, 642–659. [CrossRef]
24. Xu, J.; Cai, X.; Cai, S.; Shao, Y.; Hu, C.; Lu, S.; Ding, S. High-Energy Lithium-Ion Batteries: Recent Progress and a Promising Future in Applications. *Energy Environ. Mater.* **2023**, *6*, e12450. [CrossRef]
25. Yan, C.; Li, H.R.; Chen, X.; Zhang, X.Q.; Cheng, X.B.; Xu, R.; Huang, J.Q.; Zhang, Q. Regulating the Inner Helmholtz Plane for Stable Solid Electrolyte Interphase on Lithium Metal Anodes. *J. Am. Chem. Soc.* **2019**, *141*, 9422–9429. [CrossRef]
26. Lin, X.; Huang, J.; Zhang, B. Correlation between the microstructure of carbon materials and their potassium ion storage performance. *Carbon* **2019**, *143*, 138–146. [CrossRef]
27. Han, Y.; Guo, Y.; Zhou, J.; Ding, X.; Zhang, Y.; Li, W.; Liu, Y.; Chen, Y.; Jie, Y.; Lei, Z.; et al. Riveting Nucleation Enabled Long Cycling Life Calcium Metal Anodes. *Adv. Mater.* **2025**, *37*, e2415657. [CrossRef]
28. Zhao, S.; Che, H.; Chen, S.; Tao, H.; Liao, J.; Liao, X.-Z.; Ma, Z.-F. Research Progress on the Solid Electrolyte of Solid-State Sodium-Ion Batteries. *Electrochem. Energy Rev.* **2024**, *7*, 3. [CrossRef]
29. Hwang, J.-Y.; Myung, S.-T.; Sun, Y.-K. Sodium-ion batteries: Present and future. *Chem. Soc. Rev.* **2017**, *46*, 3529–3614. [CrossRef] [PubMed]
30. Pomerantseva, E.; Bonaccorso, F.; Feng, X.; Cui, Y.; Gogotsi, Y. Energy storage: The future enabled by nanomaterials. *Science* **2019**, *366*, eaan8285. [CrossRef]
31. Zhang, W.; Yin, J.; Sun, M.; Wang, W.; Chen, C.; Altunkaya, M.; Emwas, A.H.; Han, Y.; Schwingenschlögl, U.; Alshareef, H.N. Direct Pyrolysis of Supermolecules: An Ultrahigh Edge-Nitrogen Doping Strategy of Carbon Anodes for Potassium-Ion Batteries. *Adv. Mater.* **2020**, *32*, e2000732. [CrossRef] [PubMed]
32. Li, Y.; Liu, M.; Feng, X.; Li, Y.; Wu, F.; Bai, Y.; Wu, C. How Can the Electrode Influence the Formation of the Solid Electrolyte Interface? *ACS Energy Lett.* **2021**, *6*, 3307–3320. [CrossRef]
33. Cao, B.; Zhang, Q.; Liu, H.; Xu, B.; Zhang, S.; Zhou, T.; Mao, J.; Pang, W.K.; Guo, Z.; Li, A.; et al. Graphitic Carbon Nanocage as a Stable and High Power Anode for Potassium-Ion Batteries. *Adv. Energy Mater.* **2018**, *8*, 1801149. [CrossRef]

inorganics

Article

Simplified Preparation of N-Doped Carbon Nanosheets Using EDTA Route

Denghu Wei [1,*], Zongfu Sun [2] and Leilei Xu [1]

[1] School of Materials Science and Engineering, Liaocheng University, Liaocheng 252059, China
[2] School of Materials Science and Physics, China University of Mining and Technology, Xuzhou 221116, China
* Correspondence: dhweilcu@163.com

Abstract: Amorphous carbon and its heteroatom-doped derivatives often exhibit wrinkled, defective, porous structures, and find wide applications in the fields of energy storage and catalysis. To date, although many methods for preparing doped carbon materials have been reported, the preparation process is relatively complex, and there are still few simple methods available. Therefore, it is necessary to further develop simple and feasible preparation methods. In this study, we employed commercially available manganese disodium ethylenediaminetetraacetate (EDTA-Na_2Mn, serving as both carbon and nitrogen sources) as the precursor. Through thermal decomposition under a nitrogen atmosphere, a nitrogen-doped carbon composite embedded with manganese monoxide (MnO) was initially obtained. Subsequently, hydrochloric acid etching was applied to remove the MnO phases, yielding the final product: nitrogen-doped carbon, denoted as C-N-Mn. Notably, the carbonization and nitrogen-doping processes were simultaneously accomplished during pyrolysis, thereby streamlining the synthesis route for nitrogen-doped carbons. To demonstrate the versatility of this approach, we extended the methodology to two additional metal–organic salts (EDTA-Na_2Zn and EDTA-NaFe), successfully synthesizing nitrogen-doped carbon materials (C-N-Zn and C-M-Fe) in both cases. The phase composition, morphology, microstructure, specific surface area, and pore volume of the products were systematically characterized using X-ray diffraction (XRD), scanning/transmission electron microscopy (SEM/TEM), Raman spectroscopy, X-ray photoelectron spectroscopy (XPS), and nitrogen adsorption/desorption analysis. These nitrogen-doped carbons exhibit high specific surface areas and tunable pore volumes, suggesting their potential applicability in energy storage systems.

Keywords: N-doped carbon; thermolysis; EDTA metal salt; Synchronous carbonization and doping

1. Introduction

Amorphous carbon, including soft carbon, hard carbon, and soft/hard composite carbon, has drawn continuous and widespread attention in the fields of energy storage [1–5], CO_2 adsorption [6–9], energy conversion [10–12], and photocatalysis [13–15] due to its porous structure, high specific surface area, stable physicochemical properties, and environmental friendliness. Therefore, the synthesis of porous carbon materials is particularly important. Currently, there are mainly two preparation methods, namely template and template-free methods [16–18]. Template methods include soft template [19,20] and hard template [21–23]. Soft template mainly utilizes block copolymers and surfactants as structure-directing agents to synthesize porous carbon materials. The steps of preparing

 https://doi.org/10.3390/inorganics13050148

porous carbon via hard template method can be summarized as follows: 1. composite of carbon precursor and template; 2. high-temperature carbonization process; 3. removal of the template. For example, Zhang et al. [23] used monodisperse Fe_3O_4 nanospheres as a template and glucose as a carbon source to obtain a Fe_3O_4-C composite through high-temperature carbonization. After corrosion of the thermal nitric acid, nitrogen-doped porous carbon nanospheres were obtained and used as negative electrode materials for lithium storage. In addition, porous carbon materials can also be obtained by direct carbonization of the organometallic compounds (template-free method). For instance, Li et al. [24] first synthesized rod-shaped copper tartrate (micrometer scale) through hydrothermal synthesis, then carbonized the precursor to obtain Cu-C composites, before finally corroding the Cu-C product with thermal nitric acid to remove the copper. As a result, nitrogen-doped porous carbon materials were obtained, and their lithium storage performance was tested.

The two researchers first synthesized carbon materials and then doped nitrogen using hot nitric acid corrosion. Recent reports showed that nitrogen-doped carbon can be produced via high-temperature pyrolysis of nitrogen-containing biomass (lotus seed pot biomass) [8] or polymers (polyacrylonitrile fiber) [9], proving the feasibility of simultaneous carbonization and nitrogen doping. However, our research finds few reports on using cheap, stable nitrogen-containing organic small molecules (e.g., disodium ethylenediaminetetraacetate) as precursors for nitrogen-doped carbon synthesis. Thus, in this study, commercial and relatively inexpensive ethylenediaminetetraacetic acid disodium manganese (serving as C, N, Mn, and O sources simultaneously) was used as the raw material, mixed with sodium chloride in a 1:1 ratio, and then subjected to high-temperature pyrolysis under an inert atmosphere in a tube furnace to obtain MnO-C-N composites. After hydrochloric acid corrosion, nitrogen-doped porous carbon materials were obtained. We have extended this method, for example, to carbonization of ethylenediaminetetraacetic acid disodium zinc, ethylenediaminetetraacetic acid sodium iron, etc., which can all yield nitrogen-doped porous carbon materials, and the potassium storage performances of these three products were tested. The innovation of this article is mainly reflected in the preparation of materials as follows: (i) abandoning the complex preparation process; (ii) choosing carbon and nitrogen homologs of organic metal compounds with moderate prices and stable properties as precursors; (iii) pyrolysis of the commercialized precursors can be carried out to achieve the simultaneous completion of carbonization and nitrogen doping, thereby simplifying the synthesis steps; (iv) the metal compounds in the product obtained after pyrolysis can be easily corroded with dilute hydrochloric acid, which will add more porous structures to the carbon materials. Of course, this simple and controllable synthesis method can also be used inthe preparation of other carbon materials.

2. Results

Figure 1 illustrates the synthesis of nitrogen-doped carbon. Ethylenediaminetetraacetic acid disodium manganese was mixed with sodium chloride in a 1:1 ratio, followed by uniform ball milling. Subsequently, the mixture was subjected to 700 °C heat treatment under a nitrogen atmosphere for 4 h, resulting in a black product (composite of carbon-coated MnO, denoted as MnO-C-N). After mild hydrochloric acid corrosion, a nitrogen-doped amorphous carbon film (denoted as C-N-Mn) was obtained. The preparation procedure for the other two carbon materials (C-N-Fe, C-N-Zn) was similar to that of C-N-Mn. Sodium chloride acted as a thermal conductivity medium, a carrier for the melted organometallic salt, and a hard template for pore formation. To calculate the yield of the N-doped carbon, we repeated the experiment (Figures S7 and S8). Based on the experimental results, we can calculate the actual yield of carbon. The theoretical yield of carbon is calculated using the formula: (12*10*0.95)/389.13. Here, 389.13 is the molecular weight of sodium manganese(II) ethylenediaminetetraacetate, 12 is the molar mass of carbon, 10 is the number of carbon atoms in each molecule of sodium manganese(II) ethylenediaminetetraacetate, and 0.95 is the purity of the reagent. Calculations show that the theoretical yield of carbon is 0.293. Therefore, when 3 g of the reagent is weighed, the theoretical yield of carbon is 0.879 g. The actual mass of carbon obtained was 0.19 g, so the actual yield of carbon is: (0.19/0.879)*100% = 21.6%.

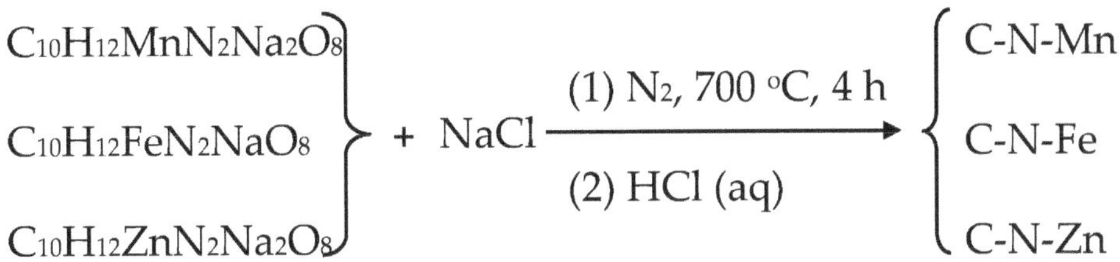

Figure 1. Synthetic route to prepare N-doped carbon materials (C-N-Mn, C-N-Fe, and C-N-Zn).

Figure S1 shows the XRD diffraction patterns of the product after the thermal decomposition of ethylenediaminetetraacetic acid disodium manganese. The diffraction peaks correspond well to the standard card (PDF no. 07-0230), indicating the cubic phase of MnO without significant impurity peaks. The diffraction peaks of amorphous carbon are not obvious, suggesting much lower intensity compared to MnO. The product after the thermal decomposition of ethylenediaminetetraacetic acid disodium iron is a composite of iron, iron nitride, and carbon (denoted as $Fe/Fe_3N/Fe_4N$-C-N), as shown in Figure S2 (PDF no. 06-0696 for Fe, PDF no. 49-1663 for Fe_3N, and PDF no. 06-0627 for Fe_4N). According to the metal activity series (as shown in Table S1), manganese is more active than zinc, which is more active than iron. Thus, the stability of the oxides is in the order: MnO > ZnO > FeO. FeO is easier to reduce than MnO and ZnO. At 700 °C with carbon present, iron oxides are first reduced by carbon to form iron, while oxygen combines with carbon to form CO_x gas, which is then carried away by the inert gas flow. Since the precursor contains nitrogen (more electronegative than carbon), some iron combines with nitrogen to form iron nitrides (Figure S2). MnO and ZnO, being more stable, are not reduced by carbon at 700 °C and remain. The product after the thermal decomposition of ethylenediaminetetraacetic acid disodium zinc is pure ZnO phase (denoted as ZnO-C-N), as shown in Figure S3 (PDF no. 36-1451). After hydrochloric acid corrosion, all three thermal decomposition products became amorphous carbon materials, with only a weak peak observed at the (002) crystal plane.

The morphology of the samples was observed by SEM and TEM. MnO-C-N exhibited a structure of carbon-coated MnO particles (Figure 2(a_1,a_3)). After corrosion, it transformed into carbon flakes (Figure 2(a_2,a_4)). The morphologies of $Fe/Fe_3N/Fe_4N$-C-N (Figure 2(b_1,b_3)) and ZnO-C-N (Figure 2(c_1,c_3)) were similar to that of MnO-C-N. After corrosion, they also turned into amorphous carbon films (Figure 2(b_2,b_4,c_2,c_4)).

The EDS spectra and elemental mapping of each sample are shown in Figure 3a,b (C-N-Mn), Figure S4 (MnO-C-N), Figure 3c,d (C-N-Fe), Figure S5 ($Fe/Fe_3N/Fe_4N$-C-N), Figure 3e,f (C-N-Zn), and Figure S6 (ZnO-C-N). Each element is uniformly dispersed in the samples and found in the EDS spectra.

Figure 4a shows the XRD patterns of the prepared three samples. Only one broad and weak peak around 25° was found in each sample, which corresponds to (002) plane of graphite and indicates the low degree of graphitization of the samples. The Raman spectroscopy was used to analyze the crystalline structure of carbon materials. As shown in Figure 5, the Raman spectra of C-N-Mn, C-N-Fe, and C-N-Zn show D and G peaks of around 1350 cm^{-1} and 1590 cm^{-1}, representing disordered and graphitic carbon structures, respectively. Because the peak width of the D- and G-band is very different, we calculated the $(I_D/I_G)_A$ ratio based on the area under the D- and G-band, as well as using the intensity ratio I_D/I_G (Figure 5a), to analyze the defect extent of the carbon material. As shown in Figure 5b–d, the $(I_D/I_G)_A$ of C-N-Mn, C-N-Fe, and C-N-Zn were 2.75, 3.2, and 2.72, respectively. The integral area ratio confirms the presence of more defects in the carbon materials [2].

X-ray photoelectron spectroscopy (XPS) measurements of samples were performed. As shown in Figure 4b, the XPS survey spectra for the samples shows peaks at C 1s, N 1s of the doped N, and O 1s. The results show that the atomic percentage of doped N in C-N-Mn, C-N-Fe, and C-N-Zn is about 10.9%, 10.6%, and 11.6%, respectively. Deconvolution of the high-resolution C 1s spectrum (Figure 4c) gives five peaks: graphitized carbon at ~284.6 eV, C=N groups at ~285.4 eV, C-O at ~286.5 eV, C=O at ~288.1 eV, and carboxyl groups at ~290.1 eV. Additionally, the XPS spectrum of N 1s (Figure 4d) was fitted to yield four peaks ascribed to pyridinic nitrogen (~398.2 eV), pyrrolic nitrogen (~400 eV), graphitic nitrogen (~400.8 eV), and oxidized nitrogen (~404.2 eV). The proportions of various types of nitrogen are presented in Table S2.

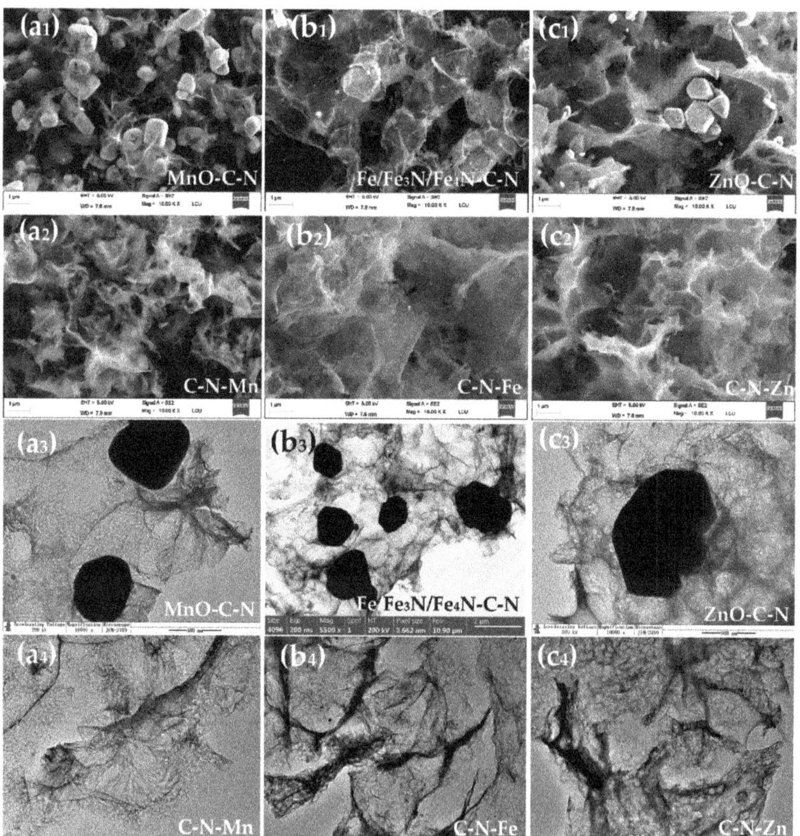

Figure 2. SEM and TEM images of the samples. (**a_1,a_3**) (MnO-C-N), (**a_2,a_4**) (C-N-Mn); (**b_1,b_3**) (Fe/Fe_3N/Fe_4N-C-N), (**b_2,b_4**) (C-N-Fe); (**c_1,c_3**) (ZnO-C-N), (**c_2,c_4**) (C-N-Mn).

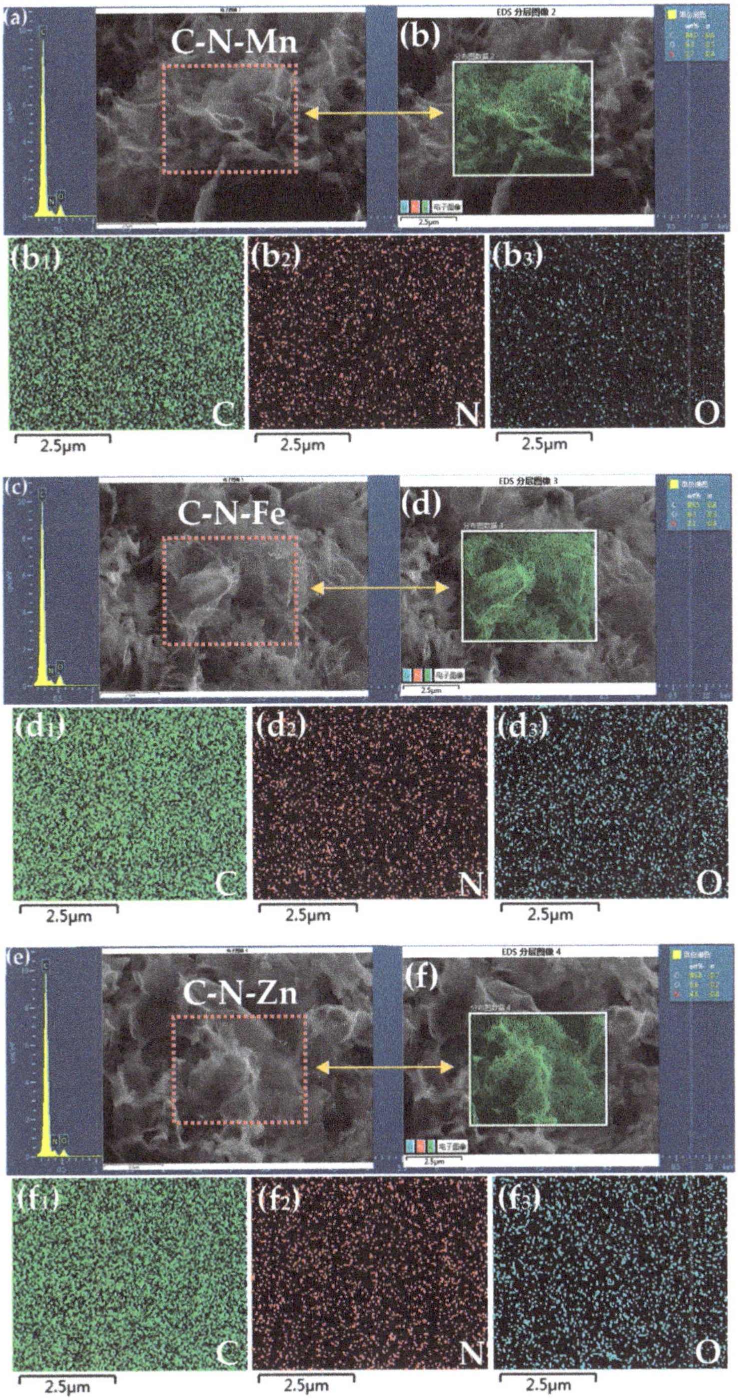

Figure 3. (**a**) EDS spectrum of the C-N-Mn sample and elemental mapping (**b**) of C ($\mathbf{b_1}$), N ($\mathbf{b_2}$), and O ($\mathbf{b_3}$); (**c**) EDS spectrum of the C-N-Fe sample and elemental mapping (**d**) of C ($\mathbf{d_1}$), N ($\mathbf{d_2}$), and O ($\mathbf{d_3}$); (**e**) EDS spectrum of the C-N-Zn sample and elemental mapping (**f**) of C ($\mathbf{f_1}$), N ($\mathbf{f_2}$), and O ($\mathbf{f_3}$).

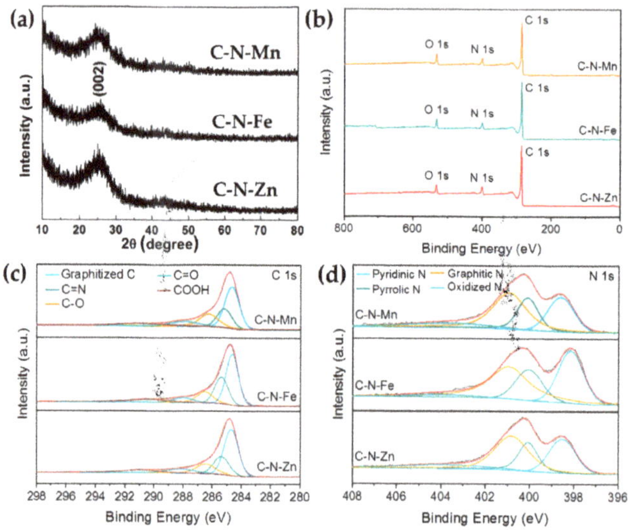

Figure 4. (**a**) XRD of the three samples; (**b**) XPS survey spectra of C-N-Mn, C-N-Fe, and C-N-Zn; (**c**) high-resolution C 1s spectrum of C-N-Mn, C-N-Fe, and C-N-Zn; (**d**) high-resolution N 1s spectrum of C-N-Mn, C-N-Fe, and C-N-Zn.

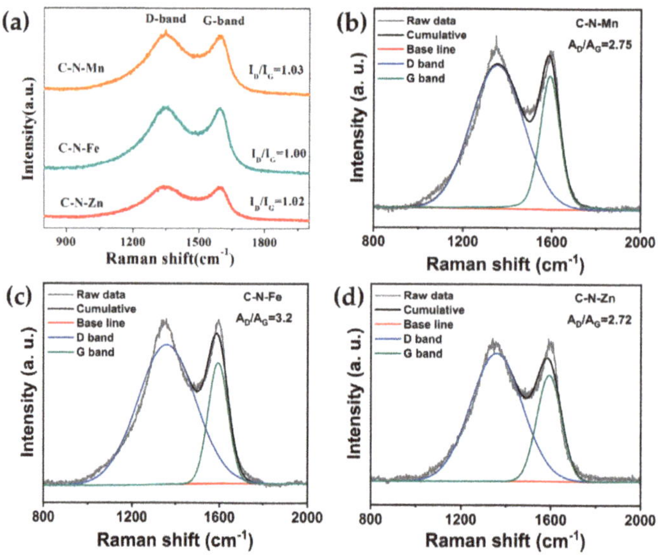

Figure 5. Raman spectra of (**a**) the three samples; (**b**) C-N-Mn; (**c**) C-N-Fe; (**d**) C-N-Zn.

The pore structures of these three carbon materials were characterized by N_2 adsorption–desorption isotherm tests, as shown in Figure 6. All of the curves (Figure 6a,c,e) of the three samples showed a typical I isotherm, and had a rapid uptake within $P/P_0 = 0.01$, which indicated the presence of micropores (<2 nm), as shown in Figure 6b,d,f. When the pressure P/P_0 rises, a hysteresis loop appears in the three curves (Figure 6a,c,e), indicating that there are a few mesopores (2~50 nm) in the samples (Figure 6b,d,f). Based on the isotherm, the calculated BET surface areas were 813.52 (C-N-Mn), 824.97 (C-N-Fe), and 998.20 $m^2\ g^{-1}$ (C-N-Zn), respectively. The pore volumes of the three carbon materials were 1.0243, 1.1307, and 1.1052 $cm^3\ g^{-1}$, respectively.

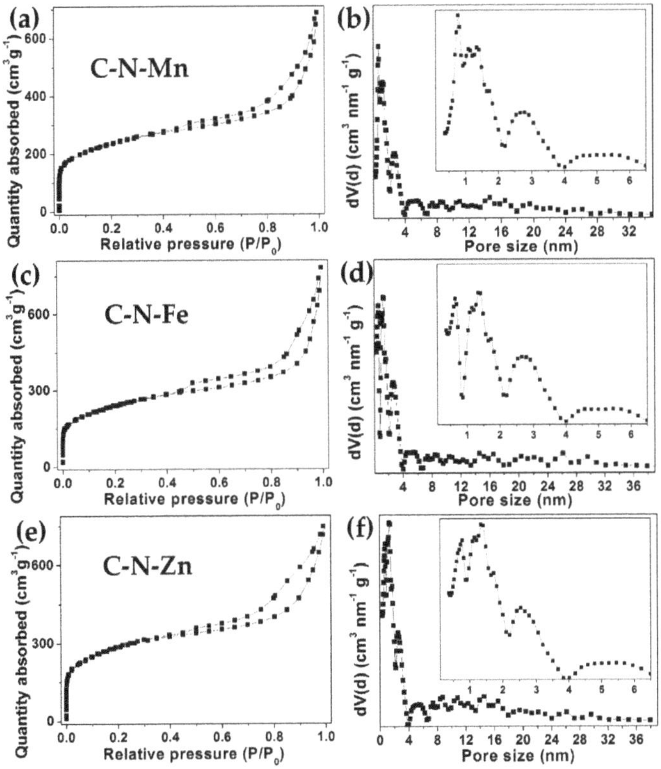

Figure 6. Nitrogen adsorption–desorption isotherms and pore size distribution of the samples: C-N-Mn (**a**,**b**); C-N-Fe (**c**,**d**); C-N-Zn (**e**,**f**).

3. Materials and Methods

All the chemical reagents used were of analytical grade without further purification.

3.1. Synthesis of C-N-Mn

C-N-Mn was obtained from the precursor MnO@C-N, which was prepared according to our previous work with a minor change [25]. Typically, $C_{10}H_{12}MnN_2Na_2O_8$ (EDTA-disodium manganese salt (3 g) and NaCl (3 g) were mixed together and ball-milled in a mortar with a rotation speed of 200 rpm for one night. Then, the mixture was pressed into several circular columns with a diameter of 15 mm (4~6 MPa for 20 s) and heated to 700 °C for 4 h at a heating rate of 5 °C min^{-1} under a N_2 atmosphere. When cooling down to room temperature, the product was dissolved in hydrochloric acid (HCl, 1 M) and stirred for 4 h to remove MnO, NaCl, and other possible sodium salt, and the final sample was collected by vacuum filtration, washed with distilled water and ethanol sequentially, and then dried at 80 °C for 5 h, after which it was named C-N-Mn.

3.2. Synthesis of C-N-Fe

C-N-Fe was synthesized the same way as the C-N-Mn sample, except $C_{10}H_{12}MnN_2Na_2O_8$ was replaced by $C_{10}H_{12}FeN_2NaO_8$ (ethylenediaminetetraacetic acid monosodium ferric salt).

3.3. Synthesis of C-N-Zn

C-N-Zn was synthesized the same way as the C-N-Mn sample, except $C_{10}H_{12}MnN_2Na_2O_8$ was replaced by $C_{10}H_{12}ZnN_2Na_2O_8$ (ethylenediaminetetraacetic acid disodium zinc salt).

3.4. Samples Characterization

X-ray diffraction (XRD diffractometer, D8 Advanced, Bruker Co., Berlin, Germany) and Raman spectroscopy (collected using a 532 nm laser, ThermoFisher DXR 2) were used to investigate the crystal structures of the samples. The morphology of the products was observed through scanning electron microscopy (SEM, Merlin Compact, Carl Zeiss AG, Jena, Germany) and transmission electron microscopy (TEM, Thermo Fischer, Talos F200x, Waltham, MA, USA). The elemental distribution of the samples was detected by energy-dispersive X-ray spectrometry (EDS) elemental mapping analysis (Merlin Compact, Carl Zeiss AG, Jena, Germany). X-ray photoelectron spectra (XPS) were acquired using an ESCALAB 250 instrument (Thermo Fischer, EACALAB 250, Waltham, MA, USA). Porosity analysis was tested by N_2 adsorption/desorption measurements at 77 K with an Autosorb-IQ-C analyzer of Quantachrome (Belsorp Max, MicrotracBEL, Osaka, Japan).

4. Conclusions

This article used commercially available ethylenediaminetetraacetic acid metal salts as raw materials, followed by thermal decomposition and acid corrosion treatment under a nitrogen atmosphere, after which three types of nitrogen-doped amorphous carbon materials were obtained. These materials are all very thin carbon films with a large surface area containing pores and wrinkled structures. The reasons why we chose such substances as raw materials are as follows: 1. They are common and chemically stable reagents that are relatively inexpensive; 2. these compounds contain both carbon and nitrogen atoms, and they are carbon–nitrogen homologous, which is conducive to the nitrogen-doping process; 3. during the carbonization process, the doping process of nitrogen atoms is simultaneously realized, thereby simplifying the synthesis route. In addition, the obtained products have a large number of micropores and mesopores, as well as a large surface area. The carbon materials prepared by this method may have potential applications in fields such as energy storage and conversion, supercapacitors, and catalyst supports.

Supplementary Materials: The following supporting information can be downloaded at: https://www.mdpi.com/article/10.3390/inorganics13050148/s1, Figure S1: XRD of the MnO-C-N sample; Figure S2: XRD of the $Fe/Fe_3N/Fe_4N$-C-N sample; Figure S3: XRD of the ZnO-C-N sample; Figure S4: Elemental mapping and EDS spectrum of the MnO-C-N sample; Figure S5: Elemental mapping and EDS spectrum of the $Fe/Fe_3N/Fe_4N$-C-N sample; Figure S6: Elemental mapping and EDS spectrum of the ZnO-C-N sample; Figure S7: Repeat experiments using disodium manganese ethylenediaminetetraacetate; Figure S8: Disodium manganese ethylenediaminetetraacetate reagent; Table S1: Metal Activity Series; Table S2. The proportion of different types of nitrogen in the three carbon materials.

Author Contributions: D.W.: conceptualization, writing, funding acquisition. Z.S. and L.X.: methodology, data curation, investigation. All authors have read and agreed to the published version of the manuscript.

Funding: This research was funded by the Natural Science Foundation of Shandong Province (ZR2017QB017).

Data Availability Statement: The authors confirm that the data supporting the findings of this study are available within the article and its Supplementary Materials.

Conflicts of Interest: The authors declare no conflict of interest.

References

1. Saju, S.K.; Chattopadhyay, S.; Xu, J.; Alhashim, S.; Pramanik, A.; Ajayan, P.M. Hard carbon anode for lithium-, sodium-, and potassium-ion batteries: Advancement and future perspective. *Cell Rep. Phys. Sci.* **2024**, *5*, 101851. [CrossRef]
2. Cherian, S.K.; Kishore, K.R.; Reddy, S.; Sharma, C.S. Candle Soot-Embedded Electrospun Carbon Nanofibers as a Flexible and Free-Standing Sulfur Host for High-Performance Lithium-Sulfur Batteries. *ACS Appl. Nano Mater.* **2023**, *6*, 15574–15587. [CrossRef]
3. Illa, M.P.; Pathak, A.D.; Sharma, C.S.; Khandelwal, M. Bacterial Cellulose-Polyaniline Composite Derived Hierarchical Nitrogen-Doped Porous Carbon Nanofibers as Anode for High-Rate Lithium-Ion Batteries. *ACS Appl. Energy Mater.* **2020**, *3*, 8676–8687. [CrossRef]
4. Illa, M.P.; Khandelwal, M.; Sharma, C.S. Modulated Dehydration for Enhanced Anodic Performance of Bacterial Cellulose derived Carbon Nanofibers. *ChemistrySelect* **2019**, *4*, 6642–6650. [CrossRef]
5. Cherian, S.K.; Nanaji, K.; Sarada, B.V.; Rao, T.N.; Sharma, C.S. Sulfur confinement into highly porous banana peduncle-derived carbon for high-rate performance lithium-sulfur battery. *J. Energy Storage* **2024**, *89*, 111803–111817. [CrossRef]
6. Tian, L.; Zhi, Y.; Yu, Q.; Xu, Q.; Demir, M.; Colak, S.G.; Farghaly, A.A.; Wang, L.; Hu, X. Enhanced CO_2 Adsorption Capacity in Highly Porous Carbon Materials Derived from Melamine-Formaldehyde Resin. *Energy Fuels* **2024**, *38*, 13186–13195. [CrossRef]
7. Lu, T.; Bai, J.; Huang, J.; Yu, Q.; Demir, M.; Kilic, M.; Altay, B.N.; Wang, L.; Hu, X. Self-Activating Approach for Synthesis of 2,6-Naphthalene Disulfonate Acid Disodium Salt-Derived Porous Carbon and CO_2 Capture Performance. *Energy Fuels* **2023**, *37*, 3886–3893. [CrossRef]
8. Xie, L.; Li, Q.; Demir, M.; Yu, Q.; Hu, X.; Jiang, Z.; Wang, L. Lotus seed pot-derived nitrogen enriched porous carbon for CO_2 capture application. *Colloids Surf. A Physicochem. Eng. Asp.* **2022**, *655*, 130226–130235. [CrossRef]
9. Ma, C.; Lu, T.; Demir, M.; Yu, Q.; Hu, X.; Jiang, W.; Wang, L. Polyacrylonitrile-Derived N-Doped Nanoporous Carbon Fibers for CO_2 Adsorption. *ACS Appl. Nano Mater.* **2022**, *5*, 13473–13481. [CrossRef]
10. Liu, J.; Chen, C.; Wu, H.; Cheng, J. Kinetics and oxidation pathways of Fe^{3+}-catalyzed carbon-assisted water electrolysis for hydrogen production. *Int. J. Hydrog. Energy* **2022**, *47*, 20432–20447. [CrossRef]
11. Jin, Q.; Xiao, L.; He, W.; Cui, H.; Wang, C. Self-supported metal(Fe, Co, Ni)-embedded nitrogen-doping carbon nanorod framework as trifunctional electrode for flexible Zn-air batteries and switchable water electrolysis. *Green Energy Environ.* **2023**, *8*, 1644–1653. [CrossRef]
12. Shen, Q.; Fu, C.; Wang, J.; Yao, W.; Wu, T.; Ding, S.; Xu, P. Exergy-cost-carbon nexus of power-to-X system from carbon dioxide/water co-electrolysis driven by solar full-spectrum energy. *Energy Convers. Manag.* **2024**, *308*, 118382. [CrossRef]
13. Yue, Y.; Yue, X.; Tang, X.; Han, L.; Wang, J.; Wang, S.; Du, C. Synergistic Adsorption and Photocatalysis Study of TiO_2 and Activated Carbon Composite. *Heliyon* **2024**, *10*, e30817. [CrossRef]
14. Asencios, Y.J.O.; Lourenço, V.S.; Carvalho, W.A. Removal of phenol in seawater by heterogeneous photocatalysis using activated carbon materials modified with TiO_2. *Catal. Today* **2022**, *388*, 247–258. [CrossRef]
15. Yu, Z.; Li, F.; Xiang, Q. Carbon dots-based nanocomposites for heterogeneous photocatalysis. *J. Mater. Sci. Technol.* **2023**, *175*, 244–257. [CrossRef]
16. Pavlenko, V.; Khosravi H, S.; Zoitowska, S.; Haruna, A.B.; Zahid, M.; Mansurov, Z.; Supiyeva, Z.; Galal, A.; Ozoemena, K.I.; Abbas, Q.; et al. A comprehensive review of template-assisted porous carbons: Modern preparation methods and advanced applications. *Mater. Sci. Eng. R Rep.* **2022**, *149*, 100682. [CrossRef]
17. Huang, J.; Liu, C.; Jin, Y.; Chen, J. Hierarchical porous carbon synthesis by carbonized polymer dots-based sacrificial template for high-performance supercapacitors. *Chem. Eng. J.* **2023**, *461*, 141930. [CrossRef]
18. Wang, C.; Yan, B.; Zheng, J.; Feng, L.; Chen, Z.; Zhang, Q.; Liao, T.; Chen, J.; Jiang, S.; Du, C.; et al. Recent progress in template-assisted synthesis of porous carbons for supercapacitors. *Adv. Powder Mater.* **2022**, *1*, 100018. [CrossRef]
19. Chen, L.; Yuan, Y.; Zhang, D.; Lin, Z.; Lin, J.; Li, S.; Guo, S. Construction of 3D hierarchical honeycomb macro/meso/micro-porous carbon with soft and hard templates for high-performance sodium-ion batteries. *Mater. Lett.* **2023**, *334*, 133737. [CrossRef]
20. Young, C.; Chen, H.-T. Supercapacitor application of a three-dimensional carbon sphere-intercalated porous carbon fabricated using a hard template and a biomass material. *Diam. Relat. Mater.* **2022**, *130*, 109528. [CrossRef]
21. Guan, L.; Hu, H.; Teng, X.-L.; Zhu, Y.-F.; Zhang, Y.-L.; Chao, H.-X.; Yang, H.; Wang, X.-S.; Wu, M.-B. Templating synthesis of porous carbons for energy-related applications: A review. *New Carbon Mater.* **2022**, *37*, 25–45. [CrossRef]
22. Zhang, W.; Cheng, R.-R.; Bi, H.-H.; Lu, Y.-H.; Ma, L.-B.; He, X.-J. A review of porous carbons produced by template methods for supercapacitor applications. *New Carbon Mater.* **2021**, *36*, 69–81. [CrossRef]
23. Zhang, K.; Li, X.; Liang, J.; Zhu, Y.; Hu, L.; Cheng, Q.; Guo, C.; Lin, N.; Qian, Y. Nitrogen-doped porous interconnected double-shelled hollow carbon spheres with high capacity for lithium ion batteries and sodium ion batteries. *Electrochim. Acta* **2015**, *155*, 174–182. [CrossRef]

24. Li, X.; Zhu, X.; Zhu, Y.; Yuan, Z.; Si, L.; Qian, Y. Porous nitrogen-doped carbon vegetable-sponges with enhanced lithium storage performance. *Carbon* **2014**, *69*, 515–524. [CrossRef]
25. Wei, D.; Xu, L.; Jiao, R.; Zhong, Z.; Ji, X.; Zeng, S. One-pot thermal decomposition of commercial organometallic salt to Fe_2O_3@C-N and MnO@C-N for lithium storage. *Dalton Trans.* **2021**, *50*, 6867–6877. [CrossRef]

inorganics

MDPI

Article

Hollow-Structured Carbon-Coated $Co_xNi_ySe_2$ Assembled with Ultrasmall Nanoparticles for Enhanced Sodium-Ion Battery Performance

Chao Wang [1,2], Weijie Si [3,*] and Xiongwu Kang [3,*]

1 National Institute of Guangdong Advanced Energy Storage, Guangzhou 510000, China; wangchaomly@163.com
2 China Southern Power Grid Technology Co., Ltd., Guangzhou 510080, China
3 New Energy Research Institute, School of Environment and Energy, South China University of Technology, Guangzhou 510006, China
* Correspondence: 202221047646@mail.scut.edu.cn (W.S.); esxkang@scut.edu.cn (X.K.)

Abstract: Transition metal selenides are considered one of the most promising materials for sodium-ion battery anodes due to their excellent theoretical capacity. However, it remains challenging to suppress the volume variation and the resulted capacity decay during the charge–discharge process. Herein, hollow-structured $CoNiSe_2$ dual transition metal selenides wrapped in a carbon shell (HS-$Co_xNi_ySe_2$@C) were deliberately designed and prepared through sequential coating of polyacrylonitrile (PAN), ion exchange of ZIF-67 with Ni^{2+} metal ions, and carbonization/selenization. The hollow structure was evidenced by transmission electron microscopy, and the crystalline structure was confirmed by X-ray diffraction. The ample internal space of HS-$Co_xNi_ySe_2$@C effectively accommodated volume expansion during the charge and discharge processes, and the large surface area enabled sufficient contact between the electrode and electrolyte and shortened the diffusion path of sodium ions for a feasible electrochemical reaction. The surface area and ionic conductivity of HS-$Co_xNi_ySe_2$@C were strongly dependent on the ratio of Co to Ni. The synergistic effect between Co and Ni enhanced the conductivity and electron mobility of HS-$Co_xNi_ySe_2$@C, thereby improving charge transfer efficiency. By taking into account the structural advantages and rational metal selenide ratios, significant improvements can be achieved in the cycling performance, rate performance, and overall electrochemical stability of sodium-ion batteries. The optimized HS-$Co_xNi_ySe_2$@C demonstrated excellent performance, and the reversible capacity remained at 334 mAh g^{-1} after 1000 cycles at a high current of 5.0 A g^{-1}.

Keywords: sodium-ion battery; hollow structure; binary transition metal selenides; synergistic effect

1. Introduction

With the increasing share of intermittent renewable energy sources, advancing high-performance energy storage technologies has become a critical focus for the scientific community. Due to the abundance and low cost of sodium resources, sodium-ion batteries (SIBs) have attracted significant attention as an alternative to conventional lithium-ion battery systems (LIBs). However, the larger ionic radius of Na^+ compared to that of Li^+ leads to substantial volume changes in electrode materials and slower ion diffusion kinetics, resulting in poor cycling stability and rate performance for the well-developed electrode materials in LIBs [1]. Thus, it is desirable to design and develop novel electrode materials for the practical application of SIBs [2,3]. Many advanced anode and cathode materials have been developed through strategies such as nanostructure engineering and alloy design to enhance structural stability and ionic conductivity, thereby improving cycling performance and rate capability [4,5].

Transition metal selenides (TMSs) have demonstrated significant potential as anode materials for sodium-ion batteries owing to their excellent electrical conductivity, high theoretical specific capacity, favorable electrochemical properties, and structural stability [6]. The weak M–Se bonds facilitate conversion reactions, enhancing cycling performance. Men et al. [7] have proposed novel porous $FeSe_2$ composites wrapped in an *N*-doped porous carbon framework (*P*–$FeSe_2$/NCF) through a scalable pyrolysis method, which retains abundant porosity and a dodecahedral structure embedded with carbon nanotubes. Even at relatively high current densities of 5 and 10 A g^{-1}, the porous $FeSe_2$ anode maintains capacities of 548 and 466 mAh g^{-1}, respectively. Such excellent rate performance has been ascribed to the unique porous structure and large surface area. Compared to monometallic selenides, bimetallic selenides exhibit electronic coupling and synergistic effects between their components, which may produce more ion defects and an ion diffusion path, thus further enhancing ion diffusion rates. Weng et al. [8] synthesized a bimetallic heterojunction selenide ($CoSe_2$/$NiSe_2$@N–C) using a Co/Ni-ZIF template, with the selenide grown within N-doped carbon nanofibers. The enhanced performance of this binary transition metal selenide was ascribed to the built-in electric field at the interface of the two TMSs.

In addition, nanostructure engineering has demonstrated an excellent strategy in the development of high-performance anode materials. Among the alloy anodes for sodium-ion batteries, yolk-shell and hollow structures often outperform the solid-state counterpart due to the enhanced large surface area of the contacting electrolyte and offering more abundant reactive sites [9]. Additionally, hollow and yolk-shell structures effectively mitigate the volume changes induced by the insertion and extraction of sodium ions during the charge–discharge process and enhance the robustness against the fragmentation of the nanostructure of the electrode towards a long lifespan. Furthermore, unlike in the solid-state counterpart, the formation of a hollow structure offers shorter ion transport pathways and facilitates the rapid insertion and extraction of sodium ions, improving the rate performance and enabling the battery to maintain high capacity and efficiency under high-rate conditions. Zhou et al. [8] employed an ion-exchange method to anchor hollow- and hetero-structured ZnSe/$NiSe_2$ bimetallic selenides (V-ZnSe/$NiSe_2$@H-NC). Owing to the unique hollow and hetero structure, it exhibited excellent performance as an anode material for sodium-ion batteries, achieving a capacity of 314 mAh g^{-1} at a current density of 2.0 A g^{-1} after 2000 cycles.

Currently, most studies focus on structural design and the introduction of metals ions into the parent metal selenide to enhance the overall electrochemical performance. However, the impact of varying the metal ratios on material performance has received limited attention, especially in the porous structure counterpart. Therefore, the simultaneous introduction of multiple metals with varying ratios during the synthesis of hollow-structured TMSs along with a systematic study of how variations in metal content influence performance, remains an under-explored research area. Such investigations could not only deepen our understanding of metal ratio optimization in enhancing electrochemical behavior but also offer new insights and guidance for the design of advanced energy storage materials.

Inspired by the aforementioned studies, we developed a hollow-structured, carbon-encapsulated Co/Ni bimetallic transition metal selenide (CoNi-Se@C) as a multifunctional anode material for sodium-ion batteries, using ZIF-67 as a sacrificial template through a straightforward PAN encapsulation and ion-exchange process. Although binary metal selenides of Co with Mn, Cu, and Fe were also explored, achieving a hollow structure remains challenging. The experimental results revealed that with increasing Ni content, the electrochemical performance initially improved but then declined, and the best performance was achieved at the ratio of Co:Ni = 1.05:0.95, possibly ascribing to the highest electrochemical surface area and ion conductance due to the optimized synergistic effects between the two metal components [10]. The hollow CoNi-Se@C with optimized Ni/Co ratio exhibited a specific discharge capacity of 334 mAh g^{-1} after 1000 cycles at a high current of 5.0 A g^{-1}, and a high discharge capacity of 428 mAh g^{-1} at a high current of 5.0 A g^{-1}, demonstrating excellent commercial potential.

2. Results

Figure 1a illustrates the synthesis scheme of the CoNi-Se@C material. First, a ZIF-67 precursor was prepared by coordinating Co^{2+} ions and 2-methylimidazole in methanol. Then, a simple "phase separation method" was employed to uniformly coat ZIF-67 with polyacrylonitrile (PAN) [11], which served as the primary carbon source, resulting in the ZIF-67@PAN precursor. Subsequently, the obtained ZIF-67@PAN was dispersed in ethanol and subjected to ion exchange with Ni^{2+} for 2 h. This process partially etched the inner ZIF-67, transforming it into a hollow layered double hydroxide (LDH) structure [12], yielding Co_xNi_y-LDH@PAN. Afterward, the dried Co_xNi_y-LDH@PAN was mixed with Se powder at a 1:1 mass ratio, placed in an alumina boat, and annealed under an Ar/H_2 atmosphere at 500 °C for 6 h. Following this high-temperature selenization process, hollow-structured $Co_xNi_ySe_2$@C was obtained and denoted as CoNi-Se@C. The ratio of Co to Ni was determined to be 0.51:0.49 by inductively coupled plasma optical emission spectroscopy (ICP-OES), and the chemical formula was denoted as $Co_{0.51}Ni_{0.49}Se_2$, as shown in Table S1, which was consistent with that determined by XPS (Table S2). The counterparts with varied ratios of Ni to Co were obtained via ion exchange for 1 h and 3 h, and are denoted as CoNi-1-Se@C and CoNi-3-Se@C. The formulas of $Co_xNi_ySe_2$ were determined to be $Co_{0.59}Ni_{0.42}Se_2$, $Co_{0.56}Ni_{0.5}Se_2$, and $Co_{0.59}Ni_{0.59}Se_2$ for CoNi-1-Se@C, CoNi-Se@C, and CoNi-3-Se@C, respectively, indicating that Ni content was consistently increasing with the prolonged Ni^{2+} exchange course.

As shown in Figure 1b,c, after high-temperature selenization, the obtained CoNi-Se@C material showed a polyhedral shape, smooth surface, and intact structure. The size of the polyhedron was approximately from 500 to 700 nm. The hollow structure of CoNi-Se@C is observed in Figure 1d. This hollow structure not only helps mitigate mechanical stress caused by volume expansion during charge and discharge but also increases the specific surface area, providing more active sites for sodium-ion storage, thereby enhancing the electrochemical performance of the material.

In addition, the high-resolution transmission electron microscopy (HRTEM) images in Figure 1e indicate that the shell of CoNi-Se@C was assembled with nanoparticles from 3 to 7 nm. The nano-sized nanoparticles display large surface area and offer abundant active sites for reaction and the pores among them favor the penetration of electrolytes and promote electrochemical reaction. Figure 1f displays the distinct lattice fringes with spacing of 0.271 and 0.265 nm, corresponding precisely to the (1,0,1) and (0,0,2) reflection of $CoNiSe_2$. Meanwhile, the selected area electron diffraction (SAED) pattern in Figure 1g presents multiple clear diffraction rings, which match well with the (0,0,2), (0,1,2), (0,0,3), and (−1,2,2) planes of $CoNiSe_2$, further verifying the crystalline structure of the material. Furthermore, the elemental mapping in Figure 1h–m shows a uniform distribution of Co, Ni, Se, N, and C elements within the CoNi-Se@C material, indicating the formation of uniform binary Co/Ni diselenide.

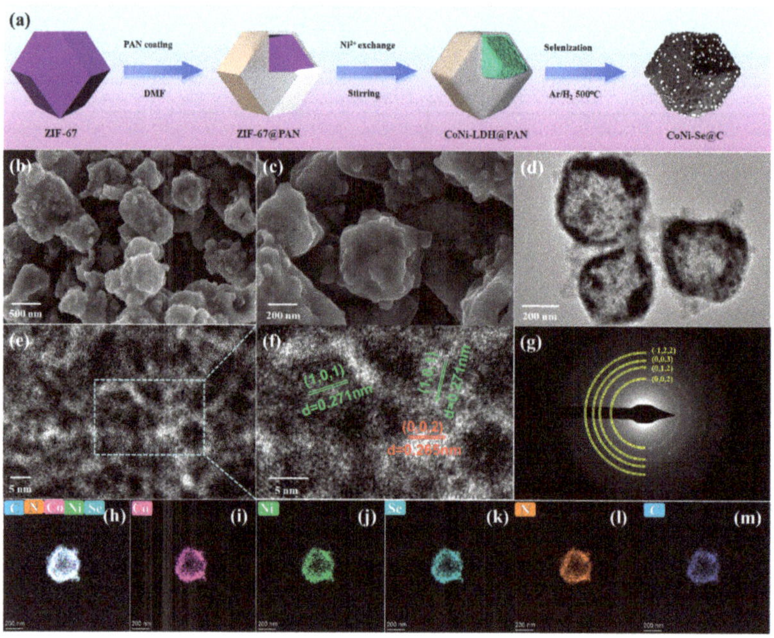

Figure 1. (**a**) Schematic illustration CoNi-Se@C synthesis. (**b**,**c**) SEM images, (**d**) TEM image, (**e**,**f**) HRTEM images, and (**g**) SAED pattern of CoNi-Se@C. (**h**–**m**) Element mapping images of Co, Ni, Se, N, and C.

Figure 2a presents the XRD pattern of $CoSe_2$@C, revealing that without ion exchange, the direct selenization of pure ZIF-67@PAN at 500 °C yielded $CoSe_2$. In contrast, the XRD patterns of CoNi-Se@C, CoNi-1-Se@C, and CoNi-3-Se@C showed a high degree of correspondence with the reference pattern of $CoNiSe_2$ (PDF#89-7162). This indicates that selenization at 500 °C resulted in the formation of a $Co_xNi_ySe_2$ (Figure 2b), and the change of the ratio of Co^{2+} to Ni^{2+} did not change the lattice parameter of $Co_xNi_ySe_2$. Additionally, the Raman spectra of CoNi-Se@C showed an apparent defect and graphitic bands with an I_D/I_G ratio of 1.15, indicating the presence of defects and disordered carbon structures, which contributed to enhanced electron transport and electrochemical activity (Figure S1).

The N_2 adsorption/desorption isotherms were employed to analyze the changes in the specific surface area of the samples and to further investigate the impact of Ni^{2+} incorporation on the structure. All four samples exhibited typical type IV adsorption behavior, with a pore size of approximately 4.5 nm (Figure 2c,d) [13]. The surface areas for $CoSe_2$@C, CoNi-1-Se@C, CoNi-Se@C, and CoNi-3-Se@C were calculated to be 93.72, 120.4, 168.15, and 131.87 m^2 g^{-1}, respectively, using the Brunauer–Emmett–Teller (BET) method. It indicates that the progressive introduction of Ni^{2+} promoted the formation of hollow structures, leading to an initial increase and further decline in surface area. The initial increase of the surface area of CoNi-Se@C with progressive ion exchange could be attributed to the creation of the hollow structure and increased cavity. However, when the ion exchange time extend to 3 h, collapse and contraction of the hollow structure may occur during selenization, which may ultimately decrease the specific surface area [14].

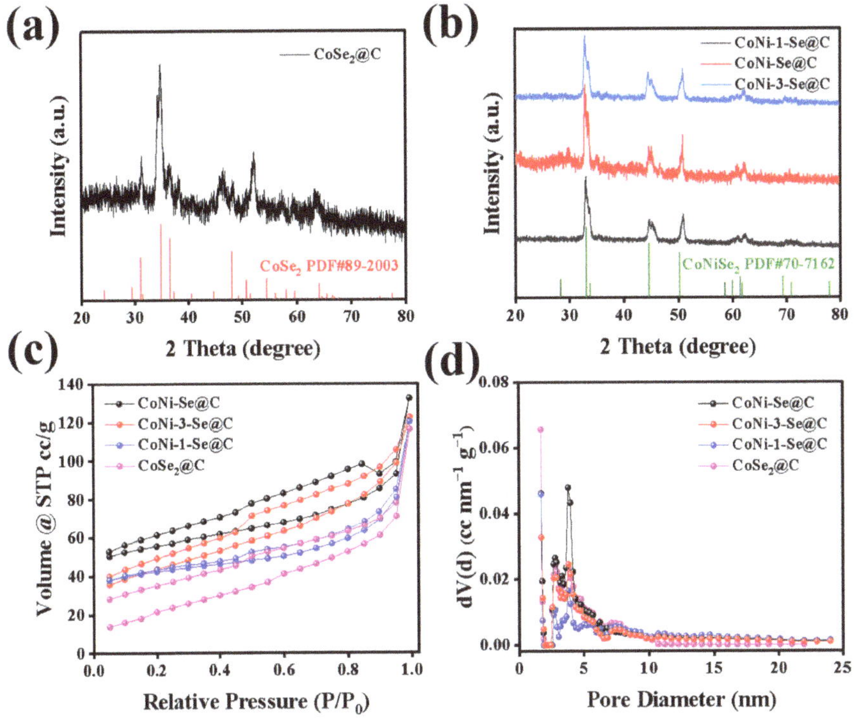

Figure 2. XRD patterns of (**a**) $CoSe_2$@C, (**b**) CoNi-Se@C, CoNi-1-Se@C, and CoNi-Se-3-@C. (**c**) Nitrogen adsorption/desorption isotherm and (**d**) pore size distributions of CoNi-Se@C, CoNi-1-Se@C, CoNi-Se-3-@C, and $CoSe_2$@C.

As shown in Figure 3a, the XPS survey spectrum clearly revealed the presence of Co, Ni, Se, N, and C. First, Figure 3b presents the high-resolution spectrum of Co 2p. The two strong spin-orbit coupling peaks located at 794.5 eV and 779.3 eV were indexed to Co^{2+} $2p_{1/2}$ and Co^{2+} $2p_{3/2}$. The two satellite peaks (Sat.) were located at 803.9 eV and 784.9 eV, individually. Additionally, two Co–O peaks were observed at binding energies of 797.9 eV and 781.4 eV, which can be attributed to surface oxidation of the sample [15,16]. In Figure 3f, the characteristic peaks of Ni^{2+} $2p_{1/2}$ and Ni^{2+} 2p3/2 were observed at 870.27 eV and 852.0 eV, and the satellite peaks were located at 859.5 eV and 878.5 eV, respectively. Similarly, corresponding Ni–O peaks are also present in the Ni 2p spectrum (873.9 eV and 854.4 eV) [17,18]. Furthermore, Figure 3d shows the high-resolution spectrum of Se 3d. Se exhibited two peaks at 53.8 eV and 52.8 eV, which were indexed to Se^{2-} $3d_{3/2}$ and Se^{2-} $3d_{5/2}$, respectively. The characteristic peak at 57.8 eV is attributed to SeO_2 [19–21]. The high-resolution N 1s spectrum (Figure 3e) identifies three nitrogen types: pyridinic N (397.1 eV), pyrrolic N (397.7 eV), and graphitic N (398.9 eV) [22]. Meanwhile, the C 1s spectrum (Figure 3f) displays three characteristic peaks at 289.1, 286.7, and 284.8 eV, corresponding to C–O, C–N, and C–C/C=C bonds, respectively [23].

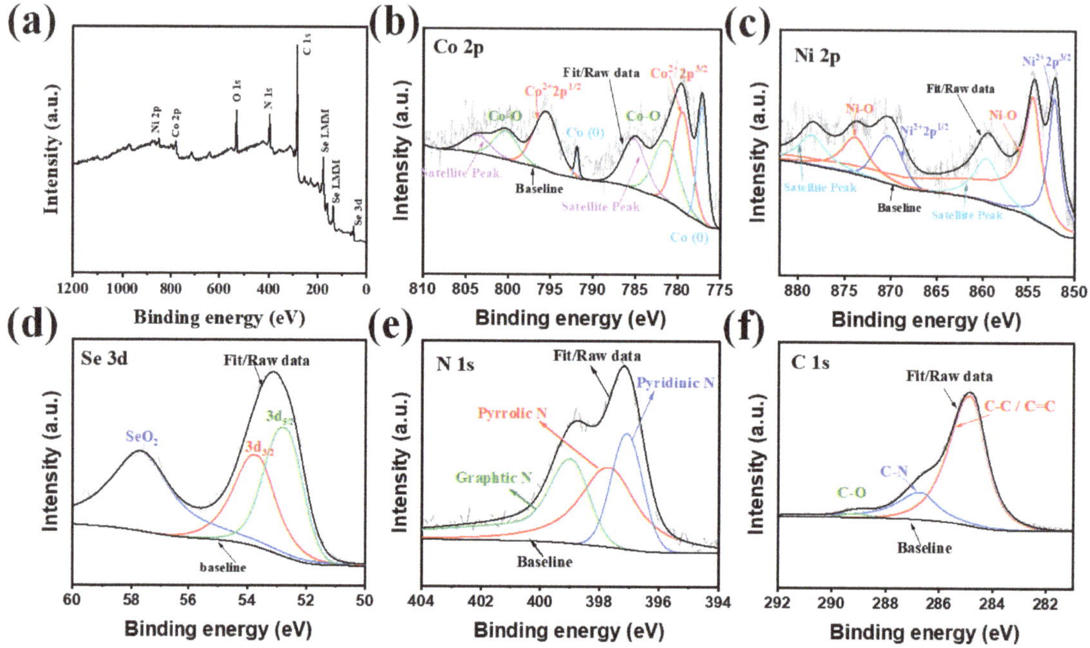

Figure 3. (**a**) Full survey XPS spectrum and (**b–f**) high-resolution XPS spectra of Co 2p, Ni 2p, Se 3d, N 1s, and C 1s of CoNi-Se@C.

The cyclic voltammetry (CV) curves of the CoNi-Se@C material with varied ratios of Co to Ni and $CoSe_2$ assembled with a sodium metal anode, measured over the voltage range from 0.01 to 3.0 V for the first three cycles, are shown in Figures 4a and S2. In the first cycle of CV scan for CoNi-Se@C (Figure 4a), a distinct reduction peak was observed at around 0.9 V, ascribing to the formation of the solid electrolyte interface (SEI) layer and the initial insertion of Na^+ ions [7]. In the subsequent second and third cycle, the high degree of overlap between the CV curves demonstrated the excellent cycling stability of the CoNi-Se@C electrode after initial activation [1,24]. These results highlight the potential of the CoNi-Se@C material for sodium-ion batteries, particularly its ability to maintain superior electrochemical performance and stability over extended cycling. In the cathodic scan, the reduction peaks were sequentially observed at 1.357, 1.071, and 0.677 V, and a single oxidation peak was observed at 1.925 V, which were possibly attributed to the specific sodiation and desodiation reactions or pseudocapacitive behavior [8,25]. To this end, XRD analysis was performed on CoNi-Se@C in the fully discharged state to investigate the phase evolution of the material. The results showed the appearance of characteristic diffraction peaks of Na_2Se and Ni/Co (Figure S3), indicating a phase transition of $Co_xNi_ySe_2$ to metallic Co/Ni and Na_2Se by Na^+ insertion [26]. The possible reactions (Equations (1)–(4)) contributed to the capacity are shown below:

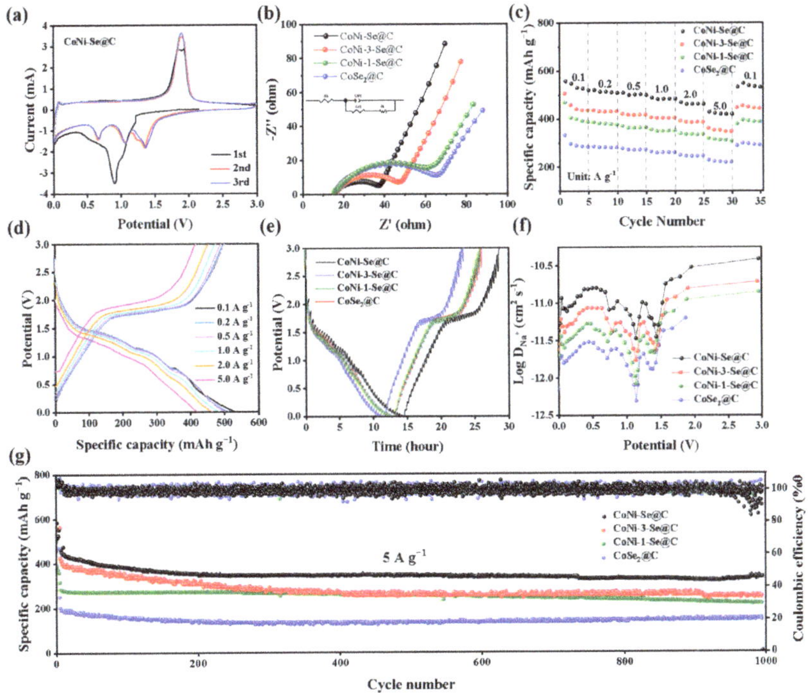

Figure 4. (**a**) CV curves for the first 3 cycles of CoNi-Se@C at 1.0 mV s^{-1}. (**b**) EIS Nyquist plots and equivalent circuit in the inset, (**c**) rate performances of $CoSe_2$@C, CoNi-1-Se@C, CoNi-Se@C, and CoNi-3-Se@C. (**d**) Discharge–charge curves at various rates of CoNi-Se@C. (**e**) GITT curves, (**f**) Na^+ diffusion coefficient in discharge process and (**g**) long-life cycling performance at 5 A g^{-1} of $CoSe_2$@C, CoNi-1-Se@C, CoNi-Se@C, and CoNi-3-Se@C.

Discharge

$$Co_{0.51}Ni_{0.49}Se_2 + xNa^+ + xe^- \rightarrow Na_xCo_{0.51}Ni_{0.49}Se_2 \quad (1)$$

$$Na_xCo_{0.51}Ni_{0.49}Se_2 + 2yNa^+ + 2ye^- \rightarrow Na_xCo_{0.51}Ni_{0.49}Se_{2\text{-}y} + yNa_2Se \quad (2)$$

$$Na_xCo_{0.51}Ni_{0.49}Se_{2\text{-}y} + (4 - 2y - x)Na^+ + (4 - 2y - x)e^- \rightarrow 0.51Co + 0.49Ni + (2 - y)Na_2Se \quad (3)$$

Charge

$$0.51Co + 0.49Ni + 2Na_2Se \rightarrow 4Na^+ + 4e^- + Co_{0.51}Ni_{0.49}Se_2 \quad (4)$$

Similar redox peaks were also observed for other anode materials of $CoSe_2$@C, CoNi-1-Se@C, and CoNi-3-Se@C. The related redox potentials and polarization derived are listed in Table S3. In the first scan, the cathodic peak potential for CoNi-1-Se@C was the highest among the three samples. This redox peak is related to the formation of SEI, and the highest potential indicates the earliest formation of an SEI film and protection of the electrode materials during the initial discharging process. With the increasing content of Ni in $Co_xNi_ySe_2$@C, the potential of the cathodic peak1 in the second cycle increased first, achieved the highest for CoNi-Se@C, and then declined considerably for CoNi-3-Se@C. The anodic peak potential in the second cycle increased consistently with the increasing of Ni content, which resulted in the consistent increase of the polarization.

Electrochemical impedance spectroscopy (EIS) with a bias of 2.4 V and amplitude of 0.05 V was employed to evaluate the impedance of different materials. Figures 4b and S4 show the Nyquist plots and fitting curves of the four materials using the equivalent circuit (inset in Figure 4b), respectively, where R_{ct}, R_s, W, and CPE represent the interfacial charge transfer resistance, solution resistance, Warburg impedance, and impedance of the constant phase element, respectively [27]. The detailed fitting parameters are shown in Table S4. R_{ct} exhibited a trend of first decreasing and then increasing with the introduction of Ni^{2+} (R_{ct}: 50.39 → 44.05 → 21.21 → 31.47 Ω), consistent with the variation of the surface area. The increases in the surface area of material resulted in more active sites, which enhanced the interfacial contact between the electrolyte and electrode, facilitating Na^+ transport and charge transfer.

The impedance of the constant phase element describes non-ideal capacitive behavior related to a rough or porous interface, varied thickness of film layer, or non-uniform reaction dynamics [28], and the variation of CPE-T and CPE-P might be caused by the roughness and porosity of the electrode materials with varied ratios of Co to Ni. It was also observed that CoNi-Se@C exhibited the highest CPE-T. The former might be related to the highest psudocapacitance contribution of CoNi-Se@C among the three samples. The Warburg element models ion transport impedance across an electrode layer to the surface of a current collector [29]. The lowest W-R indicates the highest Na^+ diffusion rate and most accessible of Na^+ to and from the electrode surface. Additionally, the synergistic effect between Ni and Co improves electronic conductivity, further optimizing charge transport during cycling [30,31]. However, excessive Ni^{2+} content causes contraction of the hollow space and condensation of the shell, which reduces the electrochemical surface area and impairs ion transport efficiency. Therefore, precise control of Ni^{2+} doping is crucial to achieving optimal electrochemical performance by balancing improved conductivity and stability [32,33].

Figure 4c compares the rate performance of $CoSe_2$@C, CoNi-1-Se@C, CoNi-Se@C, and CoNi-3-Se@C across a current range of 0.1–5.0 A g^{-1}. Among them, CoNi-Se@C exhibited the best electrochemical performance, maintaining a high discharge capacity of 428 mAh g^{-1} even at a high current of 5.0 A g^{-1}, indicating excellent rate capability and fast charge–discharge ability. Figures 4d and S5 further present the charge–discharge curves of CoNi-Se@C under different current densities. The material consistently maintains stable charge–discharge plateaus at various rates, demonstrating superior structural stability and good reversibility, making it more suitable for high-rate charge–discharge applications compared to other materials.

To further investigate the effect of Ni content on the performance of those materials, galvanostatic intermittent titration technique (GITT) testing was conducted to explore the Na^+ diffusion rate in each material (Figure 4e). Based on the results from GITT testing, the diffusion coefficient of Na^+ can be calculated using Fick's second law with the following equation [34,35]:

$$D_{Na^+} = \frac{4}{\pi\tau}(\frac{m_B V_m}{M_B S})^2(\frac{\delta E_s}{\delta E_\tau})^2 \qquad (5)$$

where τ is the relaxation time (s), m_B is the mass of the electrode material, V_m is the molar volume of the active material, M_B is the molar mass of the electrode material, S is the electrode area, ΔEs represents the potential change between two consecutive relaxation periods, and $\Delta E\tau$ denotes the potential change between the static and equilibrium states. The results in Figure 4f show that CoNi-Se@C, with a Co to Ni ratio close to 1:1, exhibited a higher sodium-ion diffusion rate compared to that of other samples. For instance, the diffusion coefficient of Na^+ (D_{Na+}) of $CoSe_2$@C, CoNi-1-Se@C, CoNi-Se@C, CoNi-3-Se@C at 0.5 V vs. Na^+/Na was approximately 3.0×10^{-12}, 4.3×10^{-12}, 1.58×10^{-11}, 8.7×10^{-12} $cm^2\ s^{-1}$, which was in the order of CoNi-Se@C > CoNi-3-Se@C > CoNi-1-Se@C > $CoSe_2$@C. The trend of Na^+ diffusion rate aligned with the previously discussed findings—initially increasing and then decreasing with the increasing ratio of Ni to Co. This phenomenon can be attributed to the partial substitution of Co^{2+} by an appropriate amount of Ni^{2+}, which exhibits the optimized synergistic interactions between the two metals, thereby achieving the highest electronic and ionic conductivity [36].

Figure 4g presents the long-term cycling performance of $CoSe_2$@C, CoNi-1-Se@C, CoNi-Se@C, and CoNi-3-Se@C at a current density of 5.0 A g^{-1}. It is evident that CoNi-Se@C exhibited the best performance, maintaining a discharge capacity of 335 mAh g^{-1} at 5.0 A g^{-1} after prolonged cycling of 1000 cycles. As shown in TEM images of CoNi-Se@C after 1000 cycles at 5.0 A g^{-1} (Figure S6), the hollow structure remained but fractures appeared, which may be related to the decline of the charge–discharge capacity. Table S5 presents a comparison of the long cycling performance of CoNi-Se@C with that of other reported metal selenides. It is evident that CoNi-Se@C exhibited excellent cycling stability and high-capacity retention over prolonged cycling. This superior performance further confirms the importance of optimizing the metal ratio for enhancing sodium storage capabilities. The synergy between Co and Ni contributes to improved structural stability, enhanced electronic conductivity, and faster ion diffusion, all of which are critical for achieving high-capacity retention and stable cycling. Conversely, inappropriate metal ratios may disrupt this synergy, leading to compromised performance.

CV tests at different scan rates ranging from 0.2 to 1.0 mV s^{-1} were conducted to investigate the contribution of the pseudo capacitance to the charge capacity of the material, and Equations (6) and (7) were utilized to calculate the contributions of pseudocapacitance and diffusion (Figures 5a,b and S7). Here, i represents the current at each peak of the CV curve, and v denotes the corresponding scan rate. In these equations, a and b serve as empirical parameters. When b approaches 0.5, the charge/discharge process is primarily controlled by diffusion, whereas when b approaches 1, pseudocapacitance becomes the dominant mechanism [37,38].

$$i = av^b \tag{6}$$

$$\log(i) = b \times \log(v) + \log(a) \tag{7}$$

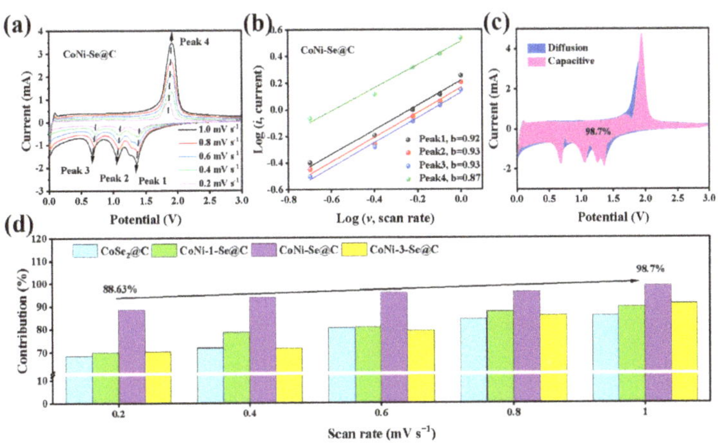

Figure 5. (**a**) CV curves at various scan rates. (**b**) The corresponding log (i) versus log (v) at each redox peak. (**c**) Capacitive contribution at 1.0 mV s^{-1}. (**d**) Contribution rates of capacitive-controlled and diffusion-controlled capacities at different scan rates.

Additionally, the contribution of pseudocapacitance can be calculated using Equation (8), where k_1v and $k_2v^{1/2}$ represent the contributions of pseudocapacitance and ion diffusion, respectively [2].

$$i(\mathrm{V}) = k_1v + k_2v^{1/2} \quad (8)$$

As shown in Figure 5c, it was clearly demonstrated that at a scan rate of 1.0 mV s^{-1}, the contribution of pseudocapacitance predominated, highlighted in pink. Figure 5d shows the contribution of pseudocapacitance of those four samples. For CoNi-Se@C, the contribution of pseudocapacitance increased from 88.63% to 98.7% as the scan rate increased from 0.2 to 1.0 mV s^{-1}. This trend indicates that as the scan rate increased, the role of surface reactions in charge storage capacity became increasingly significant.

3. Materials and Methods

3.1. Chemicals

Cobalt nitrate hexahydrate ($Co(NO_3)_2{\cdot}6H_2O$), 2-methylimidazole (2-MeIM), nickel nitrate hexahydrate ($Ni(NO_3)_2{\cdot}6H_2O$), polyacrylonitrile (PAN), and selenium powder (Se) were all purchased from Energy Chemical Co., Ltd, Shanghai, China. N,N-Dimethylformamide (DMF), methanol, and ethanol were purchased from Damao Chemical Reagent Factory, Tianjing, China.

3.2. Synthesis of ZIF-67

In a typical synthesis, 0.04 mol of $Co(NO_3)_2{\cdot}6H_2O$ was dissolved in 500 mL of methanol to prepare solution A. Simultaneously, 0.16 mol of 2-methylimidazole was dissolved in 500 mL of methanol to prepare solution B. Solution A was then gradually added to solution B, and the mixture was stirred continuously for 24 h. The resulting suspension was centrifuged, and the precipitate was washed three times with methanol. Finally, the precipitate was dried at 60 °C for 12 h to obtain ZIF-67.

3.3. Synthesis of ZIF-67@PAN

First, 0.5 g of ZIF-67 was evenly dispersed in 20 mL of DMF, followed by the addition of 0.5 g of PAN white powder. After stirring for 12 h, the solution was drawn into a syringe and added dropwise into stirred deionized water. The sample was collected by vacuum filtration and rinsed three times with deionized water. Finally, it was dried at 60 °C for 12 h to obtain ZIF-67@PAN.

3.4. Synthesis of CoNi-1-LDH@PAN, CoNi-LDH@PAN and CoNi-3-LDH@PAN

First, 0.8 g of ZIF-67@PAN was added to 400 mL of ethanol and subjected to ultrasound for uniform dispersion, referred to as solution A. Meanwhile, 0.8 g of $Ni(NO_3)_3{\cdot}6H_2O$ was dissolved in 100 mL of ethanol to prepare solution B. Solution B was then added to solution A, and the mixture was stirred before centrifuging to collect the sample. The sample was washed three times with ethanol and dried at 60 °C for 12 h to obtain CoNi-LDH@PAN. Additionally, the samples with stirring times of 1, 2, and 3 h were designated as CoNi-1-LDH@PAN, CoNi-LDH@PAN, and CoNi-3-LDH@PAN, respectively.

3.5. Synthesis of CoNi-Se@C, CoNi-1-Se@C, CoNi-3-Se@C and $CoSe_2$@C

The obtained CoNi-LDH@PAN was mixed with Se powder at a 1:1 mass ratio and placed in a porcelain boat. It was then annealed at 60 °C under an Ar/H_2 atmosphere for 6 h (at a heating rate of 2 °C min^{-1}) to obtain the CoNi-Se@C sample. Additionally, the preparation procedures for CoNi-1-Se@C, CoNi-3-Se@C, and $CoSe_2$@C followed the same method as described above, with the only difference being that the precursors used were CoNi-1-LDH@PAN, CoNi-3-LDH@PAN, and ZIF-67@PAN, respectively.

3.6. Materials Characterization

X-ray diffraction (XRD) analysis was performed on a Bruker D8 Advance (Berlin, Germany) using Cu-Kα radiation (λ = 1.5406 Å). A Phi X-tool XPS instrument from ULVAC-PHI (Kanagawa-ken, Chigasaki-shi, Japan) was used, with the standard carbon peak C1s (284.8 eV) as the calibration reference. The sample morphology was characterized by field-emission scanning electron microscopy (FESEM, Hitachi S-4800, Tokyo, Japan) and transmission electron microscopy (TEM, FEI Tecnai G2 F30, Hillsboro, OH, USA). Nitrogen adsorption–desorption isotherms were measured at 77 K using an Autosorb-iQ automatic volumetric analyzer (Quantachrome, Boynton Beach, FL, USA).

3.7. Coin Cell Assembly

CoNi-Se@C, Super P, and sodium alginate were thoroughly mixed with deionized water at a mass ratio of 8:1:1 to form a black slurry. The slurry was coated onto copper foil and dried at 60 °C for 24 h in a convection oven. The resulting electrode was then cut into 13 mm diameter discs and assembled into CR-2032 coin cells. Metallic sodium was used as the counter electrode, glass fiber (Whatman GF/D, Maidstone, UK) as the separator, and 1.0 M $NaCF_3SO_3$ in diethylene glycol dimethyl ether (DIGLYME, dodochem, Suzhou, China) was used as the electrolyte. The assembly was conducted entirely in a glovebox (Vigor-LG 2400/750TS, LTD, Suzhou, China) with oxygen and water contents below 0.1 ppm.

3.8. Electrochemical Measurements

A multi-channel battery testing system (CT2001A, LAND, Wuhan, China) was used for cycling and rate performance tests. Cyclic voltammetry (CV) and electrochemical impedance spectroscopy (EIS) were performed using an electrochemical workstation (CHI 660 electrochemical workstation, Shanghai Chi-Chi Instrumentation Co., Ltd., Shanghai, China). CV scanning voltage ranged from 0.01 to 3.0 V with a scanning rate of 0.2–1.0 mV s^{-1}, while EIS frequency ranged from 0.01 to 100,000 Hz.

4. Conclusions

In summary, carbon-coated hollow Co/Ni bimetallic selenides (HS-Co_xNi_ySe@C) were obtained by sequential coating of PAN onZIF-67, Ni^{2+} etching, and selenization. When used as anode materials for sodium-ion batteries, HS-Co_xNi_ySe@C showed improved performance and the performance of SIBs strongly dependent on the ratio of Co^{2+} to Ni^{2+}, which was ascribed to enhanced surface area and conductance of Na^+ due to the formation of the hollow structure, increased ion diffusion pathways, and the synergistic effect between Co and Ni. The excessive Ni^{2+} content caused contraction of the hollow space and condensation of the shell, which reduced the electrochemical surface area and impaired ion transport efficiency. The optimized HS-Co_xNi_ySe@C delivered a discharge capacity of 335 mAh g^{-1} after 1000 cycles at a high current density of 5.0 A g^{-1}, indicating its strong potential for commercial applications. This study not only demonstrates the superior electrochemical performance of the hollow CoNi-Se@C material but also provides innovative guidance for the design of multi-metal materials used for energy storage.

Supplementary Materials: The following supporting information can be downloaded at: https://www.mdpi.com/article/10.3390/inorganics13030096/s1, Figure S1: Raman spectra of CoNi-Se@C; Figure S2: CV curves for the first 3 cycles of (a) $CoSe_2$@C, (b) CoNi-1-Se@C, and (c) CoNi-3-Se@C at 1.0 mV s^{-1}; Figure S3: XRD pattern of CoNi-Se@C after discharged to 0.01 V; Figure S4: EIS spectra and the fitted result; Figure S5: Discharge–charge curves at various rates of (a) CoSe2@C, (b) CoNi-1-Se@C, and (c) CoNi-3-Se@C; Figure S6: TEM images of CoNi-Se@C after cycling at 5.0 A g^{-1} for 1000 cycles; Figure S7: (a,c,e) CV curves at various scan rates. (b,d,f) The corresponding log (i) versus log (v) plots at each redox peak. Among them, (a,b) correspond to $CoSe_2$@C, (c, d) CoNi-1-Se@C, and (e,f) CoNi-3-Se@C; Table S1: The atomic ratio of Co, Ni, and Se in CoNi-Se@C measured by ICP-OES; Table S2: The atomic percentages of Co, Ni, Se, N, and C in $CoSe_2$@C, CoNi-1-Se@C, CoNi-Se@C, and CoNi-3-Se@C measured by XPS. Electrochemistry performance of different anode materials for SIBs; Table S3: The potentials (V vs Na^+/Na) in CV scans derived from Figure 4a; Table S4: Electrochemical impedance spectroscopy equivalent circuit parameters table of $CoSe_2$@C, CoNi-1-Se@C, CoNi-Se@C, and CoNi-3-Se@C; Table S5: Comparison of the long cycle performance of CoNi-Se@C and metal selenides from published articles. References [24,39–46] are cited in the Supplementary Materials.

Author Contributions: Conceptualization, C.W., W.S. and X.K.; methodology, W.S.; formal analysis, C.W. and W.S.; investigation, W.S. and C.W.; resources, X.K. and C.W.; data curation, W.S.; writing—original draft preparation, C.W. and W.S.; writing—review and editing, C.W. and X.K.; supervision, X.K. All authors have read and agreed to the published version of the manuscript.

Funding: This research received no external funding.

Institutional Review Board Statement: Not applicable.

Informed Consent Statement: Not applicable.

Data Availability Statement: The data presented in this study are available on request from the corresponding author.

Conflicts of Interest: Author Chao Wang was employed by the company China Southern Power Grid Technology Co., Ltd. The remaining authors declare that the research was conducted in the absence of any commercial or financial relationships that could be construed as a potential conflict of interest.

References

1. Men, S.; Zheng, H.; Ma, D.; Huang, X.; Kang, X. Unraveling the stabilization mechanism of solid electrolyte interface on ZnSe by rGO in sodium ion battery. *J. Energy Chem.* **2021**, *54*, 124–130. [CrossRef]
2. Yang, J.; Gao, H.; Men, S.; Shi, Z.; Lin, Z.; Kang, X.; Chen, S. $CoSe_2$ Nanoparticles Encapsulated by N-Doped Carbon Framework Intertwined with Carbon Nanotubes: High-Performance Dual-Role Anode Materials for Both Li- and Na-Ion Batteries. *Adv. Sci.* **2018**, *5*, 1800763. [CrossRef] [PubMed]
3. Lin, J.; Zhou, Q.; Liao, Z.; Chen, Y.; Liu, Y.; Liu, Q.; Xiong, X. Steric Hindrance Engineering to Modulate the Closed Pores Formation of Polymer-Derived Hard Carbon for High-Performance Sodium-Ion Batteries. *Angew. Chem. Int. Ed.* **2024**, *63*, e202409906. [CrossRef] [PubMed]
4. Hwang, J.-Y.; Myung, S.-T.; Sun, Y.-K. Sodium-ion batteries: Present and future. *Chem. Soc. Rev.* **2017**, *46*, 3529–3614. [CrossRef]
5. Zhao, S.; Che, H.; Chen, S.; Tao, H.; Liao, J.; Liao, X.-Z.; Ma, Z.-F. Research Progress on the Solid Electrolyte of Solid-State Sodium-Ion Batteries. *Electrochem. Energy Rev.* **2024**, *7*, 3. [CrossRef]
6. Liu, L.; Xu, S.; Tang, F.; Wu, M.; Yang, W.; Xu, C.; Rui, X. Controllable fabrication of vanadium selenium nanosheets for a high-performance Na-ion battery anode. *Chem. Commun.* **2023**, *59*, 11365–11368. [CrossRef]
7. Men, S.; Lin, J.; Zhou, Y.; Kang, X. N-doped porous carbon wrapped $FeSe_2$ nanoframework prepared by spray drying: A potential large-scale production technique for high-performance anode materials of sodium ion batteries. *J. Power Sources* **2021**, *485*, 229310. [CrossRef]
8. Weng, J.; Zou, D.; Yuan, W.; Zhou, P.; Ding, M.; Zhou, J.; Cong, H.; Cheng, F. Bimetallic selenide heterostructure with directional built-in electric-field confined in N-doped carbon nanofibers for superior sodium storage with ultralong lifespan. *J. Energy Chem.* **2024**, *91*, 407–416. [CrossRef]
9. Zhang, X.; Zhang, Z.; Xu, S.; Xu, C.; Rui, X. Advanced Vanadium Oxides for Sodium-Ion Batteries. *Adv. Funct. Mater.* **2023**, *33*, 2306055. [CrossRef]
10. Ju, X.; Hou, X.; Liu, Z.; Du, L.; Zhang, L.; Xie, T.; Paillard, E.; Wang, T.; Winter, M.; Li, J. Revealing the Effect of High Ni Content in Li-Rich Cathode Materials: Mitigating Voltage Decay or Increasing Intrinsic Reactivity. *Small* **2023**, *19*, 2207328. [CrossRef]
11. Gao, H.; Ning, S.; Zhou, Y.; Men, S.; Kang, X. Polyacrylonitrile-induced formation of core-shell carbon nanocages: Enhanced redox kinetics towards polysulfides by confined catalysis in Li-S batteries. *Chem. Eng. J.* **2021**, *408*, 127323. [CrossRef]
12. He, P.; Yu, X.-Y.; Lou, X.W. Carbon-Incorporated Nickel–Cobalt Mixed Metal Phosphide Nanoboxes with Enhanced Electrocatalytic Activity for Oxygen Evolution. *Angew. Chem. Int. Ed.* **2017**, *56*, 3897–3900. [CrossRef]
13. Wang, F.; Jiang, Z.; Zhang, Y.; Zhang, Y.; Li, J.; Wang, H.; Jiang, Y.; Xing, G.; Liu, H.; Tang, Y. Revitalizing sodium-ion batteries via controllable microstructures and advanced electrolytes for hard carbon. *eScience* **2024**, *4*, 100181. [CrossRef]

14. Hou, C.-C.; Zou, L.; Xu, Q. A Hydrangea-Like Superstructure of Open Carbon Cages with Hierarchical Porosity and Highly Active Metal Sites. *Adv. Mater.* **2019**, *31*, 1904689. [CrossRef] [PubMed]
15. Yang, Y.; Zhang, W.; Xiao, Y.; Shi, Z.; Cao, X.; Tang, Y.; Gao, Q. $CoNiSe_2$ heteronanorods decorated with layered-double-hydroxides for efficient hydrogen evolution. *Appl. Catal. B Environ.* **2019**, *242*, 132–139. [CrossRef]
16. Lu, Q.; Zhou, T.; Zi, B.; Zhao, J.; Li, D.; Chen, M.; Sun, H.; Zhang, J.; Zhang, Y.; Liu, Q. Dual Supports by Cation Vacancies and Surface Optimization for $CoNiSe_2$-Based Hybrid Supercapacitors with High Energy Density. *ACS Energy Lett.* **2023**, *8*, 3420–3429. [CrossRef]
17. Shi, X.; Wang, H.; Ji, S.; Linkov, V.; Liu, F.; Wang, R. $CoNiSe_2$ nanorods directly grown on Ni foam as advanced cathodes for asymmetric supercapacitors. *Chem. Eng. J.* **2019**, *364*, 320–327. [CrossRef]
18. Zhou, P.; Zhang, M.; Wang, L.; Huang, Q.; Su, Z.; Xu, P.; Zou, R.; Wang, X.; Zeng, C.; Ba, K. MOFs-Derived Flower-Like Hierarchically Porous Zn-Mn-Se/C Composite for Extraordinary Rate Performance and Durable Anode of Sodium-Ion and Potassium-Ion Batteries. *Small* **2022**, *18*, 2203964. [CrossRef]
19. Feng, J.; Luo, S.-H.; Yan, S.-X.; Zhan, Y.; Wang, Q.; Zhang, Y.-H.; Liu, X.; Chang, L.-J. Rational Design of Yolk–Shell Zn—Co—Se@N-Doped Dual Carbon Architectures as Long-Life and High-Rate Anodes for Half/Full Na-Ion Batteries. *Small* **2021**, *17*, 2101887. [CrossRef]
20. Ma, S.; Yan, W.; Dong, Y.; Su, Y.; Ma, L.; Li, Y.; Fang, Y.; Wang, B.; Wu, S.; Liu, C.; et al. Recent advances in carbon-based anodes for high-performance sodium-ion batteries: Mechanism, modification and characterizations. *Mater. Today* **2024**, *75*, 334–358. [CrossRef]
21. Zhang, Y.-H.; Wang, J.-L.; Yan, H.-Y.; Zhao, J.-T.; Wang, D.-D.; Luo, S.-H.; Wang, Q.; Liu, X. Short rod-like $NiCoSe_2$ binary-metal selenide nanomaterials of carbon-coated as high-performance anode for sodium-ion batteries. *Ionics* **2023**, *29*, 3505–3515. [CrossRef]
22. Lin, J.; Zhou, Y.; Wen, J.; Si, W.; Gao, H.; Wang, G.; Kang, X. Pyrrole derivatives as interlayer modifier of Li-S batteries: Modulation of electrochemical performance by molecular perturbation. *J. Energy Chem.* **2022**, *75*, 164–172. [CrossRef]
23. Zhou, P.; Wang, L.; Zhang, M.; Wu, F.; Huang, Q.; Su, Z.; Xu, P.; Liao, M.; Hu, Y.; Lin, X. Controllable MOF-Derived Hierarchical Hollow $CoNiSe_2$ with Enhanced Mechanics and Kinetics for Extraordinary Rate Performance and Durable Anode of Sodium-Ion Batteries. *ACS Appl. Energy Mater.* **2023**, *6*, 7129–7137. [CrossRef]
24. Zhang, L.; Xie, P.; Zhang, X.; Zhu, B.; Liu, T.; Yu, J. Facile synthesis of $NiCoSe_2$@carbon anode for high-performance sodium-ion batteries. *J. Colloid Interf. Sci.* **2024**, *662*, 1075–1085. [CrossRef]
25. Xie, H.; Zhang, W.; Wang, C.; Zhao, S.; Hao, Z.; Huang, X.; Miao, K.; Kang, X. MOF-Derived Fe_2CoSe_4@NC and Fe2NiSe4@NC Composite Anode Materials towards High-Performance Na-Ion Storage. *Inorganics* **2024**, *12*, 165. [CrossRef]
26. Ou, X.; Li, J.; Zheng, F.H.; Wu, P.; Pan, Q.C.; Xiong, X.H.; Yang, C.H.; Liu, M.L. X-ray diffraction characterization of NiSe as a promising anode material for sodium ion batteries. *J. Power Sources* **2017**, *343*, 483–491. [CrossRef]
27. Santoni, F.; De Angelis, A.; Moschitta, A.; Carbone, P.; Galeotti, M.; Cinà, L.; Giammanco, C.; Di Carlo, A. A guide to equivalent circuit fitting for impedance analysis and battery state estimation. *J. Energy Storage* **2024**, *82*, 110389. [CrossRef]
28. Bisquert, J.; Garcia-Belmonte, G.; Bueno, P.; Longo, E.; Bulhões, L.O.S. Impedance of constant phase element (CPE)-blocked diffusion in film electrodes. *J. Electroanal. Chem.* **1998**, *452*, 229–234. [CrossRef]
29. Kemp, N.T. A Tutorial on Electrochemical Impedance Spectroscopy and Nanogap Electrodes for Biosensing Applications. *IEEE Sens. J.* **2021**, *21*, 22232–22245. [CrossRef]
30. Ali, Z.; Asif, M.; Huang, X.; Tang, T.; Hou, Y. Hierarchically Porous Fe_2CoSe_4 Binary-Metal Selenide for Extraordinary Rate Performance and Durable Anode of Sodium-Ion Batteries. *Adv. Mater.* **2018**, *30*, 1802745. [CrossRef]
31. Muralee Gopi, C.V.V.; Reddy, A.E.; Kim, H.-J. Wearable superhigh energy density supercapacitors using a hierarchical ternary metal selenide composite of $CoNiSe_2$ microspheres decorated with $CoFe_2Se_4$ nanorods. *J. Mater. Chem. A* **2018**, *6*, 7439–7448. [CrossRef]
32. Vivekanantha, M.; Sundhar Arul Saravanan, R.; Kumar Nayak, P.; Prakash, R.; Kamala Bharathi, K. Synergistic-effect of high Ni content and Na dopant towards developing a highly stable Li-Rich cathode in Li-ion batteries. *Chem. Eng. J.* **2022**, *444*, 136503. [CrossRef]
33. Piontek, S.; Andronescu, C.; Zaichenko, A.; Konkena, B.; Junge Puring, K.; Marler, B.; Antoni, H.; Sinev, I.; Muhler, M.; Mollenhauer, D.; et al. Influence of the Fe:Ni Ratio and Reaction Temperature on the Efficiency of $(Fe_xNi_{1-x})_9S_8$ Electrocatalysts Applied in the Hydrogen Evolution Reaction. *ACS Catal.* **2018**, *8*, 987–996. [CrossRef]
34. Xie, X.; Ma, X.; Yin, Z.; Tong, H.; Jiang, H.; Ding, Z.; Zhou, L. Bimetallic heterojunction of CuSe/ZnSe@Nitrogen-doped carbon with modified band structures for fast sodium-ion storage. *Chem. Eng. J.* **2022**, *446*, 137366. [CrossRef]
35. Wang, D.; Chao, Y.; Guo, K.; Wang, Z.; Yang, M.; Zhu, J.; Cui, X.; Xu, Q. Engineering Metal Electron Spin Polarization to Regulate p-Band Center of Se for Enhanced Sodium-Ion Storage. *Adv. Funct. Mater.* **2024**, *34*, 2405642. [CrossRef]

36. Gupta, S.; Patel, N.; Fernandes, R.; Kadrekar, R.; Dashora, A.; Yadav, A.K.; Bhattacharyya, D.; Jha, S.N.; Miotello, A.; Kothari, D.C. Co–Ni–B nanocatalyst for efficient hydrogen evolution reaction in wide pH range. *Appl. Catal. B Environ.* **2016**, *192*, 126–133. [CrossRef]
37. Xia, Y.; Yang, T.; Wang, Z.; Mao, T.; Hong, Z.; Han, J.; Peng, D.-L.; Yue, G. Van der Waals Forces between S and P Ions at the CoP-C@MoS_2/C Heterointerface with Enhanced Lithium/Sodium Storage. *Adv. Funct. Mater.* **2023**, *33*, 2302830. [CrossRef]
38. Xu, J.; Xie, L.; Niu, Y.; Chen, H.; Zhang, Y.; Jiang, Y.; Han, Q.; Qiu, X.; Miao, Y.; Zhu, L.; et al. Nitrogen-doped carbon decorated 3D $NiCoSe_2$ micro-flowers as high-performance anode materials for lithium-ion batteries. *Phys. Chem. Chem. Phys.* **2023**, *25*, 11530–11544. [CrossRef]
39. Gao, L.; Cao, M.; Zhang, C.; Li, J.; Zhu, X.; Guo, X.; Toktarbay, Z. Zinc selenide/cobalt selenide in nitrogen-doped carbon frameworks as anode materials for high-performance sodium-ion hybrid capacitors. *Adv. Compos. Hybrid Mater.* **2024**, *7*, 144. [CrossRef]
40. Xu, L.-J.; Wang, X.-J.; Tang, G.-Y.; Zhu, B.-C.; Yu, J.-G.; Zhang, L.-Y.; Liu, T. NiSe nanoparticles anchored on hollow carbon nanofibers with enhanced rate capability and prolonged cycling durability for sodium-ion batteries. *Rare Met.* **2025**, *44*, 185–194. [CrossRef]
41. Wang, L.; Huang, F.; Song, X.; Li, J.; Zhu, G.; Jin, Z.; Dai, Z. Rational Design of Quasi-1D Multicore–Shell MnSe@N-Doped Carbon Nanorods as High-Performance Anode Material for Sodium-Ion Batteries. *Nano Lett.* **2024**, *24*, 11349–11357. [CrossRef] [PubMed]
42. Chong, S.; Li, T.; Qiao, S.; Yang, Y.-C.; Liu, Z.; Yang, J.; Tuan, H.-Y.; Cao, G.; Huang, W. Boosting Manganese Selenide Anode for Superior Sodium-Ion Storage via Triggering $\alpha \rightarrow \beta$ Phase Transition. *ACS Nano* **2024**, *18*, 3801–3813. [CrossRef]
43. Shi, L.; Zhao, F.; Tang, Y.; Liu, R.; Pang, J.; Cheng, G.; Hu, M.; Ding, J. Hollow porous $Co_{0.85}Se$/$MoSe_2$@MXene heterostructured anode for sodium-ion hybrid capacitors. *Chem. Eng. J.* **2024**, *500*, 157001. [CrossRef]
44. Ren, M.; Zang, H.; Cao, S.; Liu, W.; Li, M.; Yao, J.; Cai, F.; Cui, J.; Wang, Y. Fe3Se4 decorating carbon nanotubes with superior sodium storage performance for sodium-ion batteries. *J. Energy Storage* **2024**, *81*, 110486. [CrossRef]
45. Xu, C.; Yang, J.; Chen, K.; Ma, G.; Wang, Y.; Li, Z.; Zhou, Z.; Wu, Z.; Che, S.; Ding, C.; et al. CoSe2-Modified multidimensional porous carbon frameworks as high-Performance anode for fast-Charging sodium-Ion batteries. *Chem. Eng. J.* **2024**, *497*, 154875. [CrossRef]
46. Buğday, N.; Wang, H.; Hong, N.; Zhang, B.; Deng, W.; Zou, G.; Hou, H.; Yaşar, S.; Ji, X. Fabrication of a Stable and Highly Effective Anode Material for Li-Ion/Na-Ion Batteries Utilizing ZIF-12. *Small* **2024**, *20*, 2403736. [CrossRef]

inorganics

MDPI

Article

Unveiling the Electrocatalytic Performances of the Pd-MoS_2 Catalyst for Methanol-Mediated Overall Water Splitting

Aviraj M. Teli [1,†], Sagar M. Mane [2,†], Rajneesh Kumar Mishra [3,*], Wookhee Jeon [4] and Jae Cheol Shin [1,*]

1 Division of Electronics and Electrical Engineering, Dongguk University-Seoul, Seoul 04620, Republic of Korea; avteli.teli@gmail.com
2 Department of Fiber System Engineering, Yeungnam University, 280 Dehak-Ro, Gyeongsan 38541, Gyeongbuk, Republic of Korea; manesagar99@gmail.com
3 Department of Physics, Yeungnam University, Gyeongsan 38541, Gyeongbuk, Republic of Korea
4 Department of Semiconductor, Convergence Engineering, Sungkyunkwan University, Suwon 16419, Gyeonggi, Republic of Korea; wookie92@skku.edu
* Correspondence: rajneeshmishra08@gmail.com (R.K.M.); jcshin@dgu.ac.kr (J.C.S.)
† These authors contributed equally to this work.

Abstract: Herein, this work elucidates the synthesis of the Pd-MoS_2 catalyst for application in methanol-mediated overall water splitting. The scanning electron microscope (SEM) and transmission electron microscope (TEM) pictures offer an exciting nanostructured shape of the Pd-MoS_2, depicting a high surface area. Further, high-resolution TEM (HRTEM) pictures confirm the lattice plane (100), lattice spacing (0.26 nm), and hexagonal crystal structure of the Pd-MoS_2. Moreover, high-angle annular dark-field (HAADF) images and related color maps disclose the Mo, S, and Pd elements of the Pd-MoS_2. The Pd-MoS_2 catalyst exhibits lower overpotentials of 224.6 mV [methanol-mediated hydrogen evolution reaction (MM-HER)] at -10 mA cm^{-2} and 133 mV [methanol-mediated oxygen evolution reaction (MM-OER)] at 10 mA cm^{-2}. Further, the Pd-MoS_2 illustrates noteworthy stability for 15.5 h for MM-HER and 18 h for MM-OER by chronopotentiometry test. Excitingly, the Pd-MoS_2||Pd-MoS_2 cell reveals a small potential of 1.581 V compared to the MoS_2||MoS_2 cell (1.648 V) in methanol-mediated overall water splitting. In addition, the Pd-MoS_2||Pd-MoS_2 combination reveals brilliant durability over 18 h at 10 mA cm^{-2}.

Keywords: Pd-MoS_2; methanol-mediated hydrogen evolution reaction; methanol-mediated oxygen evolution reaction (MM-OER); methanol-mediated overall water splitting; outstanding stability

1. Introduction

Currently, the rising global population and fast industrialization have led to record energy consumption, blinking a keen interest in future energy sources [1]. Fascinatingly, fossil fuels are a main energy source for industry, irrigation, and transportation, contributing to critical environmental challenges, such as pollution and greenhouse gas emissions [2]. Therefore, these issues highlight the necessity to transition toward sustainable, green, and clean energy sources [3]. Remarkably, hydrogen is a hopeful clean energy source that has the potential to transform energy systems, offering a sustainable substitution for fossil fuels due to its high energy, which could considerably reach carbon neutrality and eco-friendly combustion and mitigate climate change impacts [4]. Intriguingly, electrochemical water splitting is a scalable, effective, and simple technique for hydrogen energy production involving the decomposition of water (H_2O) into oxygen (O_2) and hydrogen (H_2) [5]. Attractively, the overpotentials and Tafel slopes mainly affect the hydrogen and oxygen productivity during the electrochemical water splitting, which is very important for hydrogen and oxygen generation [6]. Recently, methanol-mediated hydrogen evolution, methanol-mediated oxygen evolution reaction (MM-OER), and methanol-mediated overall water splitting have extended significant attention due to more efficient alternatives to traditional water splitting [7]. Combining methanol (CH_3OH) into the KOH electrolyte during the electrochemical water splitting process can lower the energy barrier for hydrogen energy production [8]. Excitingly, it can considerably decrease the overpotential, thereby reducing the energy input for hydrogen energy production. Interestingly, this approach opens new pathways for using methanol, which can be produced sustainably from biomass or captured carbon dioxide, further enhancing the environmental benefits [9]. Furthermore, methanol-mediated processes present a captivating alternative that could improve efficiency and stability for sustainable energy generation. Various Pt-based nanomaterials are currently used for hydrogen evolution reactions [10,11]. However, ruthenium and iridium-based nanomaterials are utilized for oxygen evolution reactions [12,13]. Amusingly, their expensive nature and scarcity in the crust of the earth limit their possible usage as catalysts for sustainable and commercial hydrogen generation. Therefore, it is necessary to develop cutting-edge, low-cost catalysts operating at lower overpotentials while maintaining high durability, a prerequisite for a hydrogen energy-based economy, addressing energy demands and environmental sustainability for future generations.

Transition metal dichalcogenides (TMDs) illustrate a family of two-dimensional (2D) layered structures through the well-known MX_2 formula, in which X is a chalcogen (such as S, Se, and Te) and M is a transition metal (such as Mo, W, and Ti) [14]. In addition, TMDs reveal exclusive electronic, optical, and magnetic characteristics due to their typical 2D nature. The 2D layers of the TMDs are attached by weak van der Waals forces [15]. MoS_2 comprises hexagonally packed Mo atoms between two layers of S atoms [16]. Additionally, TMDs have gained significant consideration for their vibrant applications in several fields, such as electronics, optoelectronics, and catalysis, making them a crucial theme of research in materials science and nanotechnology [17,18]. Therefore, monolayer MoS_2 endows a direct optical band gap of approximately 1.8 eV, making it suitable for optoelectronic device fabrication [19]. Additionally, MoS_2 exhibits excellent mechanical flexibility, high chemical stability, and strong spin-orbit [20]. The MoS_2's fashionable characteristics offer its research in various applications, such as transistors, sensors, and flexible electronic devices, because of its high carrier mobility and scalability [21]. Also, MoS_2 is demonstrated as an exciting catalyst for hydrogen generation reactions and other electrochemical processes [22]. Additionally, MoS_2's optical properties are harnessed in photodetectors and light-emitting devices [23,24]. The MoS_2 is synthesized through various physical and chemical methods, each offering distinct advantages. Further, the high-quality MoS_2 monolayers are produced using the mechanical exfoliation method by peeling off layers from bulk crystals [25]. In addition, the MoS_2 thin films are developed on many substrates by chemical vapor deposition (CVD), which allows for large-area, scalable, and uniform thin films for various commercial applications [26]. Moreover, the Mo and S precursor solution react under controlled temperature and pressure, yielding tunable morphology and properties of the MoS_2 using the hydrothermal and solvothermal methods [27,28]. Further, the MoS_2 can be found in various morphologies depending on the synthesis method and conditions, including nanosheets [29], nanorods [30], nanotubes [31], nanowires [32], nanoflowers [33], and quantum dots [34]. Fascinatingly, two-dimensional MoS_2 nanosheets are the most common form, offering large surface areas and high aspect ratios. On the other hand, three-dimensional MoS_2 nanotubes and nanoflowers deliver unique morphological topographies, which can enhance electrocatalytic activities. Interestingly, zero-dimensional MoS_2 quantum dots are generally suitable in optoelectronic applications due to their size-tunable optical and electronic properties [35]. Moreover, the MoS_2 nanomaterial is used in batteries, fuel cells, and supercapacitors because of its high capacity and stability [36–38]. Interestingly, MoS_2 is also used in water purification and sensing in ecological applications because of its high reactivity and large surface area [39,40]. Fascinatingly, the MoS_2 enables its use in drug delivery and cancer therapy due to its biocompatibility and photothermal properties [41,42]. However, the 2D layered MoS_2 emphasizes its potential applications in advanced technologies due to its multifunctional nature. Therefore, tuning the structural and morphological properties of the MoS_2 by doping or constructing heterostructures has been significantly studied, which offers enhanced performance in different applications. Fascinatingly, Pd doping in the MoS_2 can alter the electronic properties and create more active sites, facilitating greater adsorption and dissociation of water electrolysis during methanol-mediated overall water splitting, which makes it a highly effective catalyst for sustainable hydrogen energy generation.

In this paper, we studied the methanol-mediated hydrogen evolution reaction (MM-HER), methanol-mediated oxygen evolution reaction (MM-OER), and methanol-mediated overall water splitting (MM-OWS) of the Pd-doped MoS_2 (Pd-MoS_2) catalyst, which is a promising 2D nanomaterial for sustainable hydrogen generation through electrocatalytic water splitting. Interestingly, the presence of methanol with the KOH electrolyte works as a co-reactant, significantly boosting the electrocatalytic activities of the methanol-mediated overall water splitting, facilitating the adsorption process and also supporting to alleviate the transitional active sites on the surface of the Pd-MoS_2 electrocatalyst, thereby enhancing the rate of reaction kinetics. Fascinatingly, it is observed that the Pd-MoS_2 catalyst illustrates outstanding catalytic performances and excellent stability. The Pd-MoS_2||Pd-MoS_2 cell elucidates a lower potential of 1.581 V and tremendous durability throughout 18 h at 10 mA cm^{-2}. Therefore, it is concluded that the Pd-MoS_2 catalyst can be an excellent nanomaterial for hydrogen energy generation for commercial applications.

2. Results and Discussion

Figure 1a elucidates the XRD patterns of the Pd-doped MoS_2 and MoS_2. It is observed that the XRD spectra illustrate the lattice planes (002), (101), and (110) to MoS_2 and Pd-doped MoS_2. Therefore, it is concluded that the XRD patterns suggest the fruitful preparation of the MoS_2 and Pd-doped MoS_2, which is well matched with the JCPDS card No. 37-1492 [43]. Further, an XPS study was executed to examine the chemical composition of the Pd-doped MoS_2. Figure 1b–d elucidate the XPS patterns of the Pd-doped MoS_2 nanostructure. Figure 1b depicts the XPS patterns of Mo 3d peaks, which illustrates the various peaks such as Mo^{4+} $3d_{5/2}$ (228.72 eV), Mo^{4+} $3d_{3/2}$ (231.85 eV), and Mo^{6+} $3d_{3/2}$ (234.88 eV). Figure 1c portrays the XPS spectrum of the S 2p peaks, revealing S^{2-} $2p_{5/2}$ at 161.45 eV and S^{2-} $2p_{1/2}$ at 162.83 eV. In addition, the observed Pd 3d peaks are Pd^0 $3d_{5/2}$ (335.5 eV), Pd^{2+} $3d_{5/2}$ (337.7 eV), Pd^0 $3d_{3/2}$ (340.56 eV), and Pd^{2+} $3d_{3/2}$ (342.2 eV). Similar reports on MoS_2 have been discussed in the literature, which justifies our XPS results [44–46]. Therefore, it is determined that the XPS patterns approve the successful synthesis of the Pd-doped MoS_2.

Figure 2a,b reveal the SEM of the MoS_2 at 2 μm and 1 μm scales. It shows the layered sphere-like morphology of the MoS_2. Figure 2c,d display the SEM of the Pd-doped MoS_2 at different locations and magnifications. The SEM pictures in Figure 2c,d provide detailed insights into the morphology of Pd-doped MoS_2 nanolayers grown on Ni foam. Figure 2c elucidates the SEM image at a 5 μm scale to explore the large view of the Pd-MoS_2. Further, Figure 2d depicts the SEM image at a 2 μm scale of the Pd-MoS_2. It is observed that the Pd-MoS_2 deposited on the Ni-foam shows very thin layers. Interestingly, Pd doping may significantly influence the crystallinity, uniformity, and fine-edge nanolayer of the Pd-MoS_2.

Moreover, various features such as growth conditions, synthesis temperature, time, doping concentration, and annealing process also play vital roles in shaping the morphology of the MoS_2 [47,48]. Amazingly, it is apparent from SEM morphologies that the Pd-MoS_2 offers a high surface area and wrinkled nanolayers, which can be beneficial for methanol-mediated hydrogen, oxygen evolution, and methanol-mediated overall water splitting. Likewise, the shape of the Pd-MoS_2 and MoS_2 are also discussed using the TEM. Figure 3a–c divulge the TEM morphologies of Pd-MoS_2 at 50 nm, 10 nm, and 5 nm scale bars, respectively. Figure 3d–f unveil the TEM pictures of Pd-MoS_2 at 100 nm, 50 nm, and 10 nm scale bars, respectively. Figure 3d discloses the TEM pictures at a 100 nm scale bar to discover the morphology and layer shape of the Pd-MoS_2. Figure 3e,f illustrate the even higher magnification TEM images at 50 nm and 10 nm scales to check the layer bending and layers overlapping on each other. Interestingly, it is also observed from TEM that the Pd-MoS_2 nanolayers are stacked or slightly exfoliated, which is vital for applications depending on surface interactions, such as methanol-mediated overall water splitting. Interestingly, these TEM images at different scales collectively enhance the understanding of the structural properties of the Pd-MoS_2, indicating the successful synthesis of the Pd-MoS_2 nanomaterial and its potential applicability in methanol-mediated oxygen, hydrogen evolution, and methanol-mediated overall water splitting.

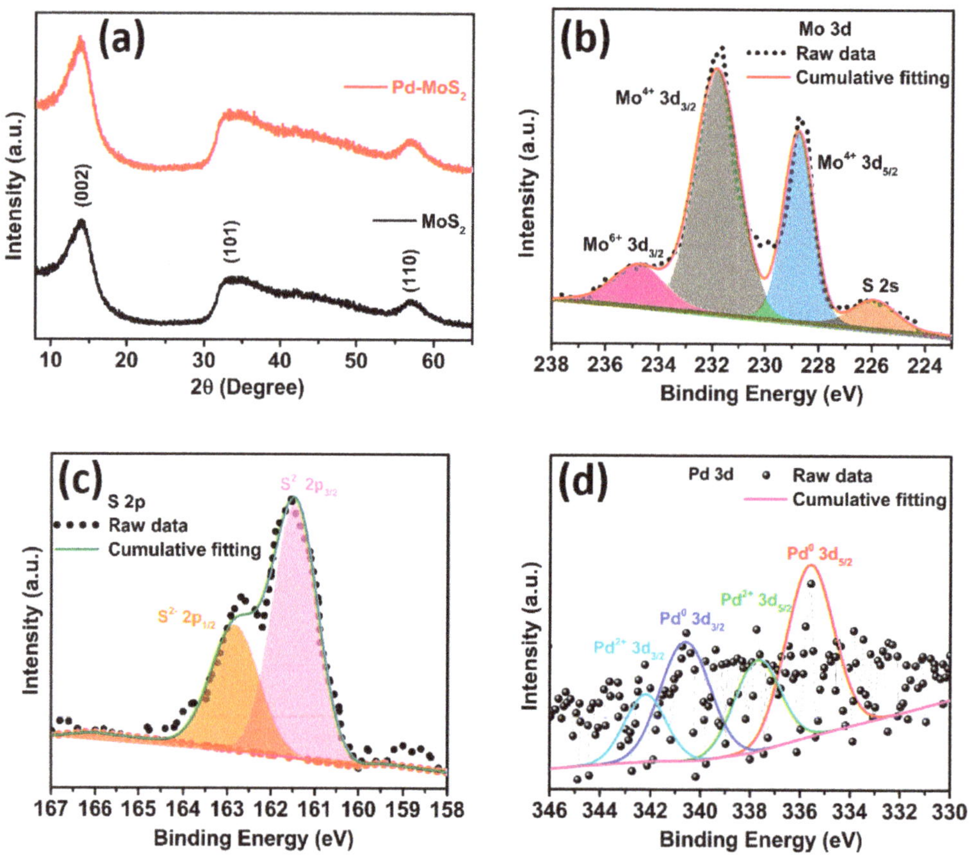

Figure 1. (**a**) XRD results of MoS_2 and Pd-MoS_2. XPS results of the (**b**) Mo 3d, (**c**) S 2p, and (**d**) Pd 3d peaks of Pd-MoS_2.

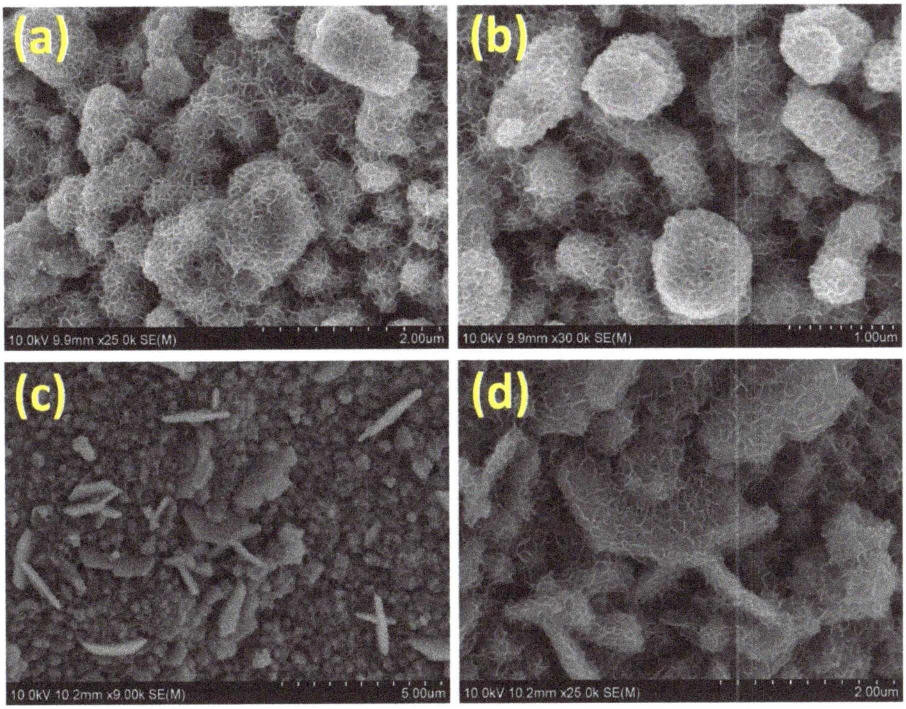

Figure 2. SEM pictures of the (**a**,**b**) MoS_2 and (**c**,**d**) Pd-MoS_2.

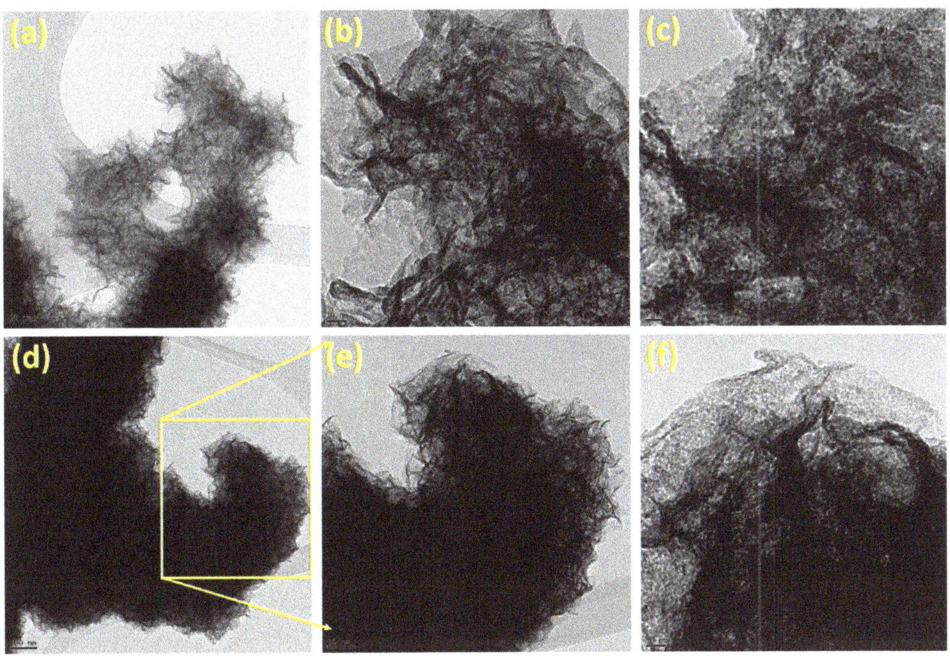

Figure 3. TEM pictures of the MoS_2 at a scale bar of (**a**) 50 nm, (**b**) 10 nm, and (**c**) 5 nm. TEM pictures of the Pd-MoS_2 at a scale bar of (**d**) 100 nm, (**e**) 50 nm, and (**f**) 10 nm.

In addition, the high-resolution transmission electron microscope (HRTEM) pictures and fast Fourier transform (FFT) pictures deliver a comprehensive indulgence of the structural features of Pd-MoS_2. Figure 4a–e unveil the HRTEM pictures, magnified HRTEM images, and FFT patterns of the Pd-MoS_2. Figure 4a portrays the HRTEM images of Pd-MoS_2, illustrating various crystallites with lattice fringes. Figure 4b,d describe the enlarged portion of the HRTEM image to measure and discuss the lattice spacings of the Pd-MoS_2. It is seen from HRTEM images that the lattice spacing of Pd-MoS_2 is 0.26 nm. Furthermore, Figure 4c,e define the FFT pattern of Pd-MoS_2 from the selected area, as revealed in Figure 4b,d. FFT patterns illustrate the lattice planes (100) of the Pd-MoS_2. Interestingly, the accuracy and uniqueness of the lattice fringes and FFT patterns suggest a high-quality synthesis of the Pd-MoS_2. Remarkably, the HRTEM and FFT results of the Pd-MoS_2 show the successful synthesis of the hexagonal crystal phase and are well coordinated with the JCPDS card no. 37-1492 [49]. In addition, we studied the elemental analysis of Pd-MoS_2 using elemental mapping. Figure 5 exhibits the HAADF result and analogous color maps of the Pd-MoS_2, which shows the elemental mappings corresponding to Mo, S, and Pd elements, providing a spatial distribution of these elements of the Pd-MoS_2. Combining the HAADF image and the elemental mappings offers compelling evidence that Pd-MoS_2 was successfully synthesized via the solvothermal method. Figure 5a designates the high-angle annular dark field (HAADF) outcome of Pd-MoS_2. Figure 5b,c portray the elementals of Mo and S of the Pd-MoS_2 from the HAADF image (Figure 5a). The Pd element mapping is vital for confirming the doping of Pd into the Pd-MoS_2 structure, as exposed in Figure 5d.

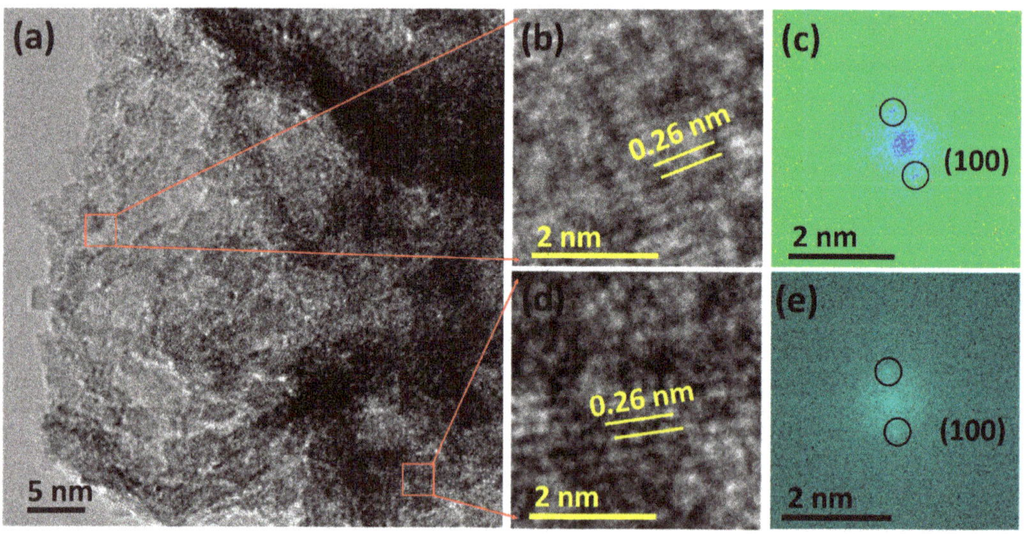

Figure 4. (**a**) HRTEM picture, (**b**,**d**) HRTEM pictures of the selected locations, and the (**c**,**e**) FFT images of the Pd-MoS_2.

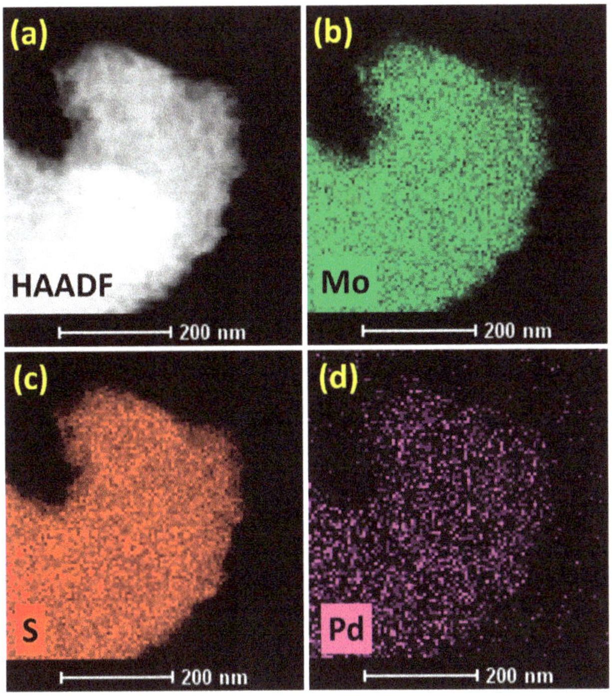

Figure 5. (**a**) HAADF picture and color maps of (**b**) Mo, (**c**) S, and (**d**) Pd of Pd-MoS_2.

The methanol-mediated oxygen evolution reaction (MM-OER) is a significant electrochemical procedure for fuel cells. Figure 6 illustrates the MM-OER of the MoS_2 and Pd-MoS_2 electrocatalysts at 5 mV s^{-1} in 1 M methanol + 1 M KOH. Figure 6a shows the LSV results of the Pd-MoS_2 and MoS_2 electrocatalysts at 5 mV s^{-1} for the MM-OER. It is seen that the Pd-MoS_2 electrocatalyst unveils superior electrocatalytic performance for the MM-OER compared to the pure MoS_2 electrocatalyst. Furthermore, the current density defines the maximum current achieved during the MM-OER for the MoS_2 and Pd-MoS_2 catalysts. In addition, higher peak current densities designate higher electrocatalytic MM-OER activities, signifying that the Pd-MoS_2 electrocatalyst can facilitate a better MM-OER rate than the pure MoS_2 electrocatalyst. Further, the inset in Figure 6a illustrates the low-scale results of LSV for Pd-MoS_2 and MoS_2 catalysts for MM-OER to elucidate the trends near the 10 mA cm^{-2}. Figure 6b displays the overpotentials of MoS_2 and Pd-MoS_2 catalysts for MM-OER at 10 mA cm^{-2}. The observed overpotentials of the Pd-MoS_2 and MoS_2 are 133 mV and 160 mV at 10 mA cm^{-2}, respectively. Stimulatingly, palladium (Pd) in Pd-MoS_2 is recognized for its brilliant electrocatalytic properties in MM-OER, which can enable the adsorption and activation of methanol and other intermediates more effectively than the MoS_2 catalyst [50]. Figure 6c exhibits the Tafel curves of MoS_2 and Pd-MoS_2 catalysts for MM-OER. The Pd-MoS_2 and MoS_2 electrocatalysts divulge the Tafel slopes of 112 mV dec^{-1} and 148 mV dec^{-1}, respectively. Captivatingly, the Tafel slope delivers insight into the rate-determining process and the efficiency of Pd-MoS_2 and MoS_2 catalysts in promoting the kinetics of MM-OER. Interestingly, the Pd-MoS_2 is a more efficient catalyst for MM-OER than pure MoS_2, supported by its lower Tafel slope, as revealed in Figure 6c. In addition, the low Tafel slope of the Pd-MoS_2 catalyst elucidates faster MM-OER reaction kinetics and a more effective electrocatalytic process, attributing to the advantageous effects of Pd doping. Further, Figure 6d displays the LSV plots of the Pd-MoS_2 at 5 mV s^{-1}, which depicts 157 mV, and 133 mV overpotential without and with methanol-mediated OER at 10 mA cm^{-2}, respectively. Therefore, it is concluded that methanol played a decisive role in the reduction in overpotentials during methanol-mediated OER. Figure 6e interprets the EIS plots of the MoS_2 and Pd-MoS_2 at respective overpotentials of MM-OER, which depicts the small series and charge-transfer resistance of the Pd-MoS_2 compared with MoS_2. Figure 6f unveils the stability results of the Pd-MoS_2 catalyst during MM-OER throughout 18 h at 10 mA cm^{-2}. Further, it is perceived that the Pd-MoS_2 catalyst performs consistently better for 18 h, suggesting its resilience and competence for long-term use without degradation in MM-OER application at 10 mA cm^{-2}.

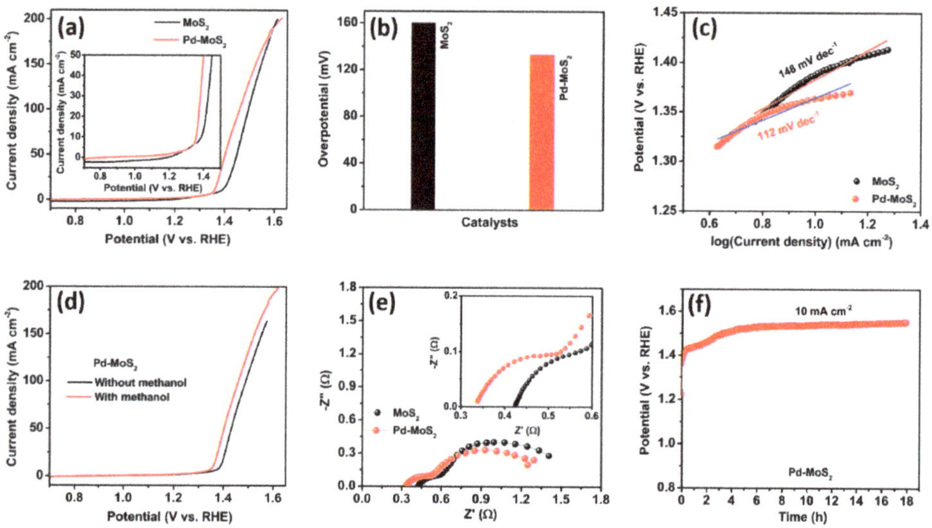

Figure 6. Electrocatalytic methanol-mediated OER. (**a**) LSV results, (**b**) overpotential graphs, and (**c**) Tafel graphs of the Pd-MoS_2 and MoS_2 electrocatalysts. (**d**) LSV curves of the Pd-MoS_2 without and with methanol. (**e**) EIS plots of the Pd-MoS_2 and MoS_2 electrocatalysts. (**f**) Durability assessment at 10 mA cm^{-2} of the Pd-MoS_2 catalyst over 18 h.

Figure 7 illustrates the methanol-mediated hydrogen evolution reaction (MM-HER) of Pd-MoS_2 and MoS_2 catalysts. Figure 7a shows the LSV results of Pd-MoS_2 and MoS_2 electrocatalysts in 1 M methanol + 1 M KOH at 5 mV s^{-1}. The MM-HER process illustrated in Figure 7a demonstrates that the Pd-MoS_2 catalyst offers superior performance compared to pure MoS_2. Further, it is also observed that nanomaterial modifications using Pd are important for developing efficient and cost-effective Pd-MoS_2 catalysts for MM-HER [51]. The LSV plots of MoS_2 and Pd-MoS_2 catalysts reveal a substantial understanding of the electrocatalytic activity during MM-HER at 5 mV s^{-1}. Figure 7b displays an overpotential plot of the Pd-MoS_2 and MoS_2 electrocatalysts at −10 mA cm^{-2} for MM-HER. It is observed that the Pd-MoS_2 electrocatalyst divulges a lesser overpotential value of 224.6 mV compared with the pure MoS_2 catalyst of 251.8 mV at −10 mA cm^{-2}. Further, the decline in overpotential from 251.8 mV of the pure MoS_2 catalyst to 224.6 mV of Pd-doped MoS_2 can attain the same reaction rate with less energy input because of a vast surface area and excellent conductivity. The improvement in electrocatalytic performance in 1 M methanol + 1 M KOH during MM-HER is due to Pd doping, which can be attributed to several factors such as increased electronic conduction, better active site availability on the Pd-MoS_2 surface, and synergistic effects between Pd and MoS_2, developing a cost-effective and efficient MM-HER catalyst for clean and green hydrogen production [52]. Figure 7c exhibits the Tafel curves of Pd-MoS_2 and MoS_2 catalysts for MM-HER. Interestingly, the Pd-MoS_2 catalyst exhibits a low Tafel curve of 113 mV dec^{-1} comparable to the MoS_2 catalyst of 115 mV dec^{-1}. It is detected that the Pd-doped MoS_2 catalyst displays a slightly lower Tafel slope compared to the pure MoS_2 catalyst, demonstrating upgraded charge transfer kinetics and overall catalytic effectiveness in the methanol-mediated hydrogen evolution reaction for hydrogen generation. On the other hand, the importance of Tafel slope enhancement lies in faster reaction kinetics, higher energy efficiency, and greater commercial feasibility, which makes the Pd-MoS_2 catalyst an auspicious candidate for MM-HER application [46]. Moreover, a low Tafel slope of the Pd-MoS_2 catalyst corresponds to the lower overpotential essential for achieving a -10 mA cm^{-2}, as exposed in Figure 7c, representing less energy consumption for hydrogen generation, which offers methanol-mediated hydrogen generation process more energy efficient. Further, Figure 7d displays the LSV plots of the Pd-MoS_2 at 5 mV s^{-1} without and with methanol-mediated HER. Interestingly, it is observed that the overpotentials are 269.6 mV, and 224.6 mV at -10 mA cm^{-2} during without and with methanol-mediated HER, respectively. Figure 7e explores the EIS spectra of the MoS_2 and Pd-MoS_2 at respective overpotentials of MM-HER, which portrays the low series and charge-transfer resistance of the Pd-MoS_2 compared with MoS_2. Figure 7f unveils a chronopotentiometry test to uncover the stability test of the Pd-MoS_2 during MM-HER at -10 mA cm^{-2}. It is observed that the stability of the Pd-MoS_2 catalyst slightly increased from 195.9 mV to 228.5 mV -10 mA cm^{-2}. Fascinatingly, the Pd-MoS_2 electrocatalyst exhibits unique nanolayered morphology, which delivers a high transport of charge carriers and plenty of active sites on the catalyst surface, facilitating effective mass transport and enhanced exposure of active sites with the 1 M KOH and 1 M methanol electrolyte ions, leading to excellent stability [53].

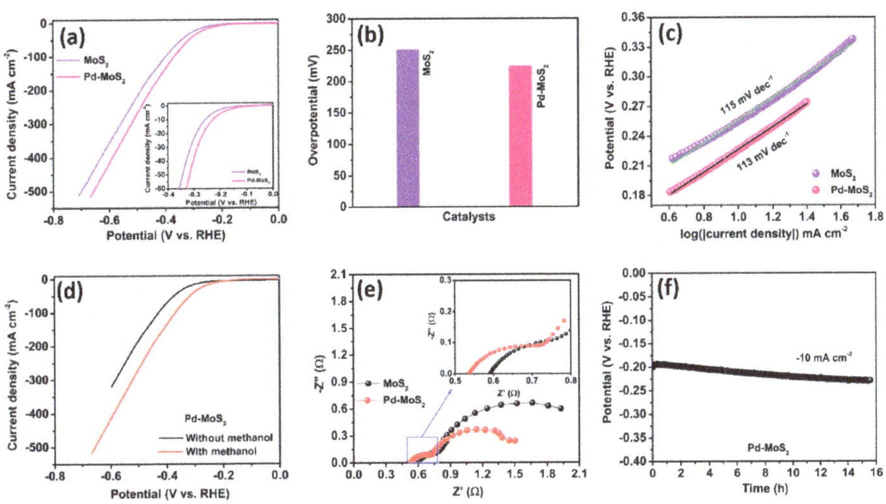

Figure 7. Electrocatalytic activities of methanol-mediated HER. (**a**) LSV results, (**b**) overpotential graphs, and (**c**) Tafel graphs of Pd-MoS_2 and MoS_2 electrocatalysts. (**d**) LSV curves of the Pd-MoS_2 without and with methanol. (**e**) EIS plots of the Pd-MoS_2 and MoS_2 electrocatalysts. (**f**) Durability examination at -10 mA cm^{-2} for the Pd-MoS_2 catalyst during 18 h.

Figure 8a–f illustrate the methanol-mediated overall water splitting (MM-OWS) of Pd-MoS_2||Pd-MoS_2 and MoS_2||MoS_2 cells. Figure 8a shows the LSV results of the MoS_2||MoS_2 cell and Pd-MoS_2||Pd-MoS_2 cell for MM-OWS in a 1 M KOH + 1 M methanol electrolyte at 5 mV s^{-1}. The catalytic performances of Pd-MoS_2||Pd-MoS_2 and MoS_2||MoS_2 are assessed by investigating the cell potential at 10 mA cm^{-2} in the LSV plot. Further, it is seen from LSV plots that the Pd doping ominously enriches the electrocatalytic activities of the Pd-MoS_2||Pd-MoS_2 cell compared with the MoS_2||MoS_2 cell in methanol-mediated overall water splitting (MM-OWS). Figure 8b displays the cell potentials of the Pd-MoS_2||Pd-MoS_2 cell and MoS_2||MoS_2 cell for MM-OWS at 10 mA cm^{-2}. Fascinatingly, the lower potential of the Pd-MoS_2||Pd-MoS_2 cell (1.581 V) compared to the MoS_2||MoS_2 cell (1.648 V) may be due to the alterations of electronic properties, synergistic, and adsorption effects introduced by Pd doping, which makes the Pd-MoS_2||Pd-MoS_2 cell a more efficient for methanol-mediated overall water splitting. Further, Figure 8c illustrates the LSV curves of the Pd-MoS_2||Pd-MoS_2 cell at 5 mV s^{-1} without and with methanol-mediated OWS. Interestingly, it is observed that the potentials of the Pd-MoS_2||Pd-MoS_2 cell are 1.656 V and 1.581 V without and with methanol-mediated HER at 10 mA cm^{-2}, respectively. Figure 8d,e show the CV plots of the non-Faradic region of the MoS_2 and Pd-doped MoS_2 catalysts to evaluate a double-layer capacitance (C_{dl}) in 1 M KOH + 1 M methanol. Figure 8f,g elucidate the C_{dl} of MoS_2 and Pd-doped MoS_2 catalysts, which were evaluated by the CV plots of the non-Faradic region (Figure 8d,e). It is detected that the Pd-MoS_2 divulges a high C_{dl} of 55.1 mF compared to the MoS_2 of 45.7 mF. Figure 8h reveals the stability test of the Pd-MoS_2||Pd-MoS_2 cell for MM-OWS throughout 18 h at 10 mA cm^{-2}. In addition, it offers a cherished understanding of the stability of the Pd-MoS_2||Pd-MoS_2 cell in MM-OWS, signifying the capability for long-term use over 18 h. The stability at 10 mA cm^{-2} supports the Pd-MoS_2||Pd-MoS_2 cell's appropriateness for enduring use in electrochemical methanol-mediated overall water splitting application. Moreover, the strength of the Pd-MoS_2||Pd-MoS_2 cell at 10 mA cm^{-2} for an extended period is decisive for hands-on applications of MM-OWS, as it suggests the cell can reliably operate under typical conditions without significant degradation. This has implications for the Pd-MoS_2||Pd-MoS_2 cell's longevity, cost-effectiveness, and commercial viability. Moreover, the excellent methanol-mediated overall water splitting performances of the Pd-MoS_2 catalyst are due to numerous factors, such as a vast surface area due to a layered structure for the adsorption and reaction of methanol and protons. Furthermore, Pd doping improves the catalytic activity by facilitating the dehydrogenation and adsorption of methanol and improving the hydrogen adsorption sites and the synergistic effects between the MoS_2 and Pd, increasing the electrical conductivity, which is crucial for efficient electron transfer during both reactions in MM-OER and MM-HER. Figure 8i portrays the schematic illustration of the operation of the Pd-MoS_2||Pd-MoS_2 cell for methanol-mediated overall water splitting. In the Pd-MoS_2||Pd-MoS_2 cell configuration, the anodic side facilitates the MM-OER reaction, generating electrons and protons; however, the cathodic electrode offers the reaction mechanism for MM-HER.

Furthermore, Table 1 shows the assessment of the electrocatalytic activities of MoS_2-based catalysts in OER, HER, and OWS. The Co@NC@MoS_2 catalyst displays a 297 mV overpotential and elucidates excellent constancy throughout 10 h in 1.0 M KOH at 10 mA cm^{-2} for oxygen evolution reaction (OER) in 1 M KOH or 0.5 M H_2SO_4 [54]. Moreover, the MoS_2@Co catalyst reveals a 370 mV overpotential, exposing brilliant reliability over 20 h at 10 mA cm^{-2} for OER [55]. Further, the CoMnCr LDH@MoS_2/NF catalyst divulges a 229 mV overpotential with outstanding stability for 24 h at 10 mA cm^{-2} for OER [56]. Also, the MoS_2@CoO catalyst discloses a 325 mV overpotential and good durability for 1000 cycles at 10 mA cm^{-2} for OER [57]. In addition, the Co-Sv-MoS_2 catalyst unveils 190 mV overpotential and robustness for 12 h at 10 mA cm^{-2} for OER [58]. Interestingly, the RuO_2 catalyst uncovers 199 mV overpotential and strength only for 5.5 h at 10 mA cm^{-2} for OER [59]. Remarkably, the LSC/MoS_2 catalyst exposes a 284 mV overpotential and long life over 1 h at −10 mA cm^{-2} for the hydrogen evolution reaction (HER) [60]. Curiously, the 2H-MoS_2 catalyst illustrates a 369 mV overpotential and robustness over 24 h at

-10 mA cm^{-2} for HER [61]. In addition, the Ni/MoS_2-1000 catalyst elucidates a 229 mV overpotential and toughness over 2000 cycles at -10 mA cm^{-2} for HER [62]. Furthermore, the MoS_2 catalyst exposes nearly a 240 mV overpotential and robustness for ~47 h at -10 mA cm^{-2} for HER [63]. Likewise, the SL-MoS_2/CP catalyst reveals the overpotential of 267 mV at -10 mA cm^{-2} and long-term stability for 12 h for HER [64]. Remarkably, the Pt/C (20 wt.%) catalyst reveals a 43 mV overpotential and long endurance for 5000 cycles at -10 mA cm^{-2} for HER [65]. Further, Fe-(NiS_2/MoS_2)/CNT||Fe-(NiS_2/MoS_2)/CNT reveals a 1.51 V potential and durability over 8 h at 10 mA cm^{-2} during overall water splitting (OWS) [66]. Additionally, $CoFe_2O_4$@MoS_2/CC||$CoFe_2O_4$@MoS_2/CC displays 1.54 V potential and cell life for 12 h at 20 mA cm^{-2} during overall water splitting (OWS) [67]. Interestingly, NiCo-MoS_2-CW||NiCo-MoS_2-CW illustrates a 1.69 V potential at 50 mA cm^{-2} during OWS [68]. Fascinatingly, Co_9S_8/MnS/MoS_2/NF-2||Co_9S_8/MnS/MoS_2/NF-2 elucidates a 1.55 V potential and feasible operation for 26 h at 2.0 V during OWS [69]. Further, 1T-MoS_2/Ni_3S_2/LDH||1T-MoS_2/Ni_3S_2/LDH elucidates a potential of 1.55 V and longstanding operation for 20 h during OWS [70]. Interestingly, the present work (Pd-MoS_2||Pd-MoS_2 catalyst) illustrates 133 mV and a 224.6 mV overpotential and stability over 18 h and 16 h during a methanol-mediated OER and methanol-mediated HER. Additionally, the present work (Pd-MoS_2||Pd-MoS_2 cell) reveals a 1.581 V potential and working strength of 18 h during methanol-mediated OWS. Therefore, it is concluded that the Pd-doped MoS_2 catalyst reveals excellent electrocatalytic properties in MM-OER, MM-HER, and MM-OWS.

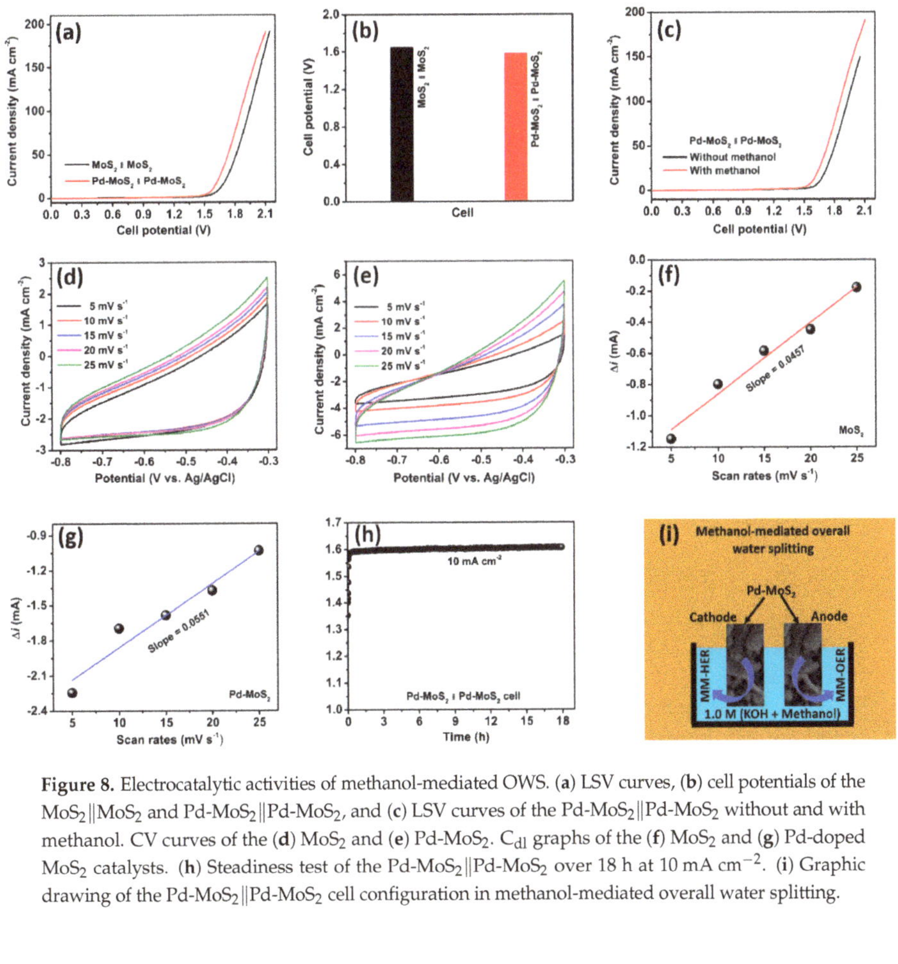

Figure 8. Electrocatalytic activities of methanol-mediated OWS. (**a**) LSV curves, (**b**) cell potentials of the MoS_2||MoS_2 and Pd-MoS_2||Pd-MoS_2, and (**c**) LSV curves of the Pd-MoS_2||Pd-MoS_2 without and with methanol. CV curves of the (**d**) MoS_2 and (**e**) Pd-MoS_2. C_{dl} graphs of the (**f**) MoS_2 and (**g**) Pd-doped MoS_2 catalysts. (**h**) Steadiness test of the Pd-MoS_2||Pd-MoS_2 over 18 h at 10 mA cm^{-2}. (**i**) Graphic drawing of the Pd-MoS_2||Pd-MoS_2 cell configuration in methanol-mediated overall water splitting.

Table 1. Assessment of the electrocatalytic properties of MoS_2-based catalysts in OER, HER, and OWS.

S. No.	Catalysts	Electrolyte	Overpotential (10 mA cm^{-2})	Stability (10 mA cm^{-2})	Ref.
	Oxygen evolution reaction (OER)/Methanol-mediated OER (MM-OER)				
1	Co@NC@MoS_2	1.0 M KOH	297 mV	10 h	[54]
2	MoS_2@Co	1.0 M KOH	370 mV	20 h	[55]
3	CoMnCr LDH@MoS_2/NF	1.0 M KOH	229 mV	24 h	[56]
4	MoS_2@CoO-coated carbon cloth	1.0 M KOH	325 mV	1000 cycles	[57]
5	Co-Sv-MoS_2	1.0 M KOH	190 mV	12 h	[58]
6	RuO_2	0.5 M H_2SO_4	199 mV	~5.5 h	[59]
7	Pd-MoS_2 (MM-OER)	1.0 M KOH + 1.0 M Methanol	133 mV	18 h	This work
	Hydrogen evolution reaction (HER)/Methanol-mediated HER (MM-HER)				
8	LSC/MoS_2	1.0 M KOH	284 mV	1 h	[60]
9	2H-MoS_2	0.5 M H_2SO_4	369 mV	24 h	[61]
10	Ni/MoS_2-1000	0.5 M H_2SO_4	229 mV	2000 cycles	[62]
11	MoS_2	0.5 M H_2SO_4	~240 mV	~47 h	[63]
12	SL-MoS_2/CP	0.5 M H_2SO_4	267 mV	12 h	[64]
13	Pt/C (20 wt.%)	1.0 M KOH	43 mV	5000 cycles	[65]
14	Pd-MoS_2 (MM-HER)	1.0 M KOH + 1.0 M Methanol	224.6 mV	16 h	This work
	Overall water splitting (OWS)/Methanol-mediated OWS (MM-OWS)				
15	Fe-(NiS_2/MoS_2)/CNT‖Fe-(NiS_2/MoS_2)/CNT	1.0 M KOH	1.51 V	8 h	[66]
16	$CoFe_2O_4$@MoS_2/CC‖$CoFe_2O_4$@MoS_2/CC	1.0 M KOH	1.54 V	12 h @20 mA cm^{-2}	[67]
17	NiCo-MoS_2-CW‖NiCo-MoS_2-CW	1.0 M KOH	1.69 V @50 mA cm^{-2}	--	[68]
18	Co_9S_8/MnS/MoS_2/NF-2‖Co_9S_8/MnS/MoS_2/NF-2	1.0 M KOH	1.55 mV	26 h @2.0 V	[69]
19	1T-MoS_2/Ni_3S_2/LDH‖1T-MoS_2/Ni_3S_2/LDH	1.0 M KOH	1.55 mV	20 h	[70]
20	Pd-MoS_2‖Pd-MoS_2 (MM-OWS)	1.0 M KOH + 1.0 M Methanol	1.581 V cell potential	18 h	This work

3. Synthesis Method

All the chemicals used were purchased from Sigma Aldrich, Seoul, Korea. This study synthesized Pd-doped MoS_2 and MoS_2 using a solvothermal procedure. Initially, 30 mL of ethanol and 30 mL of deionized water were dissolved in a 100 mL glass beaker to prepare the solvent. Subsequently, 16 mg of C_2H_5NS and 4 mmol of $Na_2MoO_2 \cdot 2H_2O$ were added to the solvent and blended using a magnetic stirrer. A Pd solution was obtained by melting 2 mg of $PdCl_2$ in 20 mL of the solvent mixture (10 mL deionized water and 10 mL ethanol) under sonication and magnetic stirring for several hours until it dissolved. While stirring the initial mixture, 1 mL of Pd solution was slowly added to the molybdenum and sulfur precursor solution. Further, the whole solution was then moved into a 100 mL autoclave. The 2 cm × 3 cm slice of cleaned Ni-foam substrate was placed into the autoclave containing the above solution. The Teflon autoclave was closed and preserved in an oven for 18 h at 180 °C to facilitate the synthesis. The Ni-foam, now deposited with Pd-doped MoS_2, was cleaned carefully with ethanol and water to exclude undesired residues. Ultimately, the cleaned Ni-foam was dehydrated in a vacuum oven at 90 °C for 15 h. Likewise, MoS_2 was prepared through the above process without a Pd precursor.

The Pd-MoS_2 shape was precisely studied by the scanning electron microscope (SEM) [S-4800 HITACHI, Ltd., Tokyo, Japan]. Furthermore, the structure, morphology, and Pd, S, and Mo elements of the Pd-MoS_2 were analyzed using TEM, HRTEM, and color mapping by JEOL [JEM-2100F, JEOL Ltd., Tokyo, Japan]. The crystal nanostructures of the Pd-doped MoS_2 and MoS_2 were analyzed by X-ray diffraction (XRD) with a PANalytical apparatus. Additionally, X-ray photoelectron spectroscopy (XPS) was employed to inspect the chemical properties of the Pd-doped MoS_2 using a Thermo Scientific system (Winsford, UK) furnished with the source of an Al K-alpha X-ray (400 μm). In addition, the methanol-mediated oxygen evolution reaction (MM-OER), methanol-mediated hydrogen evolution reaction (MM-HER), and methanol-mediated overall water splitting (MM-OWS) of the Pd-MoS_2 and MoS_2 electrocatalysts were systematically inspected via the electrochemical VersaSTAT3 (Princeton Applied Research) workstation. The MM-OER and MM-HER performances of the Pd-MoS_2 and MoS_2 catalysts were studied in a three-electrode configuration in a 1.0 M methanol + 1.0 M KOH. The MoS_2 and Pd-MoS_2 catalysts, graphite sheet, and Ag/AgCl were employed as working, counter, and reference electrodes, respectively. Further, the MM-OWS was inspected in the two-electrode configuration in 1.0 M KOH + 1.0 M methanol, in which MoS_2 or Pd-MoS_2 were utilized in both cathode and anode electrodes. The linear sweep voltammetry (LSV) of the Pd-MoS_2 and MoS_2 catalysts were examined at 5 mV s^{-1}. In addition, the Tafel graphs of the Pd-MoS_2 and MoS_2 catalysts were evaluated via LSV results, by applying $\eta = a + b \log j$, where b is the Tafel slope, j is the current density, a is the transfer coefficient, and η is the overpotential. The observed potential vs. Ag/AgCl was changed in the reversible hydrogen electrode (RHE) using $E_{RHE} = E^{\circ}_{Ag/AgCl} + E_{Ag/AgCl} + 0.059 \times pH$.

4. Conclusions

In conclusion, this paper explored the potential of palladium (Pd) doping in MoS_2 for enhancing catalytic performances in methanol-mediated overall water splitting. The Pd-MoS_2 catalyst highlights several significant advancements and insights, including enhanced electrocatalytic performances and strength. The Pd doping in MoS_2 shows small Tafel slopes and low overpotential in MM-OER and MM-HER. Further, the Pd-MoS_2||Pd-MoS_2 cell depicts low cell potential and brilliant robustness over 18 h at 10 mA cm^{-2}. Consequently, it is anticipated that this approach offers a promising route toward practical and sustainable hydrogen generation because of the unique properties of the Pd-MoS_2 electrocatalyst and the beneficial effects of methanol in KOH electrolytes.

Author Contributions: Conceptualization, R.K.M.; Methodology, A.M.T. and S.M.M.; Formal analysis, A.M.T., S.M.M. and R.K.M.; Investigation, A.M.T., S.M.M. and R.K.M.; Data curation, A.M.T. and S.M.M.; Writing—original draft, A.M.T.; Writing—review & editing, A.M.T., R.K.M., W.J. and J.C.S.; Visualization, A.M.T., R.K.M. and W.J.; Supervision, R.K.M., W.J. and J.C.S.; Project administration, W.J. and J.C.S.; Funding acquisition, J.C.S. All authors have read and agreed to the published version of the manuscript.

Funding: This work was supported by the Technology Innovation Program (RS-2023-00235844, Development of nano-structured materials and devices for super steep subthreshold swing) funded By the Ministry of Trade, Industry & Energy (MOTIE, Korea) (1415187621).

Data Availability Statement: The original contributions presented in this study are included in the article. Further inquiries can be directed to the corresponding author.

Conflicts of Interest: The authors declare no conflict of interest.

References

1. Kumari, R.; Sammi, A.; Srivastava, A.; Azad, U.P.; Chandra, P. Emerging 3D nanomaterials as electrocatalysts for water splitting reactions. *Int. J. Hydrog. Energy* **2024**, *74*, 214–231. [CrossRef]
2. Imran, S.; Hussain, M. Emerging trends in water splitting innovations for solar hydrogen production: Analysis, comparison, and economical insights. *Int. J. Hydrog. Energy* **2024**, *77*, 975–996. [CrossRef]
3. Jin, C.; Zhang, Z.; Gui, L.; Zeng, X. 2D MXene-derived 3D TiO_2-$NiCoP_x$ hierarchical heterostructures for efficient water splitting. *J. Alloys Compd.* **2024**, *1002*, 175492. [CrossRef]
4. Salah, B.; Abdelgawad, A.; Lu, Q.; Ipadeola, A.K.; Luque, R.; Eid, K. Synergistically interactive MnFeM (M = Cu, Ti, and Co) sites doped porous g-C_3N_4 fiber-like nanostructures for an enhanced green hydrogen production. *Green Chem.* **2023**, *25*, 6032–6040. [CrossRef]
5. Abdelgawad, A.; Salah, B.; Lu, Q.; Abdullah, A.M.; Chitt, M.; Ghanem, A.; Al-Hajri, R.S.; Eid, K. Template-free synthesis of M/g-C_3N_4 (M = Cu, Mn, and Fe) porous one-dimensional nanostructures for green hydrogen production. *J. Electroanal. Chem.* **2023**, *938*, 117426. [CrossRef]
6. Mushtaq, M.A.; Ahmad, M.; Shaheen, A.; Mehmood, A.; Yasin, G.; Arif, M.; Ali, Z.; Li, P.; Hussain, S.N.; Tabish, M.; et al. Advancing the development of hollow micro/nanostructured materials for electrocatalytic water splitting: Current state, challenges, and perspectives. *ACS Mater. Lett.* **2024**, *6*, 3090–3111. [CrossRef]
7. Kulkarni, R.; Lingamdinne, L.P.; Koduru, J.R.; Karri, R.R.; Chang, Y.-Y.; Kailasa, S.K.; Mubarak, N.M. Recent advanced developments and prospects of surface functionalized MXenes-based hybrid composites toward electrochemical water splitting applications. *ACS Mater. Lett.* **2024**, *6*, 2660–2686. [CrossRef]
8. Lu, L.; Gu, Y.; Guo, W.; Lu, F.; Li, Y. Extrasmall ligand-free Pt nanoparticles as dual-function catalysts for methanol-assisted water splitting systems. *ACS Appl. Nano Mater.* **2022**, *5*, 4222–4229. [CrossRef]
9. Wei, X.; Wang, S.; Hua, Z.; Chen, L.; Shi, J. Metal-organic framework nanosheet electrocatalysts for efficient H_2 production from methanol solution: Methanol-assisted water splitting or methanol reforming? *ACS Appl. Mater. Interfaces* **2018**, *10*, 25422–25428. [CrossRef] [PubMed]
10. Li, M.; Duanmu, K.; Wan, C.; Cheng, T.; Zhang, L.; Dai, S.; Chen, W.; Zhao, Z.; Li, P.; Fei, H.; et al. Single-atom tailoring of platinum nanocatalysts for high-performance multifunctional electrocatalysis. *Nat. Catal.* **2019**, *2*, 495–503. [CrossRef]
11. Li, H.; Zhang, X.; Sun, Z.; Ma, W. Rapid screening of bimetallic electrocatalysts using single nanoparticle collision electrochemistry. *J. Am. Chem. Soc.* **2022**, *144*, 16480–16489. [CrossRef] [PubMed]
12. Chen, H.; Wang, Y.; Zhou, S.; Li, N.; Liu, X.; Abdukayum, A.; Zhang, L.; Hu, G. MoS_2-decorated ordered macroporous $Fe_2Mo_3O_8$/MoO_2 with heterogeneous structure as bifunctional electrocatalyst for efficient overall water splitting. *Ceram. Int.* **2024**, *50*, 16714–16721. [CrossRef]
13. Zhu, J.; Xia, L.; Yu, R.; Lu, R.; Li, J.; He, R.; Wu, Y.; Zhang, W.; Hong, X.; Chen, W.; et al. Ultrahigh stable methanol oxidation enabled by a high hydroxyl concentration on Pt clusters/MXene interfaces. *J. Am. Chem. Soc.* **2022**, *144*, 15529–15538. [CrossRef] [PubMed]
14. Pan, J.; Liu, X.; Ji, H.; Zhu, Y.; Zhuang, Y.; Chen, K.; Sun, N.; Liu, Y.; Lei, Y.; Wang, K.; et al. The strategies to improve TMDs represented by MoS_2 electrocatalytic oxygen evolution reaction. *Chin. Chem. Lett.* **2024**, *35*, 109515. [CrossRef]
15. Jiang, R.; Xiao, M.; Zhu, H.; Zang, X.; Zhao, D.; Zhu, J.; Long, Y.; Wang, Q. Intriguing and boosting molybdenum sulfide (MoS_2)-based materials for decontamination and purification of wastewater/seawater: An upgraded review. *Sep. Purif. Technol.* **2024**, *351*, 128063. [CrossRef]
16. Tian, L.; He, C.; Hu, J.; Yang, J.; Li, X.; Chen, Z. Electronic and optical properties and electrocatalytic water splitting in a graphene/MoS_2 heterojunction. *Phys. B* **2024**, *685*, 416062. [CrossRef]
17. Anushya, G.; Benjamin, M.; Sarika, R.; Pravin, J.C.; Sridevi, R.; Nirmal, D. A review on applications of molybdenum disulfide material: Recent developments. *Micro Nanostruct.* **2024**, *186*, 207742. [CrossRef]
18. Ma, J.; Wang, X.; Li, H.; Yang, D.; Fan, J.; Liu, Y. Boosting photocatalytic overall water splitting performance by dual-metallic single Ni and Pd atoms decoration of MoS_2: A DFT study. *J. Alloys Compd.* **2024**, *991*, 174530. [CrossRef]
19. Khanlary, M.R.; Parhizkari, M.; Heydari, N. Structural and optical properties of 1T-MoS_2/MoO_3 thin films prepared by spray pyrolysis method. *Phys. B* **2024**, *680*, 415837. [CrossRef]
20. Aliyar, T.; Ma, H.; Krishnan, R.; Singh, G.; Chong, B.Q.; Wang, Y.; Verzhbitskiy, I.; Wong, C.P.Y.; Goh, K.E.J.; Shen, Z.X.; et al. Symmetry breaking and spin-orbit coupling for individual vacancy-induced in-gap states in MoS_2 monolayers. *Nano Lett.* **2024**, *24*, 2142–2148. [CrossRef]

21. Mathew, R.; Ajayan, J. Material processing, performance and reliability of MoS_2 field effect transistor (FET) technology- A critical review. *Mater. Sci. Semicond. Process.* **2023**, *160*, 107397. [CrossRef]
22. Le, K.T.; Pham, N.N.T.; Liao, Y.-S.; Ranjan, A.; Lin, H.-Y.; Chen, P.-H.; Nguyen, H.; Lu, M.Y.; Lee, S.G.; Wu, J.M. Piezoelectricity of strain-induced overall water splitting of $Ni(OH)_2/MoS_2$ heterostructure. *J. Mater. Chem. A* **2023**, *11*, 3481–3492. [CrossRef]
23. Ding, E.-X.; Liu, P.; Yoon, H.H.; Ahmed, F.; Du, M.; Shafi, A.M.; Mehmood, N.; Kauppinen, E.I.; Sun, Z.; Lipsanen, H. Highly sensitive MoS_2 photodetectors enabled with a dry-transferred transparent carbon nanotube electrode. *ACS Appl. Mater. Interfaces* **2023**, *15*, 4216–4225. [CrossRef] [PubMed]
24. Im, H.; Kim, J.; Kim, J.; Kim, S. Scalable bilayer MoS_2-based vertically inverted p-i-n light-emitting diodes. *Adv. Mater. Interfaces* **2023**, *11*, 2300319. [CrossRef]
25. Hu, B.; Wu, Y.; Wang, K.; Guo, H.; Lei, Z.; Liu, Z.; Wang, L. Gram-scale mechanochemical synthesis of atom-layer MoS_2 semiconductor electrocatalyst via functionalized graphene quantum dots for efficient hydrogen evolution. *Small* **2023**, *20*, 2305344. [CrossRef] [PubMed]
26. Man, P.; Jiang, S.; Leung, K.H.; Lai, K.H.; Guang, Z.; Chen, H.; Huang, L.; Chen, T.; Gao, S.; Peng, K.; et al. Salt-induced high-density vacancy-rich 2D MoS_2 for efficient hydrogen evolution. *Adv. Mater.* **2024**, *36*, 2304808. [CrossRef] [PubMed]
27. Chen, P.S.; Hu, Y.; Li, S.; Mazurkiewicz-Pawlicka, M.; Małolepszy, A. Preparation of a MoS_2/carbon nanotube nanocomposite by hydrothermal method for supercapacitor. *Int. J. Electrochem. Sci.* **2024**, *19*, 100523. [CrossRef]
28. Sun, K.; Guo, H.; Feng, C.; Tian, F.; Zhao, X.; Wang, C.; Chai, Y.; Liu, B.; Mintova, S.; Liu, C. One–pot solvothermal preparation of the porous $NiS_2//MoS_2$ composite catalyst with enhanced low-temperature hydrodesulfurization activity. *J. Colloid Interface Sci.* **2024**, *659*, 650–664. [CrossRef] [PubMed]
29. Wang, Y.; Zhai, W.; Ren, Y.; Zhang, Q.; Yao, Y.; Li, S.; Yang, Q.; Zhou, X.; Li, Z.; Chi, B.; et al. Phase-controlled growth of 1T-MoS_2 nanoribbons on 1H-MoS_2 nanosheets. *Adv. Mater.* **2024**, *36*, 2307269. [CrossRef] [PubMed]
30. Alrowaily, A.W.; Alyousef, H.A.; Alotaibi, B.; Alotiby, M.F.; Al-Sehemi, A.G.; Ahmad, K.; Henaish, A.; Al-Zahrani, F.A. Facile synthesis of $CuMn_2O_4$ nanoparticle supported on MoS_2 nanorods via hydrothermal route for supercapacitor application. *Mater. Chem. Phys.* **2024**, *322*, 129517. [CrossRef]
31. Liang, S.; Zheng, J.; Song, N.; Wang, X.; Tu, B.; Xu, J. Accelerated confined mass transfer of MoS_2 1D nanotube in photo-assisted metal-air batteries. *Adv. Mater.* **2024**, *36*, 2307790. [CrossRef] [PubMed]
32. Yang, L.; Yuan, X.; Dong, Y.; Qian, S.; Zhu, C. Hierarchical nanowires of MoS_2@transition metal sulfide heterostructures for efficient electrocatalytic hydrogen evolution reaction. *J. Electroanal. Chem.* **2024**, *964*, 118342. [CrossRef]
33. Wen, Y.; Li, D.; Jin, D.; Jin, H. Comparative study on gas-sensing properties of amorphous MoS_2 nanoparticles and crystalline MoS_2 nanoflowers. *Mater. Sci. Semicond. Process.* **2024**, *172*, 108077. [CrossRef]
34. Wang, S.; White, J.; Li, M.; Azam, A.; Yang, J.; Zu, X.; Qiao, L.; Reece, P.; Stride, J.; Zhang, S.; et al. Size control of MoS_2 quantum dots by varying the crystallographic orientation of sapphire substrates. *Mater. Today Chem.* **2023**, *35*, 101887. [CrossRef]
35. Mouloua, D.; LeBlanc-Lavoie, J.; Pichon, L.; Rajput, N.S.; Marssi, M.E.; Jouiad, M.; El Khakani, M.A. Tuning the optoelectronic properties of pulsed laser deposited "3D"-MoS_2 films via the degree of vertical alignment of their constituting layers. *Adv. Opt. Mater.* **2024**, *12*, 2302966. [CrossRef]
36. Zhu, W.; Zhao, J.; Tao, X. MoS_2-carbon based nanocomposites as anodes for lithium-ion batteries: A review. *J. Energy Storage* **2024**, *84*, 110934. [CrossRef]
37. Wang, W.; Tang, L.; Chen, C.; Wang, X.; Wang, Y.; Sun, W. Engineering Pt-S-Mo and Pd-S-Mo sites in hierarchical porous MoS_2 for boosted oxygen reduction activity in microbial fuel cell. *J. Power Sources* **2024**, *598*, 234143. [CrossRef]
38. Zhang, Y.; Zhang, R.; Guo, Y.; Li, Y.; Li, K. A review on MoS_2 structure, preparation, energy storage applications and challenges. *J. Alloys Compd.* **2024**, *998*, 174916. [CrossRef]
39. Cantarella, M.; Spanò, V.; Zimbone, M.; Giuffrida, F.; Lufrano, E.; Strano, V.; Franzò, G.; Sfuncia, G.; Nicotra, G.; Alberti, A.; et al. ZnO-MoS_2-PMMA polymeric nanocomposites: A harmless material for water treatment. *Mater. Today Chem.* **2024**, *36*, 101912. [CrossRef]
40. Singh, H.; Rathour, R.K.S.; Raj, S.; Bhattacharya, J. Environment friendly MOF-MoS_2 composite for Cr(VI) and Pb(II) capture from paint industry wastewater: Sequestration, regeneration and toxicity. *Inorg. Chim. Acta* **2024**, *570*, 122165. [CrossRef]
41. Ghosh, S.; Lai, J. An insight into the dual role of MoS_2-based nanocarriers in anticancer drug delivery and therapy. *Acta Biomater.* **2024**, *179*, 36–60. [CrossRef]
42. Lin, S.; Joshi, R.; Ganguly, A.; Barman, S.R.; Pal, A.; Kaswan, K.; Liu, K.; Nain, A.; Kao, F.; Lin, Z. Ultrasound-guided drug delivery system utilizing piezocatalytic MoS_2 nanomaterials for anti-inflammatory therapy. *Nano Energy* **2024**, *127*, 109732. [CrossRef]
43. Zhao, L.; Wang, Y.; Wen, G.; Zhang, X.; Huang, X. Ammonium-driven modulation of 1T-MoS_2 structure and composite with graphene: A pathway to high-performance lithium-ion battery anodes. *J. Colloid Interface Sci.* **2025**, *680*, 151–161. [CrossRef]
44. Li, B.; Jiang, L.; Li, X.; Ran, P.; Zuo, P.; Wang, A.; Qu, L.; Zhao, Y.; Cheng, Z.; Lu, Y. Preparation of monolayer MoS_2 quantum dots using temporally shaped femtosecond laser ablation of bulk MoS_2 targets in water. *Sci. Rep.* **2017**, *7*, 11182. [CrossRef]

45. Lai, B.; Singh, S.C.; Bindra, J.K.; Saraj, C.S.; Shukla, A.; Yadav, T.P.; Wu, W.; McGill, S.A.; Dalal, N.S.; Srivastava, A.; et al. Hydrogen evolution reaction from bare and surface-functionalized few-layered MoS_2 nanosheets in acidic and alkaline electrolytes. *Mater. Today Chem.* **2019**, *14*, 100207. [CrossRef]
46. Gupta, J.; Das, D.; Borse, P.H.; Sarada, B.V. In situ Pd-doped MoS_2 nanosheets as an HER electrocatalyst for enhanced electrocatalytic water splitting. *Sustain. Energy Fuels* **2024**, *8*, 1526–1539. [CrossRef]
47. Xuan, T.T.; Long, L.N.; Khai, T.V. Effect of reaction temperature and reaction time on the structure and properties of MoS_2 synthesized by hydrothermal method. *Vietnam J. Chem.* **2020**, *58*, 92–100. [CrossRef]
48. Zhang, H.; Lin, H.; Zheng, Y.; Hu, Y.; MacLennan, A. Understanding of the effect of synthesis temperature on the crystallization and activity of nano-MoS_2 catalyst. *Appl. Catal. B* **2015**, *165*, 537–546. [CrossRef]
49. Han, C.; Huang, G.; Zhu, D.; Hu, K. Facile synthesis of MoS_2/Fe_3O_4 nanocomposite with excellent Photo-Fenton-like catalytic performance. *Mater. Chem. Phys.* **2017**, *200*, 16–22. [CrossRef]
50. Wang, J.; Zhang, B.; Guo, W.; Wang, L.; Chen, J.; Pan, H.; Sun, W. Toward electrocatalytic methanol oxidation reaction: Longstanding debates and emerging catalysts. *Adv. Mater.* **2023**, *35*, 2211099. [CrossRef] [PubMed]
51. Yin, S.; Liu, S.; Wang, Z.; Xu, Y.; Li, X.; Wang, H.; Wang, L. Methanol-assisted energy-saving hydrogen production over defect-rich perforated PdIn bimetallene. *Chem. Eng. J.* **2022**, *435*, 134711. [CrossRef]
52. Li, B.; Qiao, S.; Zheng, X.; Yang, X.; Cui, Z.; Zhu, S.; Li, Z.; Liang, Y. Pd coated MoS_2 nanoflowers for highly efficient hydrogen evolution reaction under irradiation. *J. Power Sources* **2015**, *284*, 68–76. [CrossRef]
53. Yu, E.; Pan, Y. Enhancing the catalytic hydrogen evolution reaction (HER) of the defective borophene@Pt/Pd/MoS_2 heterojunction. *Int. J. Hydrog. Energy* **2024**, *50*, 920–931. [CrossRef]
54. Gaur, A.; Sachdeva, P.K.; Kumar, R.; Maruyama, T.; Bera, C.; Bagchi, V. Ultrathin MoS_2 wrapped N-doped carbon-coated cobalt nanospheres for OER applications. *Sustain. Energy Fuels* **2021**, *5*, 801–807. [CrossRef]
55. Cao, S.; Wu, W.; Liu, C.; Song, L.; Xu, Q.; Zhang, H.; Zhao, Y. Supramolecular macrocycle regulated single-atom MoS_2@Co catalysts for enhanced oxygen evolution reaction. *Energy Environ. Mater.* **2024**, *7*, e12702. [CrossRef]
56. Vedanarayanan, M.; Gopalakrishnan, S.M. Enhancement of hydrogen and oxygen evolution reaction efficiencies by coupling a MoS_2 nanoflower with a CoMnCr LDH nanocube on nickel foam. *Energy Fuels* **2023**, *37*, 12204–12214. [CrossRef]
57. Cheng, P.; Yuan, C.; Zhou, Q.; Hu, X.; Li, J.; Lin, X.; Wang, X.; Jin, M.; Shui, L.; Gao, X.; et al. Core-shell MoS_2@CoO electrocatalyst for water splitting in neural and alkaline solutions. *J. Phys. Chem. C* **2019**, *123*, 5833–5839. [CrossRef]
58. Ma, Y.; Leng, D.; Zhang, X.; Fu, J.; Pi, C.; Zheng, Y.; Gao, B.; Li, X.; Li, N.; Chu, P.K.; et al. Enhanced activities in alkaline hydrogen and oxygen evolution reactions on MoS_2 Electrocatalysts by in-plane sulfur defects coupled with transition metal doping. *Small* **2022**, *18*, 2203173. [CrossRef] [PubMed]
59. Zhao, Z.L.; Wang, Q.; Huang, X.; Feng, Q.; Gu, S.; Zhang, Z.; Xu, H.; Zeng, L.; Gu, M.; Li, H. Boosting the oxygen evolution reaction using defect-rich ultra-thin ruthenium oxide nanosheets in acidic media. *Energy Environ. Sci.* **2020**, *13*, 5143–5151. [CrossRef]
60. Zhao, Y.-N.; Sun, N.; Xu, S.; Min, S.; Dong, H.; Li, J.; Liu, C.; Chen, Z. An interface engineering strategy of MoS_2/perovskite oxide as a bifunctional catalyst to boost overall water splitting. *J. Mater. Chem. A* **2024**, *12*, 8757–8768. [CrossRef]
61. Guo, X.; Song, E.; Zhao, W.; Xu, S.; Zhao, W.; Lei, Y.; Fang, Y.; Liu, J.; Huang, F. Charge self-regulation in 1T‴-MoS_2 structure with rich S vacancies for enhanced hydrogen evolution activity. *Nat. Commun.* **2022**, *13*, 5954. [CrossRef] [PubMed]
62. Lu, X.; Sun, J.; Liu, Z.; Pan, Y.; Li, Y.; Zhang, D.; Lin, Y.; Qu, X. Scalable synthesis of hollow MoS_2 nanoparticles modified on porous Ni for improved hydrogen evolution reaction. *J. Electrochem. Soc.* **2021**, *168*, 056519. [CrossRef]
63. Kumar, M.; Nagaiah, T.C. A NiCu-MoS_2 electrocatalyst for pH-universal hydrogen evolution reaction and Zn-air batteries driven self-power water splitting. *J. Mater. Chem. A* **2023**, *11*, 18336–18348. [CrossRef]
64. Li, B.L.; Gong, C.-B.; Shen, W.; Peng, J.-D.; Zou, H.L.; Luo, H.Q.; Li, N.B. Engineering metallic MoS_2 monolayers with responsive hydrogen evolution electrocatalytic activities for enzymatic reaction monitoring. *J. Mater. Chem. A* **2021**, *9*, 11056–11063. [CrossRef]
65. Jiang, B.; Liu, S.; Cheng, L.; Zhou, L.; Cui, H.; Liu, M.; Wen, M.; Wang, C.; Wang, W.; Li, S.; et al. Mass synthesis of Pt/C catalysts with high Pt loading for low-overpotential hydrogen evolution. *Int. J. Hydrog. Energy* **2024**, *58*, 268–278. [CrossRef]
66. Li, C.; Liu, M.; Ding, H.; He, L.; Wang, E.; Wang, B.; Fan, S.; Liu, K. A lightly Fe-doped (NiS_2/MoS_2)/carbon nanotube hybrid electrocatalyst film with laser-drilled micropores for stabilized overall water splitting and pH-universal hydrogen evolution reaction. *J. Mater. Chem. A* **2020**, *8*, 17527–17536. [CrossRef]
67. Duan, Y.; Guo, Z.; Wang, T.; Zhang, J. Uniform anchoring of MoS_2 nanosheets on MOFs-derived $CoFe_2O_4$ porous nanolayers to construct heterogeneous structural configurations for efficient and stable overall water splitting. *J. Colloid Interface Sci.* **2025**, *680*, 541–551. [CrossRef] [PubMed]
68. Hu, M.; Qian, Y.; Zhang, R.; Guo, C.; Yang, L.; Li, L. Interfacial electronic modulation of NiCo decorated nano-flowered MoS_2 on carbonized wood as a remarkable bifunctional electrocatalyst for boosting overall water splitting. *J. Colloid Interface Sci.* **2024**, *677*, 729–738. [CrossRef]

69. Yan, R.; Luan, T.; Liu, Z.; Cao, Y.; Chi, K.; Yang, S.; Guo, X. $Co_9S_8/MnS/MoS_2$ heterostructure grown in situ on Ni foam as highly efficient electrocatalysts for overall water splitting. *Int. J. Hydrog. Energy* **2025**, *98*, 14–24. [CrossRef]
70. Liu, W.; Dong, J.; An, B.; Su, H.; Teng, Z.; Li, N.; Gao, Y.; Ge, L. Synergistic dual built-in electric fields in 1T-MoS_2/Ni_3S_2/LDH for efficient electrocatalytic overall water splitting reactions. *J. Colloid Interface Sci.* **2024**, *673*, 228–238. [CrossRef]

inorganics

MDPI

Article

Encapsulating Ultrafine In_2O_3 Particles in Carbon Nanofiber Framework as Superior Electrode for Lithium-Ion Batteries

Wenhe Xie [1,*], Zhe An [1], Xuefeng Li [1], Qian Wang [1], Chen Hu [1], Yuanxiao Ma [1], Shenghong Liu [1], Haibin Sun [1] and Xiaolei Sun [2]

[1] Key Laboratory of Microelectronics and Energy of Henan Province, Xinyang Normal University, Xinyang 464000, China; 13353729786@163.com (Z.A.); lxuefeng0827@163.com (X.L.); 17627269682@163.com (Q.W.); yczm100520@163.com (C.H.); 18129894906@163.com (Y.M.); liush@xynu.edu.cn (S.L.); sunhaibin@xynu.edu.cn (H.S.)

[2] School of Materials Science and Engineering, Nankai University, Tianjin 300350, China; xiaolei.sun@nankai.edu.cn

* Correspondence: xiewh@xynu.edu.cn

Abstract: Indium oxide (In_2O_3) is a promising anode material for next-generation lithium-ion batteries and is prized for its high electrical conductivity, environmental friendliness, and high theoretical capacity. However, its practical application is significantly limited by severe volume expansion and contraction during the lithium insertion/extraction process. This volume change disrupts the solid electrolyte interphase (SEI) and degrades contact with the current collector, undermining battery performance. Although the nano-structured design of In_2O_3 can mitigate the volume effect to some extent, pure In_2O_3 nanomaterials are prone to agglomeration during frequent charging and discharging. The pure In_2O_3-based electrode shows a sustained and rapid capacity degradation. In this study, we embed ultrafine In_2O_3 particles in a carbon nanofiber framework using electrospinning and thermal annealing. The 1D carbon nanofiber structure provides an effective electronic conductive network and reduces the length of lithium-iondiffusion, which enhances the reactivity of the nanocomposite and improves electrode kinetics. Additionally, the carbon nanofiber framework isolates ultrafine In_2O_3 particles, preventing their aggregation. The small volume changes due to the ultrafine size of the In_2O_3 are buffered by the carbon materials, allowing the overall structure of the In_2O_3/C composite nanofiber to remain largely intact without crushing during charging and discharging cycles. This stability helps avoid electrode fracture and excessive SEI growth, resulting in superior cycle and rate performance compared with the pure In_2O_3 nanofiber electrodes.

Keywords: In_2O_3/C nanocomposite; negative electrode; lithium-ion batteries

1. Introduction

In recent years, stimulated by various preferential policies, the global energy landscape has undergone significant changes; new renewable energy sources, such as solar energy, wind energy, and tidal energy, are rapidly developing and partially replacing the market of fossil energy. However, limited by various factors such as weather, climate, geography, etc., the power generated by these new renewable energy sources is unstable; thus, supporting energy storage technologies should also be supplied to ensure the sustainable development of the new energy industry. Among them, lithium-ion batteries play a key role in large-scale energy storage due to their high energy density, environmental friendliness, and long lifespan [1–7].

As typical secondary rechargeable batteries [8–13], lithium-ion batteries (LIBs) have been widely investigated in view of their rocking chair working principle [14]. During the charging process, lithium ions from the positive electrode migrate to the negative electrode through the separator inside the battery, while the corresponding electrons from the positive electrode are transferred to the negative electrode through the external circuit.

In the discharge process, ions and electrons return to the negative electrode through the battery's internal separator and outer circuitry, respectively. The performance of a battery is mainly determined by its positive and negative electrode materials, and the commonly used commercial positive electrode materials are mainly transition metal oxides and phosphides containing lithium. For the negative electrode, the current commercially available material is graphite, which has low and stable lithium insertion/extraction working potential; however, the theoretical capacity of graphite is only 372 mAh g^{-1}, and such a low actual capacity greatly limits the overall performance improvement of lithium-ion batteries; thus, both academia and enterprises are developing high-capacity negative electrode materials to meet urgent increasing needs.

Various novel negative electrode materials have been intensely explored, including intercalation-type materials ($Li_4Ti_5O_{12}$, TiO_2) [15–17], transition metal oxides (TiO_2, NiO, Fe_3O_4, CoO) [18–26], or sulfides (Ni_3S_2, CoS, MnS) [27–31], which can store lithium via a conversion reaction mechanism. The typical electrochemical reaction equation is $MN + Li \longleftrightarrow M + LiN$, where M is a transition metal and N is O or S, the theoretical reversible capacity > 600 mAh g^{-1}; however, these transition metal-based negative electrode materials suffer high hysteresis voltage in the charge/discharge process, and the corresponding lithium extraction working platform ranges from 1.0 to 2.5 V vs. Li/Li^+. Such a high working potential will lower the output voltage and energy density of the full battery. On the contrary, many group 3, 4, and 5 elements like Ga, In, Si, Ge, Sn, and Sb [32–36] can reversibly form alloys with lithium at a relatively low working platform (0.1–1.0 V vs. Li/Li^+) and high theoretical capacity (600–4200 mAh g^{-1}); despite the tempting advantages, these electrode materials have insufficient reactivity when alloying with lithium; as a result, the actual capacity is far below the theoretical value, and the rate performance is rather poor.

On the other hand, group 3–5 element-based oxides (GeO_2 [37], In_2O_3, SiO, SnO, SbO) [38,39] not only retain a low alloying potential, but also exhibit relatively high electrochemical reactivity; among them, In_2O_3 has aroused increasing interest in consideration of it environmental friendliness, high electrical conductivity, and high theoretical reversible capacity (580 mAh g^{-1} when In_2O_3 holds 4 Li, 965 mAh g^{-1} when In_2O_3 holds 10 Li) [40,41]. Nevertheless, despite the significant advantages mentioned above, the practical application of In_2O_3 as a negative electrode is largely limited by the severe volume expansion/contraction in the repeated lithium insertion/extraction process; such volume change cracks the formed SEI (solid electrolyte interphase, which originates from the irreversible decomposition of organic electrolytes on the surface of active materials) layer; thus, part of the active material is re-exposed in the electrolyte and induces the repeated growth of SEI, and the unstable SEI growth hinders the diffusion rate of lithium ions; even worse, some active materials are forced to detach from the current collector, leading to low Coulombic efficiency, rapid capacity degradation, and unsatisfactory rate performance of the bulk material-based electrodes.

Multiple strategies have been proposed to respond to the aforementioned challenges. The primary measure is to design various In_2O_3 nanomaterials such as nanoparticles, nanosheets, and nanospheres [42–45]; nanostructured materials with high specific surface area not only provide a large number of electrochemical reaction sites, but also greatly reduce the diffusion distance of lithium ions. Moreover, the rich pore volume of nanomaterials buffers the volume change and relieves the electrode stress. To sum up, the In_2O_3 nanomaterial-based electrodes exhibit superior electrochemical lithium storage performance compared with bulk-based electrodes. However, the free energy of the nanostructured sample is much higher than that of the bulk sample; thus, in order to maintain the stability of the system, nanostructured In_2O_3 tends to spontaneously aggregate during repeated charge and discharge processes [46]. As a consequence, the corresponding electrochemical performance significantly decreases when the initial nanostructure is gradually destroyed after dozens of cycles.

Regarding this issue, researchers have introduced a variety of foreign materials (metal, carbon, or metal oxide) [47] to composite with nanostructured In_2O_3 to further improve its electrochemical performance. Typical carbon materials like graphene and conducting polymers have a relatively high elastic modulus and appropriately accommodate stress caused by volume effects. Nevertheless, the low proportion of carbon coating on In_2O_3 nanomaterial cannot maintain its structure in long-term cycling; on the other hand, the high proportion of carbon coating will lower the overall specific capacity of the In_2O_3/C compound. Thus, it is still a challenge an achievement of highly dispersed In_2O_3/C nanocomposites to maintain a durable and stable electrode structure in the repeated lithium insertion/extraction reactions.

Herein, ultrafine In_2O_3 particles are uniformly embedded inside the carbon nanofiber framework via electrospinning and thermal annealing. Firstly, the 1D carbon nanofiber supplies a good electronic conductive network and lowers the diffusion length of lithium ions, which is beneficial for enhancing the reactivity of the nanocomposite and improving electrode kinetics. Secondly, the ultrafine In_2O_3 particles are isolated by the carbon nanofiber framework, preventing In_2O_3 nanoparticles from aggregating with each other. Thirdly, the relatively small volume change caused by the ultrafine size of indium oxide is buffered by carbon materials, the overall structure of the In_2O_3/C composite nanofibers remains largely intact without crushing during the charge and discharge processes, avoiding the electrode fracture and excessive growth of the SEI layer. Therefore, the In_2O_3/C composite nanofibers electrode exhibits superior cycle performance and rate performance compared with the pure In_2O_3/C nanofiber electrode.

2. Results and Discussion

Figure 1 presents the morphological characterization of the In_2O_3 nanofibers sample, which was prepared via calcining the precursor of the electrospinning product at a high temperature in an air atmosphere. From the SEM images (Figure 1a,b), we can see randomly distributed In_2O_3 nanofibers of 1~2 microns in length and 100~200 nanometers in diameter. From the TEM images (Figure 1c,d), one can clearly see that the rough In_2O_3 nanofiber is composed of a large number of nanoparticles, which are proved to be highly crystalline In_2O_3 by HRTEM observation in Figure 1e. Moreover, several clear selected area electron diffraction rings (SAED, Figure 1f) confirm that the nanofiber sample is polycrystalline. No carbon material can be examined inside the nanofiber because the PVP carbon source is calcined and decomposed in air.

The comprehensive morphological characterization of prepared In_2O_3/C composite nanofibers is displayed in Figure 2. From the low magnification SEM image (Figure 2a), we can see that a large number of nanofibers are interwoven together with a length exceeding 10 microns, which is much longer than pure In_2O_3 nanofibers (in Figure 1a); the relatively long fiber scale indicates that In_2O_3/C composite nanofibers have higher strength and can resist mechanical strain during the annealing process; therefore, the high stress of active materials can also resist stress caused by volume changes during the lithiation/delithiation process [48,49]. From the enlarged SEM images in Figure 2b,c, one can see that the surface of In_2O_3/C composite nanofibers is relatively smooth compared with the In_2O_3 nanofibers sample; the smooth surface is due to the fact that numerous tiny In_2O_3 particles are embedded inside the nanofibers framework during the annealing process in an argon atmosphere, and such a structure reduces the size of active In_2O_3. Moreover, the carbon framework prevents the aggregation of In_2O_3 and helps maintain the integrity of the electrode during cyclic testing. A more detailed morphological structure is observed by the TEM image in Figure 2d–f; no obvious particles can be examined inside the composite nanofibers, proving that tiny In_2O_3 particles are highly dispersed in the carbon framework. Additionally, the relatively blurry SAED rings (the insert in Figure 2e) indicate the low crystallinity of In_2O_3 particles in the composite nanofibers, which is consistent with the SEM and TEM characterization results. In addition, Figure 2g–k present the element mapping of a single composite nanofiber, from which one can see that In, C, N, and O elements are uniformly dispersed in the In_2O_3/C composite nanofibers sample.

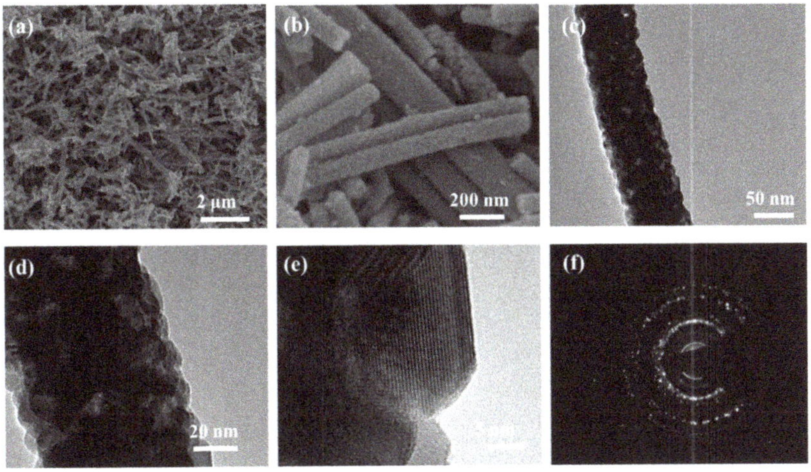

Figure 1. Morphological characterization of the In_2O_3 nanofibers sample. (**a**,**b**) SEM images, (**c**–**e**) TEM images, (**f**) SAED patterns.

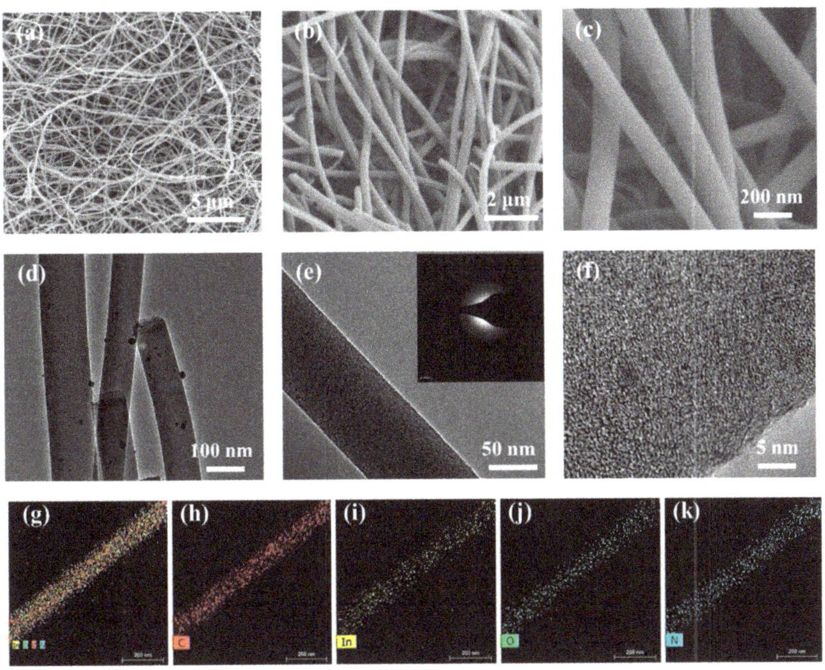

Figure 2. Morphological characterization of In_2O_3/C composite nanofibers sample. (**a**–**c**) SEM images, (**d**,**e**) TEM images, SAED patterns in the insert of (**e**), (**f**) HRTEM image, (**g**–**k**) the elemental maps.

Figure 3a records the X-ray diffraction (XRD) patterns of the prepared In_2O_3 nanofiber and In_2O_3/C composite nanofiber samples; the main diffraction peaks of the two nanofiber samples match well with cubic In_2O_3 (JCPDS 06-0416); on the other hand, the diffraction peaks of the In_2O_3/C composite nanofibers sample is wider than that of the pure In_2O_3 nanofibers sample, implying the smaller size and lower crystallinity of In_2O_3 in the composite nanofibers, which is in good agreement with the above SEM and TEM research results. Furthermore, no obvious diffraction peaks of carbon material can be examined from the XRD patterns, indicating that the carbon material is highly amorphous in the In_2O_3/C composite nanofibers. Meanwhile, the detailed surface composition of In_2O_3/C composite nanofibers is studied using X-ray photoelectron spectroscopy (XPS); from the full XPS spectrum curve in Figure 3b, we can clearly see the presence of In, O, C, and N elements in the sample; these four detected elements are completely consistent with the TEM element mapping results. The high-resolution XPS spectrum of In 3d shows two adjacent peaks at about 454.6 and 445.0 eV, which corresponds to In $3d_{3/2}$ and In $3d_{1/2}$ of In^{4+}, respectively. Moreover, according to the fitting results in Figure 3c, we can see the precise C 1s peak can be fitted into four peaks with bonding energies at 284.7, 285.4, 286.3, and 287.6 eV, which correspond to the C-C, C-N, C-O, and O-C=O groups, respectively. The precise N 1s (Figure 3e) peak consists of graphitic N (at about 400.4 eV), pyrrolic N (at about 399.0 eV), and pyridinic N (at about 398.4 eV). Significantly, the high-resolution O 1s peak is composed of a large proportion of O-In and a small amount of C-O, C=O groups. Thus, on the basis of the above characterizations and analysis, we can conclude that C, N, and a small amount of O elements originate from the PVP binder, while In and the remaining O elements originate from the indium oxide particles.

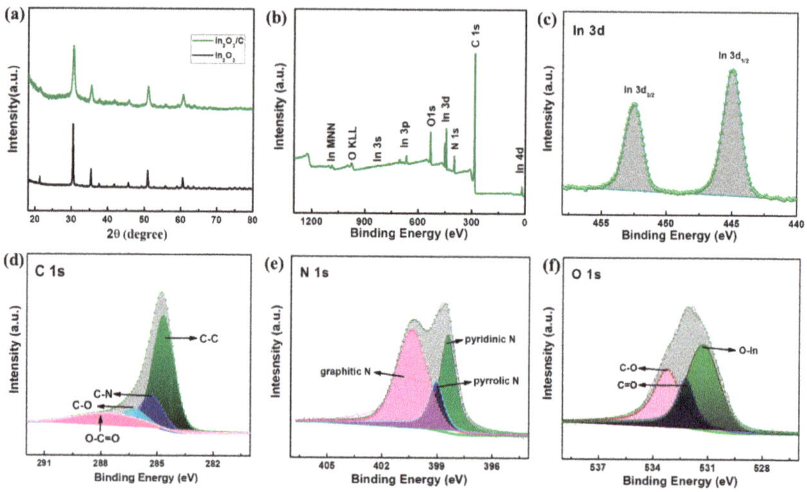

Figure 3. (**a**) XRD patterns In_2O_3 nanofibers and In_2O_3/C composite nanofibers. (**b**) Full XPS spectrum of In_2O_3/C composite nanofibers, (**c**–**f**) High-resolution In 3d, C 1s, N 1s, and O1s XPS spectrum of In_2O_3/C composite nanofibers.

A series of electrochemical tests were conducted to evaluate the lithium storage performance of the two kinds of nanofiber electrodes. Figure 4a,c present the CV curves of the In_2O_3 nanofibers and In_2O_3/C composite nanofibers electrode at a scan rate of 0.2 mV s^{-1} for the initial five cycles, respectively. These two electrodes exhibit similar but distinct profiles; they both deliver a prominent broad peak at about 0.48 V in the first reduction scan curve, which is mainly related to the generation of the SEI layer. Two obvious reduction peaks at about 0.8 and 0.4 V can be recorded in the next reduction scan curves, which correspond to the conversion reaction ($In_2O_3 + 6Li \longrightarrow 2In + 3Li_2O$) and alloying reaction ($In + xLi \longrightarrow InLix$), respectively. Notably, during the subsequent anodic scan process, the two nanofiber electrodes deliver two significant oxidation peaks at 0.45 and 0.70 V, which can be attributed to the gradual dealloying reaction of the indium lithium alloy ($InLi_x \longrightarrow In + xLi$) [42,47]. In addition, the higher anodic peaks (range from 1.0–2.0 V) correspond to the part oxidation of In to In_2O_3 ($2In + 3Li_2O \longrightarrow In_2O_3 + 6Li$) [40]. Based on the peaks position analysis above, the electrochemical lithium storage reactions process can be summarized as follows:

$$In_2O_3 + 6Li \longrightarrow 2In + 3Li_2O \quad (1)$$

$$In + xLi \longleftrightarrow InLi_x \quad (2)$$

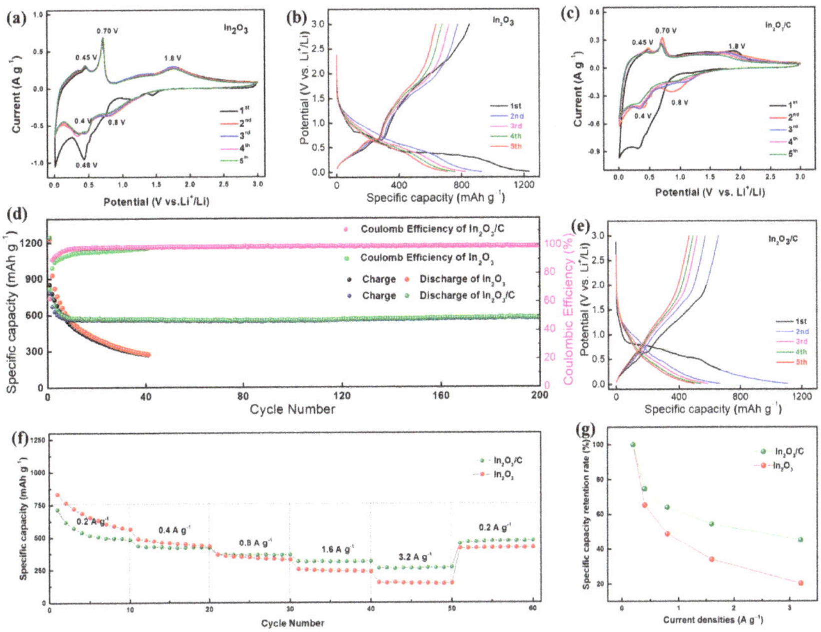

Figure 4. Electrochemical characterization of two nanofiber-based half cells; (**a**) CV curves and (**b**) charge–discharge profiles of In_2O_3 nanofibers electrode in the initial 5 cycles, (**c**) CV curves and (**e**) charge–discharge profiles of In_2O_3/C composite nanofibers electrode in the initial 5 cycles, (**d**) cyclic performance and Coulombic efficiency curves of the two nanofiber-based electrodes, (**f**) rate performance for the two nanofiber-based electrodes, (**g**) capacity retention of these two electrodes at different current densities.

It should be emphasized that the In_2O_3/C composite nanofibers electrode exhibits weaker oxidation-reduction peaks compared to the In_2O_3 nanofibers electrode; such a phenomenon is mainly due to the fact that the carbon nanofiber framework prevents direct contact between lithium and In_2O_3 in the In_2O_3/C nanofibers electrode. The corresponding typical charge–discharge profiles of the In_2O_3 nanofiber and In_2O_3/C composite nanofiber electrodes are demonstrated in Figure 4b,e, respectively. The initial discharge capacity of these two electrodes is significantly higher than their charge capacity; the large irreversible capacity is mainly due to the irreversible decomposition of the electrolyte into the SEI layer. Moreover, the In_2O_3/C composite nanofibers electrode delivers a weaker charging and discharging platform compared with the In_2O_3 composite nanofibers electrode, which is consistent with the CV testing and analysis. The detailed cycling performance and Coulombic efficiency of the two electrodes at a constant current of 0.3 A g^{-1} are plotted in Figure 4d; it can be seen that the In_2O_3/C composite nanofiber electrode's (olive dots) capacity tends to stabilize after several cycles; the delivered reversible capacity is 571 mAh g^{-1} after 200 charge/discharge cycles. On the other hand, the In_2O_3 nanofibers electrode (red dots) undergoes a sustained and rapid decay; its corresponding reversible capacity is only 269 mAh g^{-1} after 40 charge/discharge cycles.

Moreover, by observing the upper part of Figure 4d, we can see that the measured Coulombic efficiency of the In_2O_3/C composite nanofibers electrode (purple dots) is higher than that of the In_2O_3 nanofibers electrode (green dots), indicating that the In_2O_3/C composite nanofibers have a more stable SEI layer and superior reversible lithium storage capabilities compared with the In_2O_3 nanofibers sample.

In order to further compare their rate performance, the two nanofiber electrodes were tested at different current densities as shown in Figure 4f; the measured average reversible capacities of the In_2O_3/C composite nanofibers electrode are 544, 423, 365, 312, and 264 mAh g^{-1} at current densities of 0.2, 0.4, 0.8, 1.6, and 3.2 A g^{-1}, respectively. In contrast, the delivered capacities of the In_2O_3 nanofibers electrode are 663, 457, 344, 244, and 148 mAh g^{-1} at the same test current densities. The In_2O_3/C composite nanofiber electrode's capacity can restore to 470 mAh g^{-1} when the current density is set back to 0.2 A g^{-1}. By contrast, the recovery capacity of the pure In_2O_3 nanofibers electrode is only 418 mAh g^{-1}. In addition, the average capacity retention rates vs. current densities of the two nanofiber electrodes are recorded in Figure 4g; the reversible capacity retention rates of the In_2O_3/C composite nanofibers electrode are 100%, 78%, 67%, 57%, and 48% at 0.2, 0.4, 0.8, 1.6, and 3.2 A g^{-1}; the corresponding capacity retention rates of the In_2O_3 nanofibers electrode are 100%, 69%, 51%, 37%, and 22%. Thus, based on the above electrochemical testing and some recent reported In_2O_3 and carbonaceous materials in Table 1, we can conclude that the In_2O_3/C composite nanofibers electrode has better cycling stability and rate performance compared with the In_2O_3 nanofibers electrode.

Table 1. Electrochemical performance comparison table of present In_2O_3/C composite nanofibers with recent reported In_2O_3 and carbonaceous materials.

Electrode Description	Current Density (mA g^{-1})	Cycle Number	Reversible Capacity (mAh g^{-1})	High Rate Capability	Reference
In_2O_3/C composite nanofibers	300	200	571	264 mAh g^{-1} at 3.2 A g^{-1}	This work
SnO_2–In_2O_3/GNS	60	50	969	263 mAh g^{-1} at 0.6 A g^{-1}	[40]
C/In_2O_3 nanosheets	100	100	893	379 mAh g^{-1} at 2 A g^{-1}	[43]
In_2O_3/HPNC	1000	2000	623	139 mAh g^{-1} at 20 A g^{-1}	[44]
In_2O_3/C fibers	100	500	435	190 mAh g^{-1} at 1.5 A g^{-1}	[41]
In_2O_3 NC-EBA	200	100	910	316 mAh g^{-1} at 20 A g^{-1}	[38]
In_2O_3/carbon	100	150	720	140 mAh g^{-1} at 1.0 A g^{-1}	[47]

In order to better reveal the excellent lithium storage mechanism of the In_2O_3/C composite nanofibers, the cycled cells were put back into the glove box and disassembled to remove the working electrode. After washing away the residual electrolyte with propylene carbonate solvent, the cycled In_2O_3/C nanofibers sample was characterized as shown in Figure 5. From the SEM images in Figure 5a,b, we can see that the sample after cycling still maintains its typical nanofiber structure; moreover, the enlarged TEM images in Figure 5c,d demonstrate that the composite nanofibers have become rougher after cycling, which is mainly due to the formation of the SEI layer. Moreover, the element mapping demonstrates the existence of the In, C, O, N, F, and P elements of the cycled In_2O_3/C nanofiber (Figure 5e–j); compared with the pristine sample element mapping in Figure 2, the appearance F and P elements (Figure 5i,j) comes from the SEI layer, which involves the reversible decomposition of the $LiPF_6$-based electrolyte. On the whole, the In_2O_3/C nanofibers can buffer stress changes during the charging and discharging process. In addition, Electrochemical impedance spectroscopy (EIS) of In_2O_3/C composite nanofiber and pure In_2O_3 nanofiber electrodes after cycling is shown in Figure 6; the Nyquist data consist of a semicircle (at the high-medium frequency region) and a line (at the low frequency region); we can see that the R_{ct} (charge transfer, which can be obtained from the intercept of the semicircle on the *X*-axis) values of the In_2O_3/C composite nanofibers electrode and the In_2O_3 nanofibers electrode are about 58.5 and 205.5 Ω, respectively. The smaller resistance indicates that the In_2O_3/C composite nanofibers electrode possesses superior kinetics compared with the In_2O_3 nanofibers electrode.

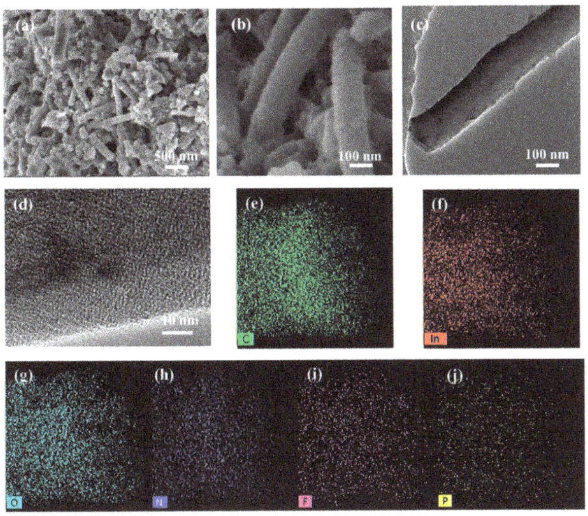

Figure 5. Morphological characterization of In_2O_3/C composite nanofibers sample after cycling. (**a**,**b**) SEM images and (**c**,**d**) TEM images. (**e**–**j**) elemental maps.

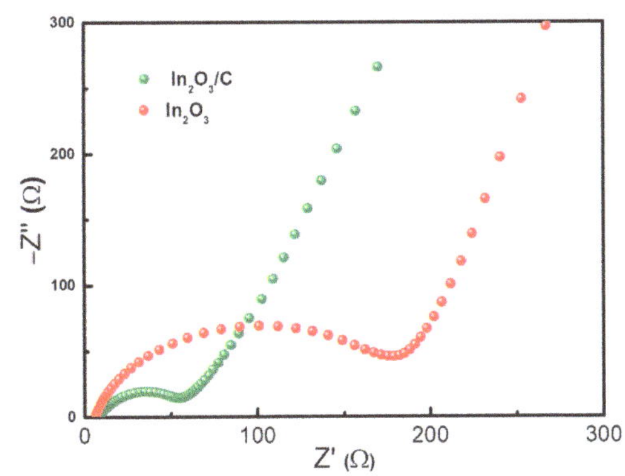

Figure 6. Electrochemical impedance spectroscopy of the In_2O_3/C composite nanofiber and the In_2O_3 nanofiber electrodes after cycling.

3. Materials and Methods

3.1. Synthesis of Sample

The present indium oxide-based nanofibers were prepared by typical electrostatic spinning technology and thermal annealing process; the first step mainly involves preparing a suitable spinning gel solution; a primary, 1.2 g polyvinylpyrrolidone (PVP, purchased from Sigma Aldrich, Steinheim, Germany) polymer with molecular weight of 1,300,000 was mixed into 8 g N,N-dimethylformamide solvent (DMF, purchased from Innochem, Beijing, China); the mixed solution was continuously stirred by magnetic force for about 4 h until the binder was completely dissolved; then, 0.6 g $In(NO_3)_3{\cdot}H_2O$ (purchase from

Innochem, Beijing, China) was placed in the mixed solution; the electrospinning precursor solution was finally prepared when the mixed solution was further stirred to form a uniform viscous gel.

The second step was to convert the precursor gel solution into nanofibers via electrospinning technology. The obtained precursor gel solution was inhaled into a 5 mL plastic injection coupling with stainless steel needle; the inner diameter of the stainless steel needle was about 0.5 mm. The plastic injection filled with precursor gel solution was loaded on an electrospinning machine (WL-2, Beijing Aibo Zhiye Ion Technology Limited Company, Beijing, China), the gel solution was evenly pushed out from the syringe by the propeller with speed about 1.5 mL per hour, and the stainless steel needle was connected with the high-voltage positive wire of the power supply with a positive voltage of 13.5 kV. The gel solution coming out of the needle was stretched into nanofibers under the action of a strong electric field during the electrospinning process; simultaneously, the rapid DMF solvent evaporation lead to nanofiber solidification, the electrospinning nanofibers product was collected by a grounded drum with a rotation speed of 40 revolutions per minute, the straight-line distance between the stainless steel needle tip and the drum was approximately 13 cm, and the temperature and humidity inside the electrospinning machine were about 30 °C and 40%, respectively.

When the gel solution was completely exhausted, the electrospinning product was carefully peeled off from the drum. Immediately, the product was placed in an 80 °C oven for 6 h to dry the adsorbed water and residual solvents; after that, the sharped electrospinning nanofibers sample was placed in a 260 °C muffle furnace in air for about 4 h; the relatively low oxidation temperature is sufficient to oxidize indium nitrate but does not completely calcine the PVP carbon source. Finally, the pre-oxidized nanofibers sample was put into a tube furnace filled with argon atmosphere at 600 °C for 4 h; such relatively high temperature is sufficient to carbonize PVP without reducing indium oxide. Thus, In_2O_3 and carbon materials were in situ compounded in the nanofibers. As a comparison, the In_2O_3 nanofibers sample was obtained by calcining the pre-oxidized nanofibers in air at 600 °C for 4 h to completely remove the PVP carbon source.

3.2. Material Characterization

The composition of the obtained sample was examined using X-ray diffraction equipment (XRD, Rigaku D/Max-2400 diffractometer, Tokoy, Japan); the wavelength of the incident electromagnetic wave is 0.15406 nm, which originates from the transition from L layer electrons to K layer electrons in metallic copper. The diffraction angle of the sample was set between 10 and 80° with a scanning speed of 5° min^{-1}. The detailed sample elemental composition, chemical valence, and bonding energy were studied by X-ray photoelectron spectroscopy (XPS, Kratos AXISUltra DLD, Manchester, UK); the photoelectrons were generated by the transition of electrons from the L layer to the K layer in aluminum metal, the corresponding energy was 1486.6 eV, the sample testing was conducted in a high vacuum environment, and XPS Peak software was applied to fit the high-resolution spectra.

The precise microstructure of the sample was characterized by field-emission scanning electron microscopy (FE-SEM, S-4800, Hitachi, Tokoy, Japan) and FEI transmission electron microscopy (TEM, FEI, Tecnai G^2 F20, Waltham, MA, USA). In order to conduct SEM observation, the prepared nanofibers sample was adhered on a metal tray with conductive adhesive. To prepare the TEM observation, a small amount of the nanofibers sample was sonicated and dispersed in an alcohol solvent for about 10 min; subsequently, a dropper was applied to extract the mixed solution and drop it onto a microgrid; after drying the residual alcohol, the microgrid was sent into the sample chamber for vacuum pumping.

3.3. Cell Preparation and Testing

The electrochemical lithium storage performance of prepared In_2O_3 nanofibers and In_2O_3/C composite nanofibers were investigated by assembling half cells. The working electrodes were fabricated via the conventional slurry approach. Briefly, the active materials of the nanofibers sample, acetylene black (as a conductive agent) and sodium alginate (as a binder), were accurately weighed with mass ratio of 8:1:1; these three components were manually ground and mixed in an agate mortar for about 1 h, and then an appropriate amount of deionized water was added to the mixture and continuously stirred to form a viscous slurry, which was evenly coated on the surface of copper foil by a scraper. After drying the deionized water solvent in a vacuum oven, the coated copper foil was sliced into small circular pieces with a diameter of 12 mm as working electrodes.

Next, the cells were assembled in a glove box (Etelux, Beijing, China) filled with inert gas atmosphere (the concentration of argon gas is higher than 99.9%); the prepared working electrode, glass fiber (as a separator), lithium metal foil (as a reference electrode and counter electrode), stainless steel gasket, and spring sheet were sequentially placed into the positive electrode shell; subsequently, approximately 120 μL of organic electrolyte (1 M $LiPF_6$ dissolved in ethylene carbonate (EC) and dimethyl carbonate (DEC)) was dropped into the cell, and then the negative electrode shell was placed on top of the positive electrode case. Finally, the two sides of the cells were sealed using a hydraulic press. In order to allow the electrolyte to fully infiltrate the battery system, the assembled cells were taken out of the glove box and aged for about 24 h; the open circuit voltage of these half cells was approximately 2.0–3.5 V. The cycling and rate performance of the batteries were evaluated on a battery testing channel (BTS-610, Neware, Shenzhen, China). The corresponding cyclic voltammetry (CV) was conducted on electrochemical workstation (CHI-660E, Chenhua, Shanghai, China), Electrochemical impedance spectroscopy (EIS) was carried out in the frequency range of 0.01–100,000 Hz with an applied perturbation voltage of 5 mV, and all of these electrochemical test voltage windows ranged from 0.01 to 3.0 V vs. Li^+/Li.

4. Conclusions

In summary, we have encapsulated ultrafine In_2O_3 particles in a carbon nanofiber framework as an advanced negative electrode material for LIBs; the carbon nanofiber framework isolating In_2O_3 particles prevents the agglomeration of active materials and the uncontrollable growth of SEI, and the volume changes in ultrafine In_2O_3 particles are buffered by the carbon materials, allowing the overall structure of the In_2O_3/C composite nanofiber to remain largely intact without crushing during charging and discharging cycles, which enables superior cycle performance of the In_2O_3/C composite nanofibers electrode (571 mAh g^{-1} after 200 cycles at 0.3 A g^{-1}). Moreover, the carbon nanofiber structure provides an effective electronic conductive network and reduces the length of lithium-iondiffusion, which enhances the reactivity of the In_2O_3/C nanocomposite and improves electrode kinetics (264 mAh g^{-1} at 3.2 A g^{-1}). In contrast, the pure In_2O_3 nanofibers electrode shows a sustained and rapid capacity decay trend. This superior design is beneficial to promote the application of the In_2O_3/C composite in the negative electrode material of next-generation lithium-ion batteries.

Author Contributions: Conceptualization, W.X.; methodology, W.X. and Z.A. validation, X.L. and Q.W.; formal analysis, C.H. and Y.M.; investigation, Y.M.; resources, S.L.; data curation, H.S.; writing—original draft preparation, W.X.; writing—review and editing, X.S.; visualization, X.S.; supervision, H.S.; project administration, W.X.; funding acquisition, H.S. All authors have read and agreed to the published version of the manuscript.

Funding: This work was supported by Key Research Projects of Henan Provincial Department of Education (20A480005), National Natural Science Foundation of China (No. U2004174), and Research Projects for College students of XYNU (24100101262).

Data Availability Statement: The data presented in this study are available on request from the corresponding author.

Acknowledgments: All authors thank the Analysis and Testing Center of Xinyang Normal University for providing assistance.

Conflicts of Interest: The authors declare no conflicts of interest.

References

1. Li, M.; Lu, J.; Chen, Z.; Amine, K. 30 Years of Lithium-Ion Batteries. *Adv. Mater.* **2018**, *30*, 1800561. [CrossRef]
2. Sun, X.; Yang, J.; Chen, Y.; Luo, F. Interconnected MoO_2/MoS_2@NC nanosheets as anodes with high-rate and long-life for lithium-ion and sodium-ion batteries. *Chem. Eng. J.* **2024**, *495*, 153418. [CrossRef]
3. Sun, X.; Jing, M.; Dong, H.; Xie, W.; Luo, F. CuO-ZnO submicroflakes with nanolayered Al_2O_3 coatings as high performance anode materials in lithium-ion batteries. *J. Alloys Compd.* **2023**, *953*, 170137. [CrossRef]
4. Kulova, T.L.; Fateev, V.N.; Seregina, E.A.; Grigoriev, A.S. A Brief Review of Post-Lithium-Ion Batteries. *Int. J. Electrochem. Sci.* **2020**, *15*, 7242–7259. [CrossRef]
5. Bai, X.; Li, D.; Zhang, D.; Yang, S.; Pei, C.; Sun, B.; Ni, S. Boosting high-rate lithium storage in Li_3VO_4 via a honeycomb structure design and electrochemical reconstruction. *J. Mater. Chem. A* **2023**, *11*, 12164–12175. [CrossRef]
6. Zhang, Z.; Yang, C.; Wu, S.; Wang, A.; Zhao, L.; Zhai, D.; Ren, B.; Cao, K.; Zhou, Z. Exploiting Synergistic Effect by Integrating Ruthenium–Copper Nanoparticles Highly Co-Dispersed on Graphene as Efficient Air Cathodes for Li–CO_2 Batteries. *Adv. Energy Mater.* **2019**, *9*, 1802805. [CrossRef]
7. Zhang, X.; Lv, B.; Peng, Y.; Li, Q.; Chen, M.; Liu, X.; Wang, Z.; Yang, J.; Chen, J.; Gu, D. Dual-confined growth of mesoporous molybdenum nitride@carbon arrays for fast pseudocapacitive lithium storage. *J. Alloys Compd.* **2024**, *1003*, 175521. [CrossRef]
8. Manthiram, A.; Goodenough, J.B. Lithium-based polyanion oxide cathodes. *Nat. Energy* **2021**, *6*, 844–845. [CrossRef]
9. Du, J.; Zhang, C.; Li, S.; Zhang, L.; Zhang, W. Two-stage prediction method for capacity aging trajectories of lithium-ion batteries based on Siamese-convolutional neural network. *Energy* **2024**, *295*, 130947. [CrossRef]
10. Zhang, X.-Q.; Zhao, C.-Z.; Huang, J.-Q.; Zhang, Q. Recent Advances in Energy Chemical Engineering of Next-Generation Lithium Batteries. *Engineering* **2018**, *4*, 831–847. [CrossRef]
11. Gao, S.; Tang, Y.; Zhao, H.; Liu, L.; Gu, Y.; Sheng, R. MoO/C hybrid synthesized by a facile molten-salt-assisted approach for high-performance lithium-ion batteries. *Int. J. Energy Res.* **2021**, *45*, 6418–6425. [CrossRef]
12. Chen, C.; Hu, Q.; Xue, H.; Li, H.; Li, W.; Cao, S.; Peng, T.; Yang, Y.; Luo, Y. Ultrafast and ultrastable $FeSe_2$ embedded in nitrogen-doped carbon nanofibers anode for sodium-ion half/full batteries. *Nanotechnology* **2024**, *35*, 055404. [CrossRef] [PubMed]
13. Nzereogu, P.U.; Omah, A.D.; Ezema, F.I.; Iwuoha, E.I.; Nwanya, A.C. Anode materials for lithium-ion batteries: A review. *Appl. Surf. Sci. Adv.* **2022**, *9*, 100233. [CrossRef]
14. Gao, S.; Liu, L.; Mao, F.; Zhang, Z.; Pan, K.; Zhou, Z. Coal-based ultrathin N-doped carbon nanosheets synthesized by molten-salt method for high-performance lithium-ion batteries. *Nanotechnology* **2022**, *33*, 425401. [CrossRef]
15. Becker, D.; Haberkorn, R.; Kickelbick, G. Mechanochemical Induced Structure Transformations in Lithium Titanates: A Detailed PXRD and 6Li MAS NMR Study. *Inorganics* **2018**, *6*, 117. [CrossRef]
16. Opra, D.P.; Sokolov, A.A.; Sinebryukhov, S.L.; Tkachenko, I.A.; Ziatdinov, A.M.; Gnedenkov, S.V. Electronic Structure, Optical and Magnetic Properties of Oxygen-Deficient Gray $TiO_{2-\delta}$(B). *Inorganics* **2022**, *10*, 184. [CrossRef]
17. Opra, D.P.; Sinebryukhov, S.L.; Modin, E.B.; Sokolov, A.A.; Podgorbunsky, A.B.; Ziatdinov, A.M.; Ustinov, A.Y.; Mayorov, V.Y.; Gnedenkov, S.V. Manganese, Fluorine, and Nitrogen Co-Doped Bronze Titanium Dioxide Nanotubes with Improved Lithium-Ion Storage Properties. *Batteries* **2023**, *9*, 229. [CrossRef]
18. Chen, M.; Liu, F.-M.; Zhao, H.; Chen, S.-S.; Qian, X.; Yuan, Z.-Y.; Wan, R. In situ encapsulation of iron oxide nanoparticles into nitrogen-doped carbon nanotubes as anodic electrode materials of lithium ion batteries. *Phys. Chem. Chem. Phys.* **2022**, *24*, 27114–27120. [CrossRef] [PubMed]
19. Chen, M.; Zhao, M.-Y.; Liu, F.-M.; Li, M.-T.; Zhang, M.-L.; Qian, X.; Yuan, Z.-Y.; Li, C.-S.; Wan, R. Self-Catalyzed Synthesis of Length-Controlled One-Dimensional Nickel Oxide@N-Doped Porous Carbon Nanostructures from Metal Ion Modified Nitrogen Heterocycles for Efficient Lithium Storage. *Langmuir* **2024**, *40*, 4852–4859. [CrossRef]
20. Guo, Y.; Zhang, D.; Bai, Z.; Yang, Y.; Wang, Y.; Cheng, J.; Chu, P.K.; Luo, Y. MXene nanofibers confining MnO_x nanoparticles: A flexible anode for high-speed lithium ion storage networks. *Dalton Tran.* **2022**, *51*, 1423–1433. [CrossRef]
21. Guo, Y.; Zhang, D.; Yang, Y.; Wang, Y.; Bai, Z.; Chu, P.K.; Luo, Y. MXene-encapsulated hollow Fe_3O_4 nanochains embedded in N-doped carbon nanofibers with dual electronic pathways as flexible anodes for high-performance Li-ion batteries. *Nanoscale* **2021**, *13*, 4624–4633. [CrossRef]
22. Cao, K.; Jia, Y.; Wang, S.; Huang, K.-J.; Liu, H. Mn_3O_4 nanoparticles anchored on carbon nanotubes as anode material with enhanced lithium storage. *J. Alloys Compd.* **2021**, *854*, 157179. [CrossRef]
23. Cao, K.; Zheng, R.; Wang, S.; Shu, J.; Liu, X.; Liu, H.; Huang, K.-J.; Jing, Q.-S.; Jiao, L. Boosting Coulombic Efficiency of Conversion-Reaction Anodes for Potassium-Ion Batteries via Confinement Effect. *Adv. Funct. Mater.* **2020**, *30*, 2007712. [CrossRef]
24. Sun, X.; Chen, Y.; Luo, F. Nanoporous ALD-modified oxygen-deficient NiO flakes as anodes for lithium-ion batteries. *Ceram. Int.* **2024**, *50*, 3480–3490. [CrossRef]
25. Li, C.; Ke, S.; Liu, S.; Wu, G.; Li, Q.; Zhang, Y.; Cao, K. Heterostructured Mn–Sn Bimetallic Sulfide Nanocubes Confined in N, S-co-Doped Carbon Framework as High-Performance Anodes for Sodium-Ion Batteries. *Langmuir* **2024**, *40*, 15815–15823. [CrossRef] [PubMed]
26. Chen, M.; Liu, F.-M.; Chen, S.-S.; Qian, X.; Zhao, Y.-J.; Sun, Y.; Li, C.-S.; Wan, R.; Yuan, Z.-Y. Low-temperature metal-catalyzed synthesis of encapsulated metal oxide nanoparticles in nitrogen-doped carbon nanotubes from carbon nitride as anodic materials of high-performance lithium-ion batteries. *New J. Chem.* **2023**, *47*, 3215–3221. [CrossRef]

27. Wang, Y.; Yang, Y.; Zhang, D.; Wang, Y.; Luo, X.; Liu, X.; Kim, J.-K.; Luo, Y. Inter-overlapped MoS_2/C composites with large-interlayer-spacing for high-performance sodium-ion batteries. *Nanoscale Horiz.* **2020**, *5*, 1127–1135. [CrossRef]
28. Liu, H.; He, Y.; Zhang, H.; Cao, K.; Wang, S.; Jiang, Y.; Jing, Q.-S.; Jiao, L. Lowering the voltage-hysteresis of CuS anode for Li-ion batteries via constructing heterostructure. *Chem. Eng. J.* **2021**, *425*, 130548. [CrossRef]
29. Xu, J.; Liu, Q.; Dong, Z.; Wang, L.; Xie, X.; Jiang, Y.; Wei, Z.; Gao, Y.; Zhang, Y.; Huang, K. Interconnected MoS_2 on 2D Graphdiyne for Reversible Sodium Storage. *ACS Appl. Mater. Interfaces* **2021**, *13*, 54974–54980. [CrossRef]
30. Guo, Y.; Zhang, Y.; Wang, Y.; Zhang, D.; Lu, Y.; Luo, R.; Wang, Y.; Liu, X.; Kim, J.-K.; Luo, Y. Vertically aligned ultrathin MoS_2 nanosheets grown on graphene-wrapped hollow carbon microtubes derived from loofah sponge as advanced anodes for highly reversible lithium storage. *Electrochim. Acta* **2019**, *296*, 989–998. [CrossRef]
31. Liu, H.; He, Y.; Zhang, H.; Cao, K.; Wang, S.; Jiang, Y.; Jing, Q.-S.; Jiao, L. Bi-continuous ion/electron transfer avenues enhancing the rate capability of SnS2 anode for potassium-ion batteries. *J. Power Sources* **2021**, *506*, 230160. [CrossRef]
32. Wang, Y.; Zhang, D.; Yang, Y.; Guo, Y.; Bai, Z.; Chu, P.K.; Luo, Y. Three-dimensional nano/micro-structured porous MoP/CNTs microspheres as high-capacity anode for lithium-ion batteries. *J. Alloys Compd.* **2021**, *872*, 159608. [CrossRef]
33. Guo, W.; Mei, L.; Li, X.; Mao, M.; Ma, J. Electrospun In@C nanofibers as a superior Li-ion battery anode. *RSC Adv.* **2015**, *5*, 92522–92525. [CrossRef]
34. Na, Z.; Yao, R.; Yan, Q.; Wang, X.; Sun, X. Metal-organic frameworks derived In-based nanoparticles encapsulated by carbonaceous matrix for highly efficient energy storage. *Appl. Surf. Sci.* **2020**, *513*, 145894. [CrossRef]
35. Zeng, Q.; Dong, Y.; Chen, Y.; Yue, X.; Liang, Z. Recent advancements and perspectives of fast-charging composite anodes for lithium-ion batteries. *Sci. Chin.-Chem.* **2024**, *67*, 3952–3963. [CrossRef]
36. Pendashteh, A.; Tomey, R.; Vilatela, J.J. Nanotextile 100% Si Anodes for the Next Generation Energy-Dense Li-ion Batteries. *Adv. Energy Mater.* **2024**, *14*, 2304018. [CrossRef]
37. Du, X.; Zhang, H.; Lan, X.; Yuan, B.; Hu, R. Sn Alloy and Graphite Addition to Enhance Initial Coulombic Efficiency and Cycling Stability of SiO Anodes for Li-Ion Batteries. *Energy Environ. Mater.* **2022**, *5*, 353–359. [CrossRef]
38. Pham, H.T.; Lee, D.-S.; Dao, T.D.; Jeong, H.-D. In_2O_3 nanocrystal–π conjugated molecule hybrid materials for high-capacity anode in lithium ion battery. *J. Ind. Eng. Chem.* **2018**, *57*, 22–27. [CrossRef]
39. Wang, Q.; Xu, J.; Shen, G.; Guo, Y.; Zhao, X.; Xia, Y.; Sun, H.; Hou, P.; Xie, W.; Xu, X. Large-scale carbon framework microbelts anchoring ultrafine SnO_2 nanoparticles with enhanced lithium storage properties. *Electrochim. Acta* **2019**, *297*, 879–887. [CrossRef]
40. Yang, H.; Song, T.; Lee, S.; Han, H.; Xia, F.; Devadoss, A.; Sigmund, W.; Paik, U. Tin indium oxide/graphene nanosheet nanocomposite as an anode material for lithium ion batteries with enhanced lithium storage capacity and rate capability. *Electrochim. Acta* **2013**, *91*, 275–281. [CrossRef]
41. Zhao, H.; Yin, H.; Yu, X.-X.; Zhang, W.; Li, C.; Zhu, M.-Q. In_2O_3 nanoparticles/carbon fiber hybrid mat as free-standing anode for lithium-ion batteries with enhanced electrochemical performance. *J. Alloys Compd.* **2018**, *735*, 319–326. [CrossRef]
42. Yue, L.; Pan, X.; Chen, S.; Song, J.; Liu, C.; Luo, G.; Guan, R.; Zhang, W. High lithium storage capacity achieved by regulating monodisperse C/In_2O_3 nanosheet composite with double phases. *Mater. Chem. Phys.* **2017**, *193*, 89–98. [CrossRef]
43. Yue, L.; Zhang, W.; Zhang, W.; Zhang, Q.; Guan, R.; Hou, G.; Xu, N. One-step solvothermal process of In_2O_3/C nanosheet composite with double phases as high-performance lithium-ion battery anode. *Electrochim. Acta* **2015**, *160*, 123–130. [CrossRef]
44. Xu, H.; Wang, L.; Zhong, J.; Wang, T.; Cao, J.; Wang, Y.; Li, X.; Fei, H.; Zhu, J.; Duan, X. Ultra-stable and High-rate Lithium Ion Batteries Based on Metal–organic Framework-derived In_2O_3 Nanocrystals/Hierarchically Porous Nitrogen-doped Carbon Anode. *Energy Environ. Mater.* **2020**, *3*, 177–185. [CrossRef]
45. Kim, M.; Park, C.; Jung, W.; Hur, J. Enabling high capacity and reversible Li storage in indium (III) oxide anode surrounded by carbon nanotube matrix. *J. Alloys Compd.* **2024**, *996*, 174796. [CrossRef]
46. Xue, Y.; Xu, T.; Guo, Y.; Song, H.; Wang, Y.; Guo, Z.; Li, J.; Zhao, H.; Bai, X.; Lai, C. Amorphous/crystalline heterostructured indium (III) sulfide/carbon with favorable kinetics and high capacity for lithium storage. *Adv. Compos. Hybrid Mater.* **2024**, *7*, 213. [CrossRef]
47. Jin, L.; Zhao, X.; Qian, X.; Wang, S.; Shen, X.; Dong, M. Synthesis of porous In_2O_3 /carbon composites derived from metal-organic frameworks for high performance Li-ion batteries. *Mater. Lett.* **2017**, *199*, 176–179. [CrossRef]
48. Li, X.; Zhang, Z.; Gong, L.; Zhang, Z.; Liu, G.; Tan, P. Revealing the mechanism of stress rebound during discharging in lithium-ion batteries. *J. Energy Storage* **2023**, *58*, 106454. [CrossRef]
49. Li, R.; Li, W.; Singh, A.; Ren, D.; Hou, Z.; Ouyang, M. Effect of external pressure and internal stress on battery performance and lifespan. *Energy Storage Mater.* **2022**, *52*, 395–429. [CrossRef]

inorganics MDPI

Article

A Novel Spinel High-Entropy Oxide $(Cr_{0.2}Mn_{0.2}Co_{0.2}Ni_{0.2}Zn_{0.2})_3O_4$ as Anode Material for Lithium-Ion Batteries

Changqing Jin [1,*], Yulong Wang [1], Haobin Dong [1], Yongxing Wei [1], Ruihua Nan [1], Zengyun Jian [1], Zhong Yang [1] and Qingping Ding [2,*]

1 Shaanxi Key Laboratory of Optoelectronic Functional Materials and Devices, School of Materials and Chemical Engineering, Xi'an Technological University, Xi'an 710021, China; wyl20000223@163.com (Y.W.); dhaobin@163.com (H.D.); weiyx1985@gmail.com (Y.W.); nanrh@xatu.edu.cn (R.N.); jianzengyun@xatu.edu.cn (Z.J.); yz750925@163.com (Z.Y.)

2 Ames National Laboratory, U.S. DOE, and Department of Physics and Astronomy, Iowa State University, Ames, IA 50011, USA

* Correspondence: eaglejin@xatu.edu.cn (C.J.); qpding@ameslab.gov (Q.D.)

Abstract: In this study, we synthesized spinel high-entropy oxide (HEO) $(Cr_{0.2}Mn_{0.2}Co_{0.2}Ni_{0.2}Zn_{0.2})_3O_4$ nanoparticles by a simple solution combustion method. These particles were investigated for their performance as anodes in lithium-ion batteries. The reversible capacity is 132 mAh·g^{-1} after 100 cycles at a current density of 100 mA·g^{-1}, 107 mAh·g^{-1} after 1000 cycles at a current density of 1 A g^{-1}, and 96 mAh·g^{-1} rate capacity at a high current density of 2 A g^{-1}. The outstanding cycle stability under high current densities and remarkable rate performance can be attributed to the stable structure originating from the high entropy of the material.

Keywords: high-entropy oxide; lithium-ion battery; spinel; nanoparticle; anode

1. Introduction

Lithium-ion batteries (LIBs) have seen swift adoption in consumer electronics and electric vehicles [1–5]. However, existing research has yet to satisfy the growing demand for batteries with higher electrochemical lithium storage performance. A key limitation lies in the lithium storage capabilities of anode materials [6,7]. Currently, the widely used commercial anode materials for LIBs—natural or synthetic graphite—have a specific capacity of only 372 mAh·g^{-1}, which can no longer meet the demands for high energy density and high-rate performance in the next generation of lithium-ion batteries. Commonly studied anode materials, including transition metal oxides [8,9] and silicon [10], undergo significant volume expansion during lithium intercalation/deintercalation. This expansion can lead to electrode structural failure and increased internal resistance, which in turn degrade lithium storage performance. Hence, the innovation of novel anode materials is essential.

A material that has shown promise in anode materials is the high-entropy oxide (HEO). Derived from the concept of high-entropy alloys (HEAs) [11–14], HEOs are composed of five or more metals in equal molar amounts, leading to notable alterations in composition, structure, and properties. HEOs leverage configurational entropy for phase stabilization [15], with higher entropy generally correlating with greater phase stability. Unlike other materials like transition metal oxides (TMOs), which depend on intrinsic energy for phase stabilization and have crystal structures determined by energy minimization under specific conditions, HEOs with five or more metal elements demonstrate potential for innovative lithium storage performance.

In 2016, Berardan [16] first introduced HEOs, emphasizing the unique properties stemming from their elevated configurational entropy. In 2018, Sarkar [17] et al. synthesized a rock salt structured HEO, $(Co_{0.2}Cu_{0.2}Mg_{0.2}Ni_{0.2}Zn_{0.2})O$, via pyrolysis, which showed a specific capacity of 770 mAh/g after 100 cycles at 200 mA/g. That year, Qiu [18] et al. created

(MgCoNiCuZn)O HEOs by a solid-phase reaction, showcasing an initial discharge capacity of 1585 mAh/g and excellent cycling stability; after 300 cycles at 100 mA/g, it retained a reversible capacity of 920 mAh/g. Even at 3000 mA/g, it maintained 490 mAh/g. In 2020, Wang [19] et al. developed a spinel-structured HEO, $(FeCoNiCrMn)_3O_4$, with outstanding cycling stability and rate performance due to its large lattice parameter facilitating 3D ion and electron diffusion. The anode, composed of five elements with diverse valences, induced lattice distortions and electronic modifications, enhancing conductivity and lithium storage. In 2021, Chen [20] et al. detailed the lithium storage mechanism of the spinel HEO $(Ni_{0.2}Co_{0.2}Mn_{0.2}Fe_{0.2}Ti_{0.2})_3O_4$, which delivered a high capacity of 560 mAh/g at 100 mA/g. Duan [21] et al. synthesized single-phase spinel HEOs, $(FeCoNiCrMnXLi)_3O_4$ (X = Cu, Mg, Zn), with seven metal elements in equal molar ratios, enhancing LIB anode performance through rapid 3D Li+ transport and entropy-driven stabilization. Rock salt-structured HEOs exhibit durable and reversible lithium storage due to the entropic stabilization effect, enhancing their cyclic stability [22,23]. In contrast, spinel-structured HEOs, which include metal cations with a broader range of oxidation states from divalent to tetravalent [24], can cause larger lattice distortions and more defects, and greatly alter the electronic structure. These effects improve electrical conductivity, ion diffusion, and lithium storage capacity [25–27]. However, despite these advantages for lithium-ion batteries, the utilization of HEOs in these applications remains limited.

Researchers in this field are predominantly engaged in material exploration and the enhancement of their electrochemical lithium storage properties. However, a universal framework for material design has not yet been established. The element selection for HEO synthesis poses a critical challenge in leveraging HEOs as anode materials in lithium-ion batteries (LIBs). There is an urgent need for a coherent strategy to design HEOs with diverse structures and high specific capacities. For instance, our strategy for spinel-structured HEOs encompasses three key aspects: (1) structural stability, ensuring that the ionic radii of the ions in the HEO, normalized by their coordination numbers, are nearly the same; (2) ionic valence selection that satisfies the chemical formula, where variable valences of ions in the HEO must align with the formula's requirements; for a spinel with AB_2O_4, suitable valences might be A^{2+} and B^{3+} or A^{4+} and B^{2+}; and (3) achieving a high specific capacity by formulating the HEO to allow for the reduction of high-valence metal cations to elemental metals or lithium alloys during lithiation, requiring the constituent simple oxides to possess a high theoretical specific capacity. By adhering to these principles, a variety of HEOs can be engineered with properties optimized for superior performance in energy storage, catalysis, and sensing applications. On this basis, we have successfully developed a novel HEO, $(Cr_{0.2}Mn_{0.2}Co_{0.2}Ni_{0.2}Zn_{0.2})_3O_4$, tailored for electrochemical lithium storage.

2. Results

The XRD pattern in Figure 1a confirms the successful synthesis of $(Cr_{0.2}Mn_{0.2}Co_{0.2}Ni_{0.2}Zn_{0.2})_3O_4$ with a single-phase spinel structure (space group Fd-3m). As illustrated in Figure 1, the diffraction peaks at 2θ of 18.71°, 30.78°, 36.27°, 37.94°, 44.10°, 54.74°, 58.37°, 64.13°, and 75.95° correspond to (111), (220), (311), (222), (400), (422), (511), (440), and (533) crystal planes of $(Cr_{0.2}Mn_{0.2}Co_{0.2}Ni_{0.2}Zn_{0.2})_3O_4$, respectively. No impurity phase was observed, indicating that $(Cr_{0.2}Mn_{0.2}Co_{0.2}Ni_{0.2}Zn_{0.2})_3O_4$ forms a single-phase spinel structure. The Rietveld method (software: FullProf(64 bits), Version: 5-May-2020) was performed based on the XRD data using the space group of Fd3m. Figure 1b shows the fitting results; the lattice constant, density, and crystallite size of the sample are determined to be 8.289 Å, 9.250 g/cm^3, and 28.0 nm, respectively.

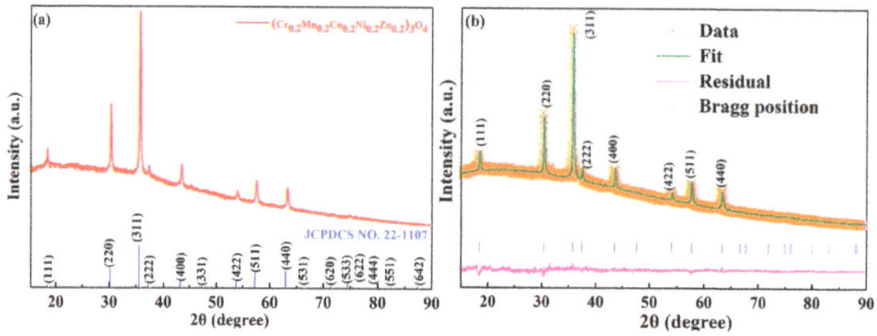

Figure 1. (**a**) XRD pattern; (**b**) Rietveld refinement pattern of the $(Cr_{0.2}Mn_{0.2}Co_{0.2}Ni_{0.2}Zn_{0.2})_3O_4$ nanoparticles.

Figure 2 displays the morphology and microstructure of the $(Cr_{0.2}Mn_{0.2}Co_{0.2}Ni_{0.2}Zn_{0.2})_3O_4$ nanoparticles. The SEM image (a) and TEM image (g) represent the overall particle structure at low and high magnification, respectively. As seen in the detailed illustration in Figure 2g, the nanoparticles exhibit a relatively narrow distribution range of particle diameter (10.9–67.7 nm), with an average particle diameter of approximately 28 nm. This is attributed to the short calcination time (40 min) and the influence of high-temperature gas during calcination, which prevents further growth and aggregation of nanoparticles. EDS mapping confirms the presence of massive metallic elements Cr (Figure 2b), Mn (Figure 2c), Co (Figure 2d), Ni (Figure 2e), and Zn (Figure 2f) within the $(Cr_{0.2}Mn_{0.2}Co_{0.2}Ni_{0.2}Zn_{0.2})_3O_4$ nanoparticles, with similar distribution densities observed. Quantitative analysis of each element's content was performed, and the results are presented in Table 1. The atomic proportions of Cr, Mn, Co, Ni, and Zn elements are approximately equal, aligning with our experimental goals. The high-resolution TEM image in the inset of Figure 2g indicates that the well-resolved interplanar spacing of the lattice fringe is about 0.285 nm, which corresponds to the (220) plane of spinel-structured $(Cr_{0.2}Mn_{0.2}Co_{0.2}Ni_{0.2}Zn_{0.2})_3O_4$.

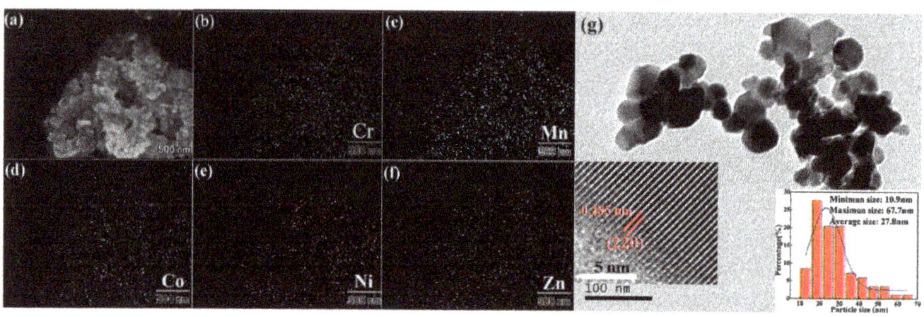

Figure 2. SEM image (**a**) and corresponding EDS elemental maps of Cr (**b**), Mn (**c**), Co (**d**), Ni (**e**), and Zn (**f**), and TEM image (**g**) of $(Cr_{0.2}Mn_{0.2}Co_{0.2}Ni_{0.2}Zn_{0.2})_3O_4$ nanoparticles. The inset in (**g**) shows the distribution of the particle size.

Table 1. The atomic fraction of each element in the HEO sample.

Element	Atomic Fraction (%)
O	63.65
Cr	7.69
Mn	7.60
Co	7.08
Ni	7.29
Zn	6.69

Figure 3 illustrates the electrochemical lithium storage performance. Figure 3a illustrates the initial three cyclic voltammetry (CV) cycles tested at a scan rate of 0.1 mV/s. The initial cathodic scan of the cell exhibits an initial reduction peak around 0.965 V, which can be attributed to the reduction of metal cations in the HEO to metals, along with the formation of Li_2O [16]. Subsequently, a significant irreversible reduction peak emerges between 0 and 0.5 V due to the formation of the SEI film [17]. The disappearance of this irreversible peak in subsequent scans suggests that the quantity of SEI film generated remains essentially stable. In the first anodic scan, oxidation peaks are observed over a broad range of 1.0–2.0 V [28,29], with similar peaks in this range observed in subsequent scans, indicating a continuous and reversible oxidation process. The reactions that occur during this redox process are shown in Equation (1):

$$\frac{1}{y}M_xO_y + 2Li \overset{2e}{\Leftrightarrow} Li_2O + \frac{x}{y}M, M = Cr, Mn, Co, Ni, Zn \quad (1)$$

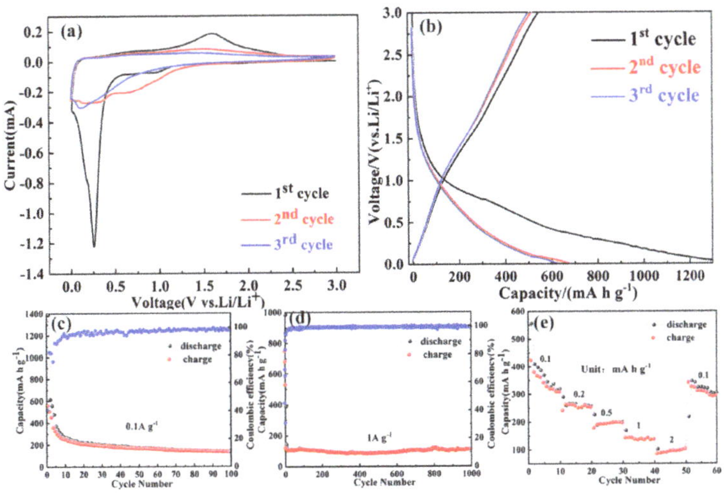

Figure 3. CV curves at a scan rate of 0. 1 mV·s^{-1} (**a**), discharge/charge profiles at 100 mA·g^{-1} (**b**), cycling performance at 100 mA·g^{-1} (**c**), long-term cycling performance at 1 A g^{-1} (**d**), and rate capability of $(Cr_{0.2}Mn_{0.2}Co_{0.2}Ni_{0.2}Zn_{0.2})_3O_4$ electrode (**e**).

Figure 3b presents the discharge/charge profiles at a current density of 100 mA·g^{-1}. Initially, a small discharge plateau around 0.965 V signifies the lithiation process of HEO. Subsequently, the discharge curve shows a gradual voltage drop until reaching the lower cutoff limit of 0.01 V, indicating the formation of SEI films due to the electrolyte decomposition and the interfacial lithium storage interaction [16]. During the charge process, the curve exhibits a continuously increasing potential slope up to approximately 1.6 V, attributed to the reversible oxidation of active metals to oxides. From Figure 3b, it can be seen that the first-turn charge-specific capacity/discharge-specific capacity of the HEO is 563/1388 mAh/g, and the first-time coulombic efficiency of the HEO electrode is low, only 40.6%.

Figure 3c,d represent the cycling performance at 100 mA·g^{-1} and 1 A g^{-1}, respectively. As shown in Figure 3c, the specific capacities initially decay rapidly during the first 10 cycles, followed by a slower degradation during the subsequent cycles until reaching a stable capacity of 132 mAh·g^{-1}. The low initial Coulombic efficiency (40.6%) is inherent to transformed anodes, primarily caused by the formation of the SEI film [30]. During subsequent cycles, particularly the second and third cycles, the charging and discharging processes induce some collapse and fragmentation of the active material, resulting in the formation of additional SEI membrane and a consequent capacity decrease. By the tenth cycle, the active material stabilizes, preventing further structural degradation and SEI formation, thus stopping rapid capacity loss. However, ongoing volume fluctuations can still cause some active material to detach from the current collector, leading to a gradual attenuation of capacity in subsequent cycles, albeit at a slow rate. Figure 3d demonstrates the stable cycling performance at 1 A g^{-1} with the specific capacity reaching 107 mAh·g^{-1} after 1000 cycles.

Figure 3e presents the rate capability at different current densities. It is evident that the specific capacity decreases rapidly at a low current density but exhibits no decay at higher current densities due to the side reaction during the lithiation/delithiation processes [16,17,31]. At low current densities, the battery charging and discharging time is longer, which allows more time for side reactions to occur in the electrolyte. This could lead to the loss of active lithium and the thickening of the SEI layer, thereby accelerating capacity decay. The discharge/charge curves show a high degree of overlap during different current densities. When the current density abruptly switched back to 0.1 A g^{-1} from 2 A g^{-1}, the specific capacity recovered to 319 mAh·g^{-1} with a capacity retention rate of 97.8%, indicating that the entropy stability effect contributes to the stability of the LIB anode. Table 2 lists the electrochemical experimental results of the spinel-structured HEO in this study compared with those of other spinel-structured HEOs when used as LIB anode materials, which demonstrates that the novel spinel-structured HEO, as synthesized in this study, exhibits potential for lithium storage capabilities in lithium-ion batteries.

Table 2. Synthesis methods and electrochemical properties of spinel-type HEOs.

Materials	Method	Rate Capability/mAh·g^{-1}	Cyclic Performance (Cycles)/ mAh·g^{-1}	Ref.
$(MgFeCoNiZn)_3O_4$	Solid-state synthesis	304 @ 0.5 A g^{-1}	360(300) @ 0.1 A g^{-1}	[32]
$(MgTiZnNiFe)_3O_4$	Solid-state reaction	93.6 @ 1 A g^{-1}	145(100) @ 0.1 A g^{-1}; 100(800) @ 1 A g^{-1}	[33]
$(CoTiZnNiFe)_3O_4$	Solid-state reaction	150.3 @ 1 A g^{-1}	290(100) @ 0.1 A g^{-1}; 130(800) @ 1 A g^{-1}	[33]
$(FeNiCrMnMgAl)_3O_4$	Solution combustion	350 @ 4 A g^{-1}	670(200) @ 0.2 A g^{-1}	[34]
$(MnFeCoNiZn)_3O_4$	Electrospinning	58 @ 2 A g^{-1}	155(550) @ 0.5 A g^{-1}	[35]
$(CrFeMnNiCo_2)_3O_4$	Sol–gel	147 @ 2 A g^{-1}	520(100) @ 0.2 A g^{-1}; 120(1000) @ 2 A g^{-1}	[36]
$(CrFeMnNiCo_3)_3O_4$	Sol–gel	101 @ 2 A g^{-1}	505(100) @ 0.2 A g^{-1}; 118(1000) @ 2 A g^{-1}	[36]
$(CrFeMnNiCo_4)_3O_4$	Sol–gel	97.2 @ 2 A g^{-1}	510(100) @ 0.2 A g^{-1}; 105(1000) @ 2 A g^{-1}	[36]
$(CrMnCoNiZn)_3O_4$	Sol–gel	96 @ 2 A g^{-1}	132(100) @ 0.1 A g^{-1}; 107(1000) @ 1 A g^{-1}	This work

We also explored the dynamic properties of the HEO electrode using electrochemical impedance spectroscopy (EIS). Initial impedance measurements were performed before cycling (HEO-0) and after 20 cycles at a current density of 0.1 A g^{-1} (HEO-20), as shown in Figure 4. The Nyquist plots featured semicircles at mid-to-high frequencies and linear trends at low frequencies. For analyzing the impedance data, we applied an equivalent circuit model that included the solution resistance (Rs), charge transfer resistance (Rct), a constant phase element (CPE) representing the mid-frequency semicircle, and Warburg impedance (W) for Li^+ diffusion representing the linear features at low frequencies. The Rct values, obtained from impedance fitting, are 282 Ω for HEO-20 and 389 Ω for HEO-0, respectively. It is evident that Rct of the HEO after 20 cycles is lower than that of the HEO before cycling.

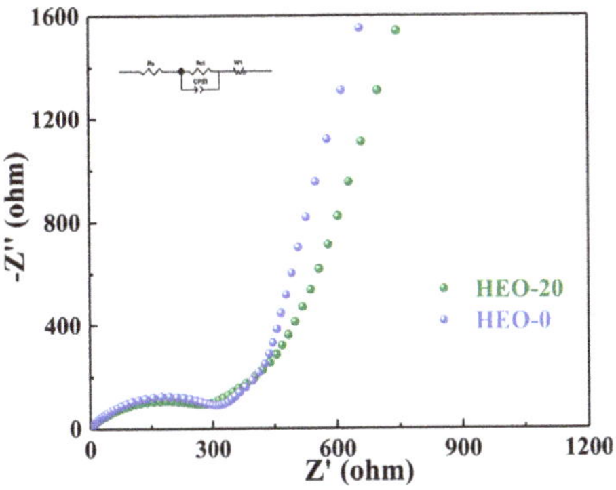

Figure 4. EIS plots of HEO electrode before (HEO-0) and after 20 cycles (HEO-20) at 0.1 A g^{-1}.

X-ray photoelectron spectroscopy (XPS) was employed to examine the elemental composition and chemical valence state changes of HEO before and after 1000 cycles at 1 A g^{-1}, which confirmed the presence of Cr, Mn, Co, Ni, Zn, and O and the corresponding valence states before and after cycling (as depicted in Figure S1). Figure 5 shows the Co 2p and Ni 2p spin–orbit peaks before and after 1000 cycles, which identify the oxidation states of the elements. The Co 2p spectrum in Figure 5a, fitted with Gaussian shapes before cycling, displays four main peaks at 779.3, 781.2, 794.6, and 796.5 eV, along with two satellite peaks at 787.9 and 803.4 eV. The peaks at 779.3 and 794.6 eV are attributed to Co^{3+}, while the peaks at 781.2 and 796.5 eV correspond to the $2p_{3/2}$ and $2p_{1/2}$ transitions of Co^{2+}, respectively. The approximately 15.5 eV energy difference between the Co $2p_{3/2}$ (781.2/779.3 eV) and Co $2p_{1/2}$ (796.5/794.6 eV) peaks is characteristic of cobalt in both +2 and +3 oxidation states. The Ni 2p XPS spectrum in Figure 5c, also fitted with Gaussian shapes before cycling, shows four principal peaks at 853.5, 855.5, 871.5, and 873.2 eV, accompanied by two satellite peaks at 860.6 and 878.3 eV. The binding energies at 853.5/855.5 eV and 871.5/873.2 eV are indicative of Ni^{2+} and Ni^{3+}, respectively. Figure 5b,d illustrate the chemical valence states of Co and Ni after cycling. After 1000 cycles, the presence of Co^{3+} and Ni^{3+} ions is minimal, with a decrease in divalent ion content, replaced by amounts of metals Co and Ni. This could be due to the irreversible reactions of Co^{3+} and Ni^{3+} during the lithium intercalation/delithiation process, leading to the formation and residue of metals Co or Ni during cycling.

Figure 6a,b present SEM images of the HEO electrodes before and after 1000 cycles at a current density of 1 A g^{-1}. After cycling, the initial nanoparticles coalesced into larger clusters, with the appearance of microcracks and voids, which are tens of nanometers in size. This indicates that volume changes are linked to the charge and discharge processes. Although microcracks form due to particle aggregation, the elimination of inter-particle gaps leads to denser packing of the aggregates, which improves electrical conductivity. This conjecture is consistent with the results obtained from the EIS measurements. The corresponding EDS images in Figure 6c–h show a uniform distribution of Cr, Mn, Co, Ni, and Zn across the electrode surface. This uniformity suggests that the HEO maintains its structural integrity after 1000 cycles, consistent with the observed stability of long cycle performance under high current.

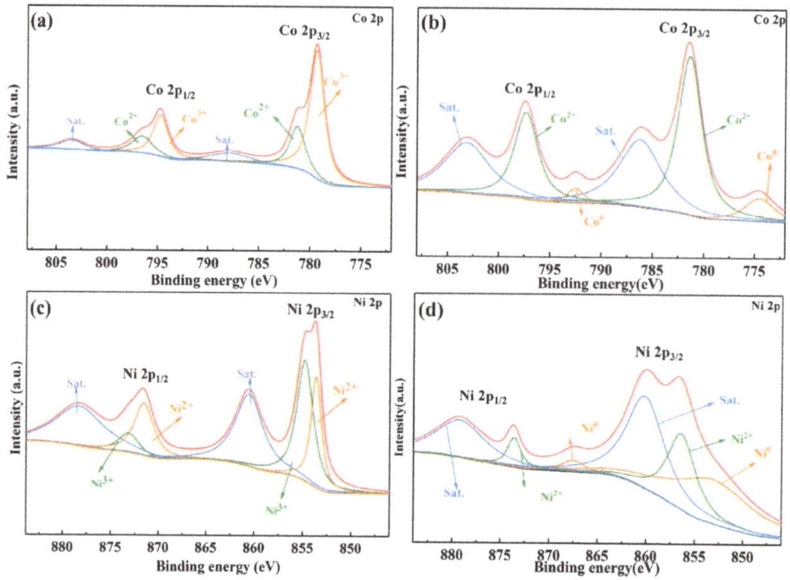

Figure 5. XPS spectra of Co 2p (a) and Ni 2p (c) of HEO before cycling and XPS spectra of Co 2p (b) and Ni 2p (d) of HEO after 1000 cycles at 1 A g^{-1}.

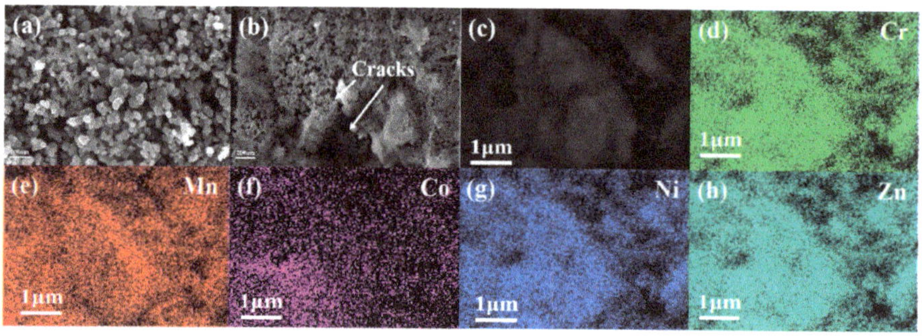

Figure 6. SEM image of HEO before (a) and after (b) 1000 cycles at a current density of 1 A g^{-1}; corresponding elemental mapping of all elements (c), and corresponding elemental mapping of Cr (d), Mn (e), Co (f), Ni (g), and Zn (h) of the HEO.

3. Conclusions

We synthesized a novel spinel-structured HEO $(Cr_{0.2}Mn_{0.2}Co_{0.2}Ni_{0.2}Zn_{0.2})_3O_4$ and thoroughly investigated its electrochemical lithium storage performance as an anode material for LIBs. This anode material demonstrates good rate capability and long-term cycling stability at elevated current densities, attributed to the entropy stabilization effect characteristic of high-entropy oxides (HEOs). However, the material's low initial Coulombic efficiency of 40.6% and the observed capacity degradation after 10 cycles at low current densities pose challenges for its development as an anode material for LIBs. To address these electrochemical issues, strategies such as material composite formation or the fabrication of artificial SEI could be employed to enhance electrochemical performance. The artificial SEI on the surface of HEO can reduce the amount of lithium ions required for SEI formation, while also decreasing the consumption of lithium ions by surface side reactions. Additionally, composites of HEO with materials that possess good electrical conductivity and flexibility can enhance the electro-transport properties of the material, mitigate volume changes during charge–discharge processes, prevent the detachment of active materials from current collectors, and thereby improve the material's electrochemical lithium storage performance.

4. Materials and Methods

Preparation of $(Cr_{0.2}Mn_{0.2}Co_{0.2}Ni_{0.2}Zn_{0.2})_3O_4$: $(Cr_{0.2}Mn_{0.2}Co_{0.2}Ni_{0.2}Zn_{0.2})_3O_4$ nanoparticles were synthesized using the solution combustion method. Firstly, equimolar amounts (0.0025 mol) of metal nitrates (Sinopharm Chemical Reagent Co., Ltd., Shanghai, China) were dissolved in 10 mL of deionized water. Next, 0.4692 g of glycine (Sinopharm Chemical Reagent Co., Ltd., Shanghai, China) was added as fuel. The resulting solution was dried at 80 °C to obtain a gel. Subsequently, the gel was calcined in a tube furnace at 750 °C for 40 min and promptly removed from the furnace. This process yielded a powder composed of $(Cr_{0.2}Mn_{0.2}Co_{0.2}Ni_{0.2}Zn_{0.2})_3O_4$ nanoparticles.

Material characterization: X-ray diffraction pattern of the samples was obtained using X-ray diffractometer (Bruker D2 Advance, Bruker, Lücken, Germany). The morphology was examined by scanning electron microscopy (SEM, Quanta 400, FEI, Waltham, Massachusetts, USA). Energy-dispersive spectroscopy (EDS) integrated into the SEM was employed to investigate the element distribution. The microstructure was observed using transmission electron microscopy (TEM, JEM-2010, JEOL, Tokyo, Japan). The composition and chemical states of the sample were investigated through X-ray photoelectron spectroscopy (XPS, Thermo Scientific K-Alpha, FEI, Waltham, Massachusetts, USA).

Electrochemical measurements: CR2032 half-cells were fabricated inside an argon-filled glove box (Dellix-S2100,Dellix Co., Ltd., Chengdu, Sichuan province, China), ensuring that oxygen and humidity levels were below 0.1 ppm. Working electrodes were prepared by casting a slurry containing 70 wt% of active materials, 20 wt% carbon black (C_{45}, Sinopharm Chemical Reagent Co., Ltd., Shanghai, China), and 10 wt% polyvinylidene (PVDF, Sinopharm Chemical Reagent Co., Ltd., Shanghai, China) fluoride onto Cu foils, followed by drying at 90 °C for 12 h under vacuum. A polypropylene membrane (Celgard 2400, Sinopharm Chemical Reagent Co., Ltd., Shanghai, China) was employed as a separator. The electrolyte used was a mixture of dimethyl carbonate, ethylene carbonate, and methyl ethyl carbonate in a volume ratio of 1:1:1 with a 1 M $LiPF_6$ concentration. Charge–discharge measurements and rate measurements were carried out using a Neware battery testing system (CT4008, Neware, Shenzhen, China). Cyclic voltammetry (CV) tests were performed within a potential window of 0.01–3 V at a scan rate of 0.1 $mV \cdot s^{-1}$ on an electrochemical workstation (CHI660E, CH Instruments, Shanghai, China). In this process, the diameter of the pole piece was 14 mm and the average mass of active substance carried on each pole piece was 0.84 mg.

Supplementary Materials: The following supporting information can be downloaded at: https://www.mdpi.com/article/10.3390/inorganics12070198/s1, Figure S1: XPS spectra of full spectra (a); Cr 2p (b); Mn 2p (c); Zn 2p (d); O 1s (e) before cycle and full spectra(f); Cr 2p(g); Mn 2p(h); Zn 2p(i); O 1s (j) after cycle.

Author Contributions: C.J. conceived the project. Y.W. (Yulong Wang), H.D., Y.W. (Yongxing Wei), R.N. and Z.J. carried out the experiments and analyzed the experimental data. Y.W. (Yulong Wang) and C.J. wrote the paper. Z.Y. participated in the analysis of experimental data and discussion of the results. C.J. and Q.D. revised the manuscript. All authors have read and agreed to the published version of the manuscript.

Funding: This work is supported by the National Natural Science Foundation of China (Grant No. 11404251). The work at Ames National Laboratory was supported by the U.S. Department of Energy, Division of Materials Sciences and Engineering. Ames National Laboratory is operated for the U.S. Department of Energy by Iowa State University under Contract No. DE-AC02-07CH11358.

Data Availability Statement: The original contributions presented in the study are included in the article/supplementary material, further inquiries can be directed to the corresponding authors.

Conflicts of Interest: The authors declare no conflicts of interest. The funders had no role in the design of the study.

References

1. Zou, C.; Xiong, B.; Xue, H.; Zhang, D.; Ge, Z.; Wang, Y.; Jing, L.Y.; Pan, S.Q. The role of new energy in carbon neutral. *Pet. Explor. Dev.* **2021**, *48*, 480–491. [CrossRef]
2. Freunberger, S.A. Batteries: Charging ahead rationally. *Nat. Energy* **2016**, *1*, 16074. [CrossRef]
3. Goriparti, S.; Miele, E.; Angelis, F.D.; Fabrizio, E.D.; Zaccaria, R.P.; Capiglia, C. Review on recent progress of nanostructured anode materials for Lion batteries. *J. Power Sources* **2014**, *257*, 421–443. [CrossRef]
4. Ji, L.W.; Lin, Z.; Alcoutlabia, M.; Zhang, X.W. Recent developments in nanostructured anode materials for rechargeable lithium-ion batteries. *Energ. Environ. Sci.* **2011**, *4*, 2682–2699. [CrossRef]
5. Wu, F.X.; Yushin, G. Conversion cathodes for rechargeable lithium and lithium-ion batteries. *Energ. Environ. Sci.* **2017**, *10*, 435–459. [CrossRef]
6. Lin, D.C.; Liu, Y.Y.; Cui, Y. Reviving the lithium metal anode for highenergy batteries. *Nat. Nanotechnol.* **2017**, *12*, 194–206. [CrossRef]
7. Sheoran, K.; Thakur, V.K.; Siwal, S.S. Synthesis and overview of carbon-based materials for high performance energy storage application: A review. *Mater. Today Proc.* **2021**, *56*, 9–17. [CrossRef]
8. Yu, S.H.; Lee, S.H.; Lee, D.J.; Sung, Y.E.; Hyeon, T.W. Conversion reaction-based oxide nanomaterials for Lithium-ion battery anodes. *Small* **2016**, *12*, 2146–2172. [CrossRef]
9. Shan, F.; Dominic, B.; Stefano, P. Transition metal oxide anodes for electrochemical energy storage in Lithium and sodium-ion batteries. *Adv. Energy Mater.* **2019**, *10*, 1902485.
10. Yoo, S.M.; Lee, J.I.; Shin, M.S.; Park, S.J. Large-scale synthesis of interconnected Si/SiO_x nanowire anodes for rechargeable lithium-ion batteries. *ChemSusChem* **2013**, *6*, 1153–1157. [CrossRef]
11. Yeh, J.-W.; Chen, S.-K.; Lin, S.-J.; Gan, J.-Y.; Chin, T.-S.; Shun, T.-T.; Tsau, C.-H.; Chang, S.-Y. Nanostructured high-entropy alloys with multiple principal elements: Novel alloy design concepts and outcomes. *Adv. Eng. Mater.* **2004**, *6*, 299–303. [CrossRef]
12. Oses, C.; Toher, C.; Curtarolo, S. High-entropy ceramics. *Nat. Rev. Mater.* **2020**, *5*, 295–309. [CrossRef]
13. Dupuy, A.D.; Chellali, M.R.; Hahn, H.; Schoenung, J.M. Multiscale phase homogeneity in bulk nanocrystalline high entropy oxides. *J. Eur. Ceram. Soc.* **2021**, *41*, 4850–4858. [CrossRef]
14. Talluri, B.; Aparna, M.L.; Sreenivasulu, N.; Bhattacharya, S.S.; Thomas, T. High entropy spinel metal oxide $(CoCrFeMnNi)_3O_4$ nanoparticles as a high-performance supercapacitor electrode material. *J. Energy Storage* **2021**, *42*, 1479–1488. [CrossRef]
15. Rost, C.M.; Sachet, E.; Borman, T.; Moballegh, A.; Dickey, E.C.; Hou, D.; Jones, J.L.; Curtarolo, S.; Maria, J.P. Entropy-stabilized oxides. *Nat. Commun.* **2015**, *6*, 8485. [CrossRef]
16. Bérardan, D.; Franger, S.; Meena, A.K.; Dragoe, N. Room temperature lithium superionic conductivity in high entropy oxides. *J. Mater. Chem. A* **2016**, *4*, 9536–9541. [CrossRef]
17. Sarkar, A.; Djenadic, R.; Usharani, N.J.; Sanghvi, K.P.; Chakravadhanula, V.S.; Gandhi, A.S.; Hahn, H.; Bhattacharya, S.S. Nanocrystalline multicomponent entropy stabilised transition metal oxides. *J. Eur. Ceram. Soc.* **2017**, *37*, 747–754. [CrossRef]
18. Qiu, N.; Chen, H.; Yang, Z.; Sun, S.; Wang, Y.; Cui, Y. A high entropy oxide $(Mg_{0.2}Co_{0.2}Ni_{0.2}Cu_{0.2}Zn_{0.2}O)$ with superior lithium storage performance. *J. Alloys Compd.* **2019**, *777*, 767–774. [CrossRef]
19. Wang, D.; Jiang, S.; Duan, C.; Mao, J.; Dong, Y.; Dong, K.; Wang, Z.; Luo, S.; Liu, Y.; Qi, X. Spinel-structured high entropy oxide $(FeCoNiCrMn)\ _3O_4$ as anode towards superior lithium storage performance. *J. Alloys Compd.* **2020**, *844*, 156158. [CrossRef]
20. Chen, T.Y.; Wang, S.Y.; Kuo, C.H.; Huang, S.C.; Lin, M.H.; Li, C.H.; Chen, H.Y.; Wang, C.C.; Liao, Y.F.; Lin, C.C.; et al. In operando synchrotron X-ray studies of a novel spinel $(Ni_{0.2}Co_{0.2}Mn_{0.2}Fe_{0.2}Ti_{0.2})_3O_4$ high-entropy oxide for energy storage applications. *J. Mater. Chem. A* **2020**, *8*, 21756–21770. [CrossRef]
21. Duan, C.; Tian, K.; Li, X.; Wang, D.; Sun, H.; Zheng, R.; Wang, Z.; Liu, Y. New spinel high-entropy oxides $(FeCoNiCrMnXLi)_3O_4$ (X = Cu, Mg, Zn) as the anode material for lithium-ion batteries. *Ceram. Int.* **2021**, *47*, 32025–32032. [CrossRef]
22. Sarkar, A.; Velasco, L.; Wang, D.I.; Wang, Q.; Talasila, G.; de Biasi, L.; Kübel, C.; Brezesinski, T.; Bhattacharya, S.S.; Hahn, H.; et al. High entropy oxides for reversible energy storage. *Nat. Commun.* **2018**, *9*, 3400. [CrossRef]
23. Triolo, C.; Xu, W.; Petrovičovà, B.; Pinna, N.; Santangelo, S. Evaluation of entropy-stabilized $(Mg_{0.2}Co_{0.2}Ni_{0.2}Cu_{0.2}Zn_{0.2})O$ oxides produced via solvothermal method or electrospinning as anodes in lithium-ion batteries. *Adv. Funct. Mater.* **2022**, *32*, 2202892. [CrossRef]
24. Wang, S.Y.; Chen, T.Y.; Kuo, C.H.; Lin, C.C.; Huang, S.C.; Lin, M.H.; Wang, C.C.; Chen, H.Y. Operando synchrotron transmission X-ray microscopy study on (MgCoNiCuZn)O high-entropy oxide anodes for lithium-ion batteries. *Mater. Chem. Phys.* **2021**, *274*, 125105. [CrossRef]
25. Liu, Z.; Yu, X.Y.; Paik, U. Etching-in-a-box: A novel strategy to synthesize unique yolk-shelled Fe_3O_4@ carbon with an ultralong cycling life for lithium storage. *Adv. Energy Mater.* **2016**, *6*, 1502318. [CrossRef]
26. Osenciat, N.; Bérardan, D.; Dragoe, D.; Léridon, B.; Holé, S.; Meena, A.K.; Franger, S.; Dragoe, N. Charge compensation mechanisms in Li-substituted high-entropy oxides and influence on Li superionic conductivity. *J. Am. Ceram. Soc.* **2019**, *102*, 6156–6162. [CrossRef]

27. Petrovičovà, B.; Xu, W.; Musolino, M.G.; Pantò, F.; Patanè, S.; Pinna, N.; Santangelo, S.; Triolo, C. High-entropy spinel oxides produced via sol-gel and electrospinning and their evaluation as anodes in Li-ion batteries. *Appl. Sci.* **2022**, *12*, 5965. [CrossRef]
28. Cao, K.Z.; Jin, T.; Yang, L.; Jiao, L.F. Recent progress in conversion reaction metal oxide anodes for Li-ion batteries. *Mater. Chem. Front.* **2017**, *1*, 2213–2242. [CrossRef]
29. Nguyen, T.X.; Patra, J.; Chang, J.K.; Ting, J.M. High entropy spinel oxide nanoparticles for superior lithiation-delithiation performance. *J. Mater. Chem. A* **2020**, *8*, 18963–18973. [CrossRef]
30. Zhou, L.M.; Zhang, K.; Hu, Z.; Tao, Z.L.; Mai, L.Q.; Kang, Y.M.; Chou, S.L.; Chen, J. Recent developments on and prospects for electrode materials with hierarchical structures for lithium-ion batteries. *Adv. Energy Mater.* **2018**, *8*, 1701415. [CrossRef]
31. Wang, Q.S.; Sarkar, A.; Li, Z.Y.; Lu, Y.; Velasco, L.; Bhattacharya, S.S.; Brezesinski, T.; Hahn, H.; Breitung, B. High entropy oxides as anode material for Li-ion battery applications: A practical approach. *Electrochem. Commun.* **2019**, *100*, 121–125. [CrossRef]
32. Marques, O.J.; Walter, M.D.; Timofeeva, E.V.; Segre, C.U. Effect of initial structure on performance of high-entropy oxide anodes for li-ion batteries. *Batteries* **2023**, *9*, 115. [CrossRef]
33. Liu, C.; Bi, J.; Xie, L.; Gao, X.; Meng, L. Preparation and electrochemical properties of two novel high entropy spinel oxides $(MgTiZnNiFe)_3O_4$ and $(CoTiZnNiFe)_3O_4$ by solid state reaction. *Mater. Today Commun.* **2023**, *35*, 106315. [CrossRef]
34. Zheng, Y.; Wu, X.; Lan, X.; Hu, R. A spinel $(FeNiCrMnMgAl)_3O_4$ high entropy oxide as a cycling stable anode material for Li-ion batteries. *Processes* **2021**, *10*, 49. [CrossRef]
35. Triolo, C.; Maisuradze, M.; Li, M.; Liu, Y.; Ponti, A.; Pagot, G.; DiNoto, V.; Aquilanti, G.; Pinna, N.; Giorgetti, M.; et al. Charge Storage Mechanism in Electrospun Spinel-Structured High-Entropy $(Mn_{0.2}Fe_{0.2}Co_{0.2}Ni_{0.2}Zn_{0.2})_3O_4$ Oxide Nanofibers as Anode Material for Li-Ion Batteries. *Small* **2023**, *19*, 2304585. [CrossRef]
36. Liu, C.; Bi, J.; Xie, L.; Gao, X.; Rong, J. High entropy spinel oxides $(CrFeMnNiCo_x)_3O_4$ (x = 2, 3, 4) nanoparticles as anode material towards electrochemical properties. *J. Energy Storage* **2023**, *71*, 108211. [CrossRef]

inorganics

MDPI

Article

Electrochemical Investigation of Lithium Perchlorate-Doped Polypyrrole Growing on Titanium Substrate

Yibing Xie [1,2,*], Jing Xu [3], Lu Lu [4] and Chi Xia [5]

1 School of Chemistry and Chemical Engineering, Southeast University, Nanjing 211189, China
2 Suzhou Research Institute, Southeast University, Suzhou 215123, China
3 College of Ecology and Resource Engineering, Wuyi University, Wuyishan 354300, China
4 School of Intelligent and Automotive Engineering, Wuxi Vocational Institute of Commerce, Wuxi 214153, China
5 Institute of Shandong Non-Metallic Materials, Jinan 250031, China
* Correspondence: ybxie@seu.edu.cn

Abstract: Lithium perchlorate-doped polypyrrole growing on titanium substrate ($LiClO_4$-PPy/Ti) has been fabricated to act as electroactive electrode material for feasible electrochemical energy storage. A theoretical and experimental investigation is adopted to disclose the conductivity, electroactivity properties and interfacial interaction-dependent capacitance of $LiClO_4$-PPy/Ti electrode. The experimental measurement results disclose that $LiClO_4$-PPy/Ti reveals lower ohmic resistance (0.2226 Ω cm^{-2}) and charge transfer resistance (2116 Ω cm^{-2}) to exhibit higher electrochemical conductivity, a more reactive surface, and feasible ion diffusion to present higher double-layer capacitance (0.1930 mF cm^{-2}) rather than $LiClO_4$/Ti (0.3660 Ω cm^{-2}, 65,250 Ω cm^{-2}, 0.0334 mF cm^{-2}). $LiClO_4$-PPy/Ti reveals higher Faradaic capacitance caused by the reversible doping and dedoping process of perchlorate ion on PPy than the electrical double-layer capacitance of $LiClO_4$/Ti caused by the reversible adsorption and desorption process of the $LiClO_4$ electrolyte on Ti. Theoretical simulation calculation results prove that a more intensive electrostatic interaction of pyrrole N$\cdots$Ti (2.450 Å) in $LiClO_4$-PPy/Ti rather than perchlorate O$\cdots$Ti (3.537 Å) in $LiClO_4$/Ti. $LiClO_4$-PPy/Ti exhibits higher density of states (57.321 electrons/eV) at Fermi energy and lower HOMO-LUMO molecule orbital energy gap (0.032 eV) than $LiClO_4$/Ti (9.652 electrons/eV, 0.340 eV) to present the enhanced electronic conductivity. $LiClO_4$-PPy/Ti also exhibits a more declined interface energy (-1.461×10^4) than $LiClO_4$/Ti (-5.202×10^3 eV) to present the intensified interfacial interaction. $LiClO_4$-PPy/Ti accordingly exhibits much higher specific capacitances of 0.123~0.0122 mF cm^{-2} at current densities of 0.01~0.10 mA cm^{-2} rather than $LiClO_4$/Ti (0.010~0.0095 mF cm^{-2}, presenting superior electroactivity and electrochemical capacitance properties. $LiClO_4$-PPy/Ti could well act as the electroactive supercapacitor electrode for feasible energy storage.

Keywords: lithium perchlorate doping polypyrrole; titanium substrate; electrochemical capacitance; interfacial interaction; simulation calculation

1. Introduction

The electrochemical behaviors and performances of energy storage devices are related to the electrode and electrolyte materials as well as their interfacial interaction [1–3]. In addition, the electrode substrate materials are also related to the properties of functional electrodes. The electrochemical capacity and mechanical stability are most concerned for the electrodes of energy storage devices. These electrode materials mainly involve the various graphite carbon, conductive polymers, transition metal sulfates or oxides, and their composites [4–7]. In particular, the conductive polymers become suitable supercapacitor electrode materials for flexible energy storage [8,9]. The electrical conductive materials, such as carbon fiber, graphene, graphite carbon, titanium, and so on, have been used to prepare the energy storage electrodes [10–12]. They could act as the multifunctional roles, such as the electrode substrate, current collector, and electrode. However, the contradictory behaviors of these substrate materials also emerged in respect to the electrochemical activity and electrochemical stability. For example, titanium metal could resist the chemical and electrochemical corrosion in neutral, alkaline, and even acidic conditions [13]. Titanium keeps the chemical and electrochemical stability in a certain degree, and is applied to electrode material as well as electrode-supporting material [14–17]. The structure-adjusted titanium, such as titanium sheet, titanium mesh, and titanium foam with high electrochemical stability, can be used to support various electroactive materials [18–21]. On the other hand, titanium metal also behaves the relatively low electrochemical activity due to the passivation layer of titanium oxides on the surface [22]. Surface modification using conductive polymers is a useful method to Improve the electrochemical activity of titanium. The Ti metal sheet can act as the electrode substrate. The conductive polymers usually includes polyaniline, polypyrrole, and polythiophene, which have been used to act as electroactive materials for the supercapacitor electrode application [23]. The different electroactive materials can be coated on Ti substrate to form functional electrodes. The polypyrrole (pPy) has been electrodeposited on Ti substrate to form pPy/Ti electrode through cyclic voltammetry polymerization process [24]. The polyaniline (PANI) has been electrodeposited and coated on Ti substrate to form PANI/Ti electrode through a pulse potentiostatic polymerization process [25]. These conductive polymers could conduct the reversible ion doping–dedoping process to provide superior capacitance [26–28]. The interfacial bonding strength between the doped conductive polymers and metal substrates is related to the electrochemical energy storage performance [29]. In addition, the reactive electrolytes also influence the electrical double-layer capacitance due to the interfacial adsorption on metal substrates and the Faradaic capacitance due to the ion doping state of conductive polymers. In previous studies, the system of a lithium perchlorate electrolyte and polypyrrole has been attempted for various applications in different areas [30,31]. However, the interfacial interaction between lithium perchlorate doping polypyrrole and metal titanium substrate is still sparsely investigated [32].

This study focuses on disclosing the promotion effect of electroactive polypyrrole on the electrochemical properties of lithium perchlorate interacting with titanium substrate. The polypyrrole is used as the typical conductive polymer material, and a metal titanium sheet is used as the stable substrate material. The titanium sheet has chemical and electrochemical stability to resist the electrochemical corrosion to act as a suitable current collector. The polypyrrole growing on titanium substrate (PPy/Ti) is designed as the supercapacitor electrode in the application of electrochemical energy storage, which is prepared through the electrochemical polymerization process. The lithium perchlorate ($LiClO_4$) is used as a neural electrolyte to study the electrical double-layer capacitance of titanium metal substrate and the Faradaic capacitance of polypyrrole. The interfacial interaction is intensively considered between the $LiClO_4$ electrolyte and the PPy/Ti electrode when the electrochemical behavior is investigated for $LiClO_4$-Ppy/Ti. The electrochemical performances are fully evaluated through cyclic voltammetry, electrochemical impedance spectrum, and galvanostatic charge/discharge measurements. The theoretical simulation calculation has been applied to disclose the atomic charge distribution, electrostatic potential, and interfacial affinity of $LiClO_4$-PPy/Ti, which could lead to a better understanding of capacitance performance.

2. Results and Discussion

2.1. Microstructure and Morphology Characterization

Figure 1A–D shows SEM images of Ti and PPy/Ti. The Ti sheet substrate reveals the uniquely complete surface structure without any crack. The chemical polishing treatment could lead to forming the uniform metal surface. Comparatively, PPy/Ti revels the compact covering layer of PPy film supported on the Ti sheet. The PPy also is composed of the aggregating particles, forming uniquely complete film structure without any crack. Concerning the surface microstructure, the Ti sheet shows smooth surface and PPy/Ti shows rough surface. The insert in Figure 1D shows the cross-section SEM image of PPy/Ti. The PPy covered on the surface of the Ti substrate exhibits a close-packing film. The morphology, microstructure, and thickness of such PPy film could be adjusted by modulating the electropolymerization reaction parameters in the CV electrodeposition process. This study focuses on the interfacial interaction between PPy film and Ti substrate, which is related to the microstructure of PPy. It is believed that PPy/Ti could conduct more feasible ion diffusion and doping reaction in the presence of lithium perchlorate in comparison the ion adsorption of PPy/Ti.

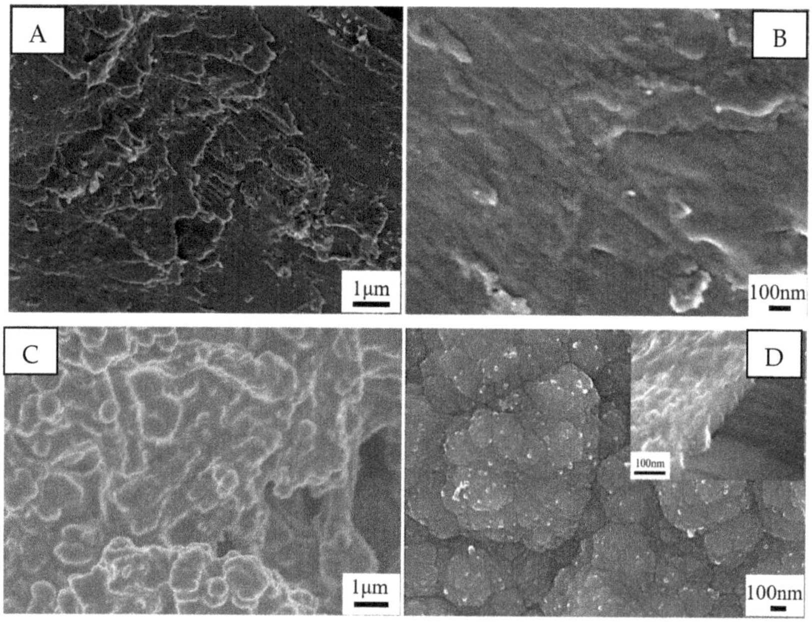

Figure 1. (**A**) SEM image and (**B**) magnified SEM image of Ti; (**C**) SEM image of PPy/Ti; and (**D**) magnified SEM image and cross-section SEM image of PPy/Ti.

Figure 2A shows the XRD pattern of the titanium sheet substrate. Concerning the titanium sheet, the distinct characteristic diffraction peaks at 2θ of 35.1°, 38.5°, 40.3°, 53.1°, 63.1°, 70.8°, 76.3°, 77.5°, and 82.3° correspond to the respective crystal planes of (100), (002), (101), (102), (110), (103), (112), (201), and (202) [33]. The titanium sheet involves the close-packed hexagonal crystal phase structure, which is acted as stable metal substrate. Figure 2B shows the Raman spectrum of the Ti substrate. The Ti substrate does not reveal any characteristic Raman peaks in full wavenumber range due to the absence of electronic transition-relaxation of crystal lattice Ti atoms at 785 nm excitation photon energy. Figure 2C shows the Raman spectrum of PPy/Ti synthesized in this study. The characteristic Raman peaks of PPy/Ti are observed at 983 and 1050 cm^{-1}, with a broad band of 1340–1410 cm^{-1} and 1580 cm^{-1}, which are assigned to C–H ring deformation vibration, C–N stretching vibration, and C=C aromatic stretching vibration, respectively. The characteristic Raman peak position of the literature-reported PPy is located at 980, 1040, 1314–1410 cm^{-1}, and 1597 cm^{-1} [34]. The characteristic Raman peaks of PPy are well-matched between this study-formed PPy/Ti and the literature-reported PPy. Therefore, the Raman spectrum analysis result could well confirm the formation of PPy in the as-prepared PPy/Ti electrode in this study. Figure 2D shows the FTIR spectrum of PPy/Ti. The characteristic peaks at 1540, 1480, 1190, and 1040 cm^{-1} were ascribed to C=C double bond stretching vibration, C-C single bond stretching vibration, C-N single bond stretching vibration, and N-H in-plane bending vibration. The characteristic peaks at 912, 768, 742, and 675 cm^{-1} are ascribed to the C-H out-of-plane bending vibration. The observation of the C=C double bond, C-C single bond, and C-N single bond could be ascribed to the formation of PPy. Accordingly, PPy fully grew on the Ti substate to form PPy/Ti through the electrochemical polymerization and deposition process.

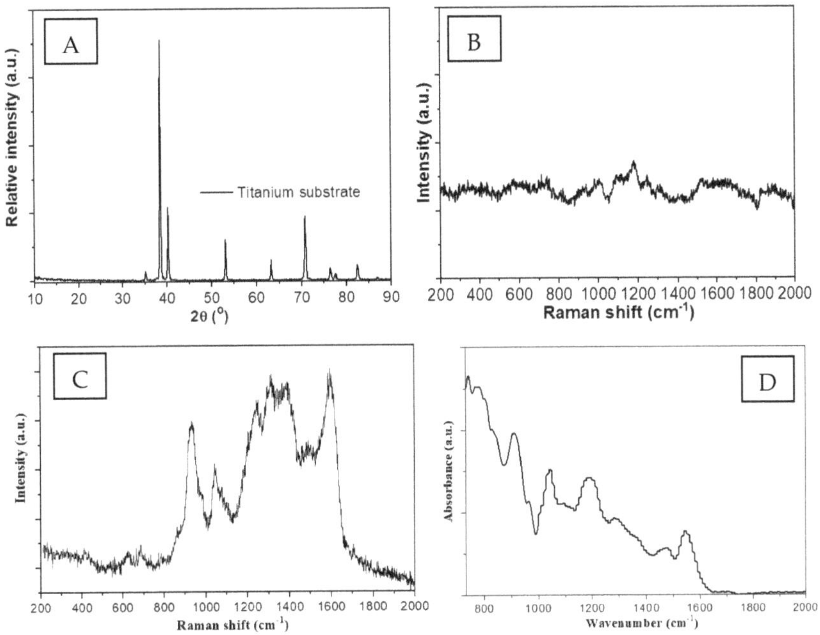

Figure 2. (**A**) XRD pattern and (**B**) Raman spectrum of Ti substrate; (**C**) Raman spectrum of PPy/Ti; and (**D**) FTIR spectrum of PPy/Ti.

2.2. Electrochemical Properties

The CV curves can be used to calculate the CV-based capacitance (C_{CV}, mF cm^{-2}) using Equation (1). The GCD curves can be used to calculate the GCD-based capacitance (C_{GCD}, mF cm^{-2}) using Equation (2).

$$C_{CV} = i/v = [(\int_{V_c}^{V_a} i(v) * dv)/(V_a - V_c)]/v = [\frac{1}{2}(\oint i(v) * dv)/\Delta V]/v \quad (1)$$

$$C_{GCD} = \frac{I * t}{\Delta V} \quad (2)$$

where: V_a, upper potential; V_c, lower potential; $i(v)$, response current density (mA cm^{-2}); v, potential sweep rate (V s^{-1}); I, discharge current density (mA cm^{-2}); t, discharge time (s); and ΔV, potential window (V).

Figure 3A,B shows the CV curves of $LiClO_4$/Ti and $LiClO_4$-PPy/Ti at scan rates of 5~200 mV s^{-1} and the potential range of 0~0.8 V versus saturated calomel electrode (SCE) in 1.0 M $LiClO_4$ electrolyte solution. The CV curves of $LiClO_4$/Ti demonstrate the rectangle-like shape to present electrical double-layer capacitance due to the reversible adsorption–desorption of $LiClO_4$ electrolyte ion on the Ti substrate. The oxidation potential of Ti^{2+}/Ti is −1.39 V versus SCE. This means that the electrochemical oxidation of Ti is unlikely to occur in the potential range of 0~0.8 V versus SCE, presenting the high electrochemical stability of the Ti substrate in $LiClO_4$ electrolyte. Comparatively, the CV curves of $LiClO_4$-PPy/Ti exhibit an obviously deviated rectangle shape in the potential range of 0~0.8 V, presenting the polarization effect. The redox potential of PPy is around 0.5 V versus SCE [35]. $LiClO_4$-PPy conducts reversible ClO_4^- doping-dedoping reaction which contributes to the Faradaic capacitance performance.

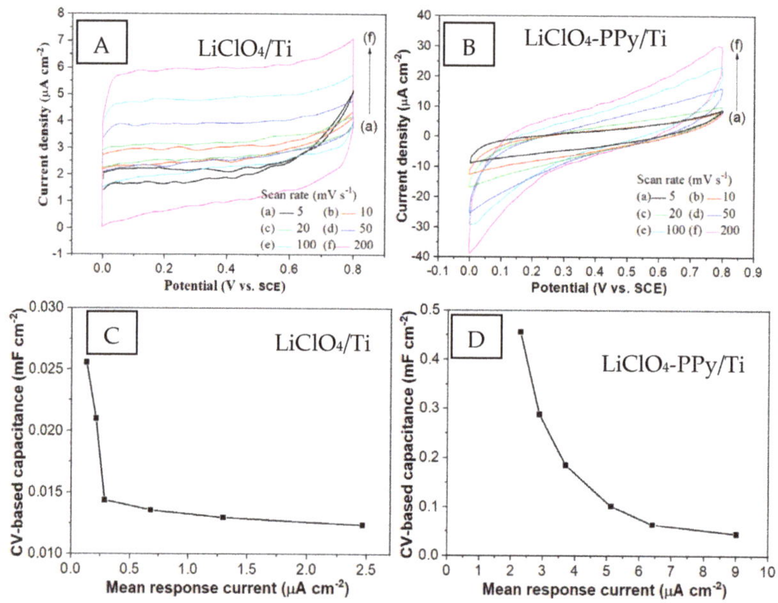

Figure 3. (**A**,**B**) CV curves and (**C**,**D**) CV-based capacitance curves of $LiClO_4$/Ti and $LiClO_4$-PPy/Ti at scan rates of 5~200 mV s^{-1} in 1.0 M $LiClO_4$ electrolyte solution.

Figure 3C,D shows CV-based capacitance curves of $LiClO_4$/Ti and $LiClO_4$-PPy/Ti. When the CV cures of $LiClO_4$/Ti and $LiClO_4$-PPy/Ti conduct the scan rate increasing from 5 to 200 mV s^{-1}, their mean response current increases from 0.13 to 2.47 μA cm^{-2} and from 2.28 to 9.011 μA cm^{-2}, respectively. The CV-based capacitance decreases from 0.0256 to 0.0124 mF cm^{-2} and from 0.456–0.045 mF cm^{-2}, respectively. In comparison with $LiClO_4$/Ti, $LiClO_4$-PPy/Ti exhibits a much higher mean response current and specific capacitance performance at the same scan rate, presenting its superior electrochemical activity in the neutral $LiClO_4$ electrolyte. The oxidation potential of Ti^{2+}/Ti is 1.39 V versus SCE ($Ti - 2e \rightarrow Ti^{2+}$). The oxidation potential of PPy at the doping process is approximate 0.5 V versus SCE ($PPy - e + ClO_4^- \rightarrow PPy^+ClO_4^-$). PPy/Ti could conduct reversible doping–dedoping process of PPy rather than the electrochemical oxidation of the Ti substrate in the potential range of 0~0.8 V. Additionally, the PPy covering layer could also resist such electrochemical corrosion reaction of Ti substrate. Meanwhile, PPy could conduct the redox reaction through the reversible doping–dedoping ClO_4^- anion, contributing to improving the electrochemical activity and, accordingly, the Faradaic capacitance performance.

Figure 4A,B shows the GCD curves of $LiClO_4$/Ti and $LiClO_4$-PPy/Ti in the potential range of 0~0.8 V (vs. SCE) and 1.0 M $LiClO_4$ electrolyte solution. The GCD curves of $LiClO_4$/Ti show complete symmetric characteristics. This shows the good reversible properties of $LiClO_4$ ion adsorption–desorption during the charge–discharge process. Such $LiClO_4$/Ti exhibits ideal electrical double-layer capacitance behavior. Comparatively, the GCD curves of $LiClO_4$-PPy/Ti shows the sloping plateau. The less-symmetric characteristic is ascribed to the redox reaction in the charge–discharge process. Such $LiClO_4$-PPy/Ti exhibits Faradaic capacitance behavior. Therefore, $LiClO_4$-PPy/Ti demonstrates much higher capacitance performance than $LiClO_4$/Ti in the neutral $LiClO_4$ electrolyte. Furthermore, $LiClO_4$-PPy/Ti conducts the doping–dedoping reaction of PPy, and $LiClO_4$/Ti conducts the ion adsorption–desorption process of Ti. The GCD curve shape of $LiClO_4$-PPy/Ti shows fewer identical characteristics than that of $LiClO_4$/Ti. Figure 4C,D shows GCD-based capacitance curves of $LiClO_4$/Ti and $LiClO_4$-PPy/Ti. According to the GCD curves at current densities of 0.01~0.10 mA cm^{-2}, the corresponding GCD-based specific capacitances achieve 0.010~0.0095 mF cm^{-2} and 0.123~0.0122 mF cm^{-2}. Therefore, $LiClO_4$-PPy/Ti exhibits much higher capacitance than $LiClO_4$/Ti, since electroactive $LiClO_4$-PPy mostly contributes to Faradaic capacitance rather than electrical double layer capacitance of Ti substrate in $LiClO_4$ electrolyte. These GCD measurement results agree with the CV measurement results. Furthermore, this $LiClO_4$-PPy/Ti electrode keeps the similar capacitance performance in comparison with the ion doped PPy reported in the literature [36,37]. Figure 4E shows the cycling capacity retention curve of $LiClO_4$-PPy/Ti. The capacity retention achieves 76.1% after 1000 cycles during the cycling charge–discharge process. Therefore, the PPy/Ti electrode exhibits the reasonable cycling stability in the electrochemical Faradaic reaction process.

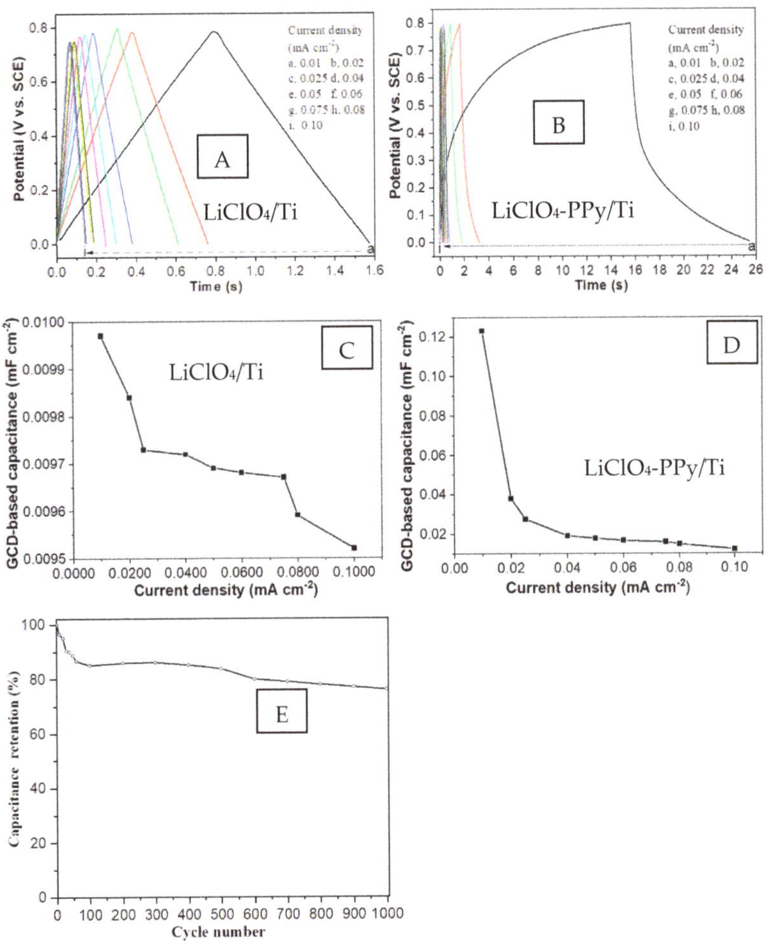

Figure 4. (**A**,**B**) GCD curves and (**C**,**D**) specific capacitance curves of $LiClO_4$/Ti and $LiClO_4$-PPy/Ti at current densities of 0.01~0.1 mA cm^{-2} in 1.0 M $LiClO_4$ electrolyte solution; and (**E**) the cycling capacity retention curve of $LiClO_4$-PPy/Ti.

An electrochemical impedance spectroscopy (EIS) is adopted to investigate the electrolyte ion diffusion and interfacial charge transfer properties of electrode materials. Figure 5A,B shows the Nyquist plots of the EIS spectra of $LiClO_4$/Ti and $LiClO_4$-PPy/Ti. Figure 5C shows equivalent circuit model for fitting curves of EIS spectra. The equivalent circuit includes charge-transfer resistance (R_{ct}), ohmic resistance (R_o), double layer capacitance (C_{dl}), constant phase element (CPE), and CPE exponent (n). n = 1 indicates ideal capacitor behavior, and n = 0 indicates pure resistor behavior. Table 1 lists the fitting parameter values of these elements used in an equivalent circuit model. In comparison with $LiClO_4$/Ti, the slightly lowered R_o value of $LiClO_4$-PPy/Ti indicates higher electronic conductivity. The ohm resistance of $LiClO_4$-PPy/Ti is mostly dependent on the Ti substrate. The conducting $LiClO_4$-PPy layer prevents the formation of semi-conducting titanium oxide. The quite lowered R_{ct} value indicates the more electroactive interface of $LiClO_4$-PPy for the charge transfer. Its higher C_{dl} value indicates higher double-layer capacitance due to the rougher surface of PPy/Ti rather than Ti substate. However, its higher CPE value indicates the higher Warburg impedance in view of electrolyte diffusion. The Warburg impedance of Ti substrate keeps relatively low value. The lithium ion has relatively high diffusion coefficient and plays a more predominant role than the electrode materials in the $LiClO_4$ electrolyte. $LiClO_4$/Ti carries out a shorter ion diffusion length than $LiClO_4$-PPy/Ti. Therefore, $LiClO_4$/Ti reveals a relatively lower CPE value than $LiClO_4$-PPy/Ti. Moreover, $LiClO_4$-PPy/Ti exhibits a larger CPE exponent (n value) than $LiClO_4$/Ti. This indicates that $LiClO_4$-PPy/Ti more likely approaches capacitor behavior rather than resistance behavior. The $LiClO_4$-PPy layer provides more active interface for electrolyte ion adsorption and charge transfer. Therefore, $LiClO_4$-PPy/Ti exhibits higher electroactivity than $LiClO_4$/Ti, contributing to a higher specific capacitance in neutral $LiClO_4$ electrolyte solution.

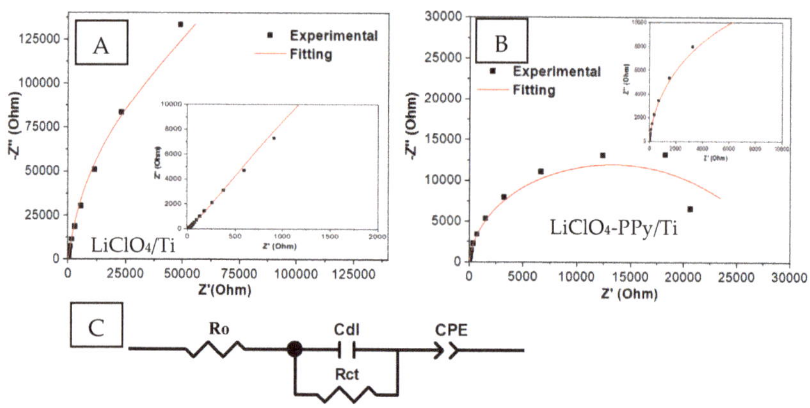

Figure 5. (**A**,**B**) Nyquist plots of EIS spectra of $LiClO_4$/Ti and $LiClO_4$-PPy/Ti.; and (**C**) equivalent circuit model used for fitting curves of EIS spectra.

Table 1. Fitting parameter values of elements used in equivalent circuit model of $LiClO_4$/Ti and $LiClO_4$-PPy/Ti.

Substance	C_{dl} (mF cm^{-2})	R_{ct} (Ω cm^{-2})	R_o (Ω cm^{-2})	CPE (S s^n cm^{-2})	n
$LiClO_4$/Ti	0.0334	65,250	0.3660	0.00263	0.746
$LiClO_4$-PPy/Ti	0.1930	2116	0.2226	0.03427	0.912

2.3. DFT Simulation Calculation of $LiClO_4$/Ti and $LiClO_4$-PPy/Ti

The $LiClO_4$ molecule conducts the adsorption–desorption to establish interfacial interaction between $LiClO_4$ and Ti substrate through Van der Waals forces in the electrochemical process. The $LiClO_4$ molecule also conducts the doping–dedoping process to establish electrostatic interaction between $LiClO_4$-PPy and Ti substrate. The atomic charge distribution could affect the interaction of $LiClO_4$/Ti and $LiClO_4$-PPy/Ti. The DFT simulation calculation is used to investigate electronic properties of $LiClO_4$/Ti and $LiClO_4$-PPy/Ti [38].

Figure 6A,B shows molecule the structure models of $LiClO_4$/Ti and $LiClO_4$-PPy/Ti. $LiClO_4$/Ti involves an interfacial electrostatic interaction between the $LiClO_4$ and Ti substrate. $LiClO_4$-PPy/Ti involves an interfacial electrostatic interaction between the $LiClO_4$-PPy and Ti substrate. $LiClO_4$-PPy presents the perchlorate ion doping state of PPy.

Figure 6C,D shows the molecular electrostatic potentials of $LiClO_4$/Ti and $LiClO_4$-PPy/Ti. Concerning $LiClO_4$/Ti, the negative charge of ClO_4^- ion could induce the charge dissociation of the Ti substrate, which leads to forming positive and negative crystal lattice Ti atoms. Concerning $LiClO_4$-PPy/Ti, the doping state of $LiClO_4$-PPy reveals a positive charge of hydrogen atoms and negative charge of conjugated pyrrole rings, which leads to the forming of a positive and negative crystal lattice of Ti atoms. The intermolecular interaction distance is determined to be 3.537 Å for perchlorate O...Ti of $LiClO_4$/Ti and 2.450 Å for pyrrole N...Ti of $LiClO_4$-PPy/Ti, respectively. Therefore, $LiClO_4$-PPy/Ti presents stronger electrostatic interaction than $LiClO_4$/Ti. The electrostatic interaction is allowed to occur for $LiClO_4$/Ti and $LiClO_4$-PPy/Ti. The intensified interaction has been established between the $LiClO_4$-PPy molecule chain and the Ti substrate, contributing to the activation of the $LiClO_4$-PPy/Ti electrode.

Figure 6E,F shows the density of states curves of $LiClO_4$/Ti and $LiClO_4$-PPy/Ti. Both $LiClO_4$/Ti and $LiClO_4$-PPy/Ti demonstrate metal conductivity characteristics, since the characteristic density of states at Fermi energy 0 eV cross from the valance band to the conduction band. Comparatively, $LiClO_4$-PPy/Ti demonstrates more orbit energy levels than the $LiClO_4$/Ti at valence band and conduction band regions, which are ascribed to the interfacial interaction between the $LiClO_4$-PPy and the Ti substrate. Furthermore, $LiClO_4$-PPy/Ti shows a higher DOS level (57.321 electrons/eV) than $LiClO_4$/Ti (9.652 electrons/eV) at Fermi energy 0 eV. $LiClO_4$-PPy/Ti exhibits higher electron states and higher electronic conductivity.

Figure 6G,H shows the band gap structure curves of $LiClO_4$/Ti and $LiClO_4$-PPy/Ti. The band gaps of $LiClO_4$/Ti and $LiClO_4$-PPy/Ti approach 0.03 and 0 eV, presenting similar superior conductivity of metal substrates. Comparatively, $LiClO_4$-PPy/Ti demonstrates much greater energy levels at the valance band and conduction band regions than $LiClO_4$/Ti, which presents superior electrical conductivity of $LiClO_4$-PPy/Ti. Therefore, the band gap analysis results agree with the DOS analysis results, proving the superior electronic conductivity of $LiClO_4$-PPy/Ti rather than $LiClO_4$/Ti. The declined interfacial energies are -5.202×10^3 eV for $LiClO_4$/Ti and -1.461×10^4 for $LiClO_4$-PPy/Ti, which presents the intensified interfacial interaction of $LiClO_4$-PPy/Ti. Therefore, $LiClO_4$-PPy/Ti shows higher electroactivity than $LiClO_4$/Ti. $LiClO_4$-PPy has more electronic distribution than $LiClO_4$-PPy/Ti, which leads to stronger interaction, lower interfacial energy, and, accordingly, higher capacitance performance.

Figure 6I shows the HOMO and LUMO HOMO and LUMO frontier molecular orbital charge distribution of $LiClO_4$/Ti and $LiClO_4$-PPy/Ti. Concerning HOMO and LUMO electron distribution in $LiClO_4$/Ti and $LiClO_4$-PPy/Ti, their molecular orbitals are mostly dependent on Ti rather than $LiClO_4$ and $LiClO_4$-PPy. In Ti atoms, there occurs an obvious electronic transition, which causes heterogeneous electron distribution. Comparatively, different O and Cl atoms in $LiClO_4$ occur insignificant electronic transition due to weak electron transfer. Similarly, the electron transition of different atoms in $LiClO_4$-PPy also becomes relatively neglective because of weak electronic transfer. The HOMO-LUMO molecule orbital energy gaps of $LiClO_4$/Ti and $LiClO_4$-PPy/Ti are determined to be 0.340 eV and 0.032 eV, respectively. Therefore, $LiClO_4$-PPy/Ti with a low HOMO-LUMO gap presents its high conductivity and high electroactivity, also. The HOMO-LUMO gap analysis results agree with the DOS and band gap analysis results. The DFT simulation calculation results support the CV and GCD measurement results, proving the higher capacitance of $LiClO_4$-PPy/Ti.

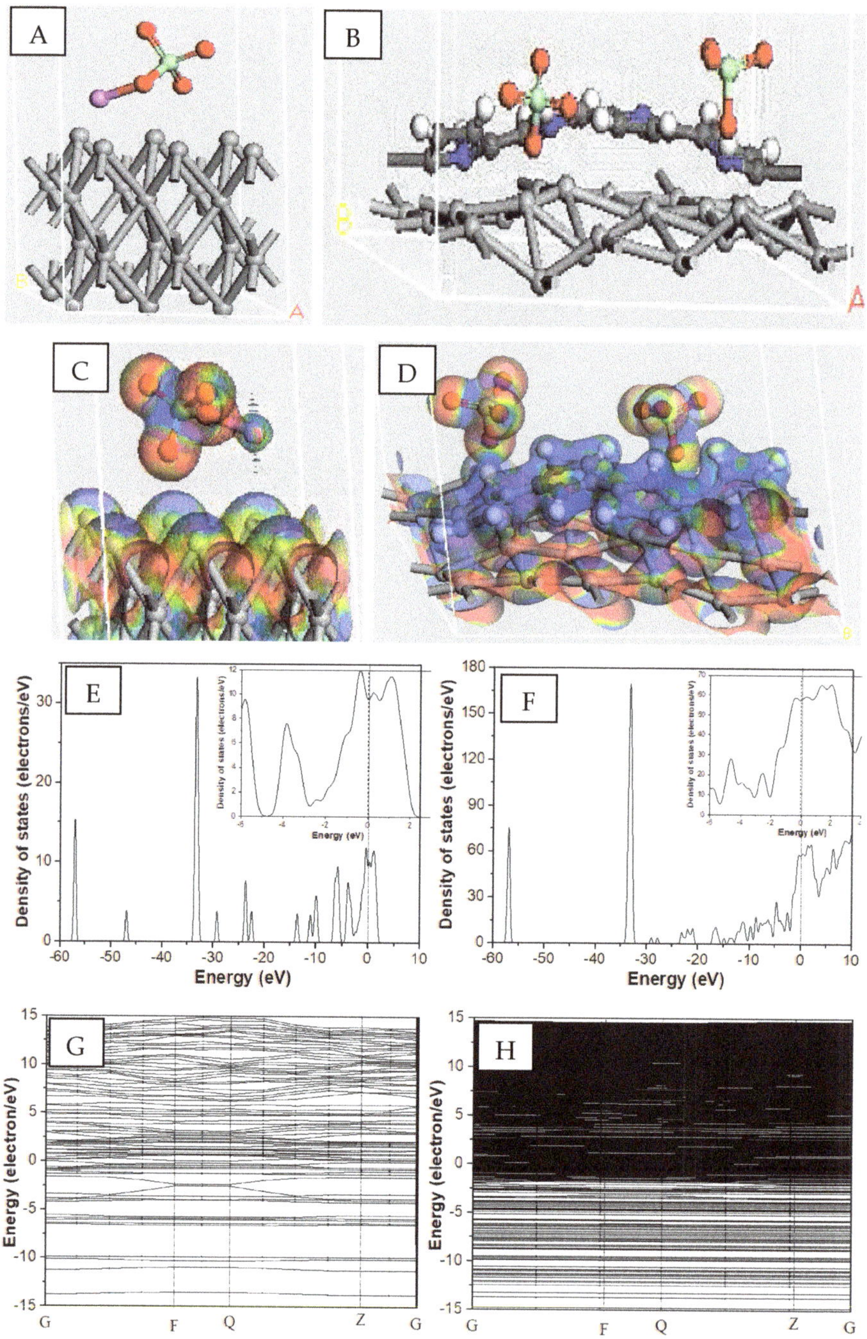

Figure 6. *Cont.*

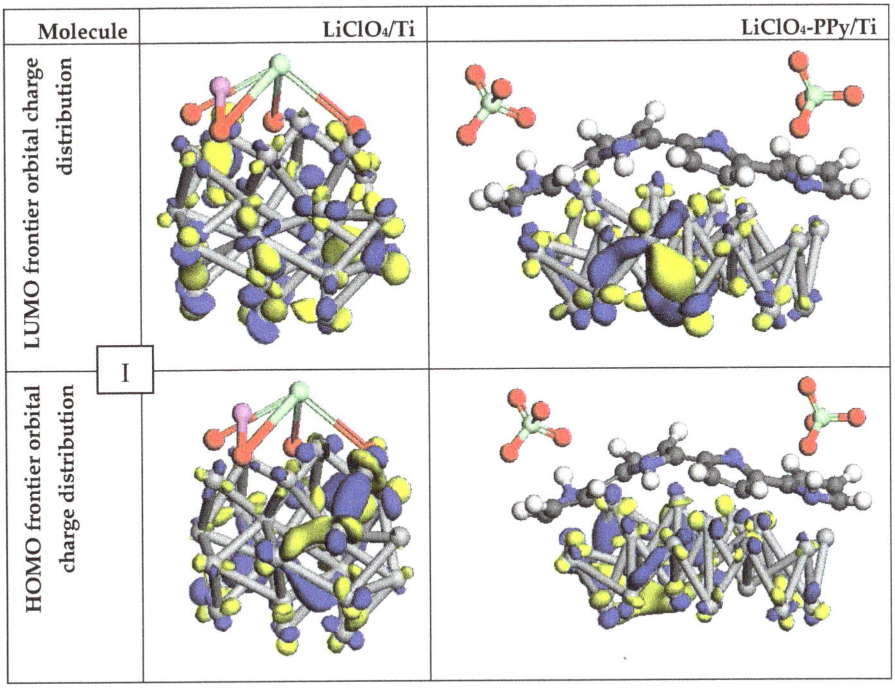

Figure 6. (**A**,**B**) Molecule structure models, (**C**,**D**) molecular electrostatic potentials (difference charge density), (**E**,**F**) total density of states curves, (**G**,**H**) band gap structure curves, (**I**) HOMO and LUMO frontier molecular orbital charge distribution of $LiClO_4$/Ti and $LiClO_4$-PPy/Ti.

Figure 7 shows the Mulliken charge population of C, O, N, Cl, and Ti atoms of $LiClO_4$/Ti and $LiClO_4$-PPy/Ti. The interfacial interaction in $LiClO_4$/Ti and $LiClO_4$-PPy/Ti leads to the center charge offset of the Ti lattice atoms and the negative charge density distribution of oxygen atoms as well as nitrogen atoms. Accordingly, Ti atoms involve the dissociation distribution of negative and positive charge in $LiClO_4$/Ti and $LiClO_4$-PPy/Ti. Table 2 lists the Mulliken charges of C, O, N, Cl, and Ti atoms in $LiClO_4$/Ti and $LiClO_4$-PPy/Ti. The negative charge densities of oxygen atoms are changed from $LiClO_4$/Ti (−0.78, −0.80, −0.82, −0.95) to $LiClO_4$-PPy/Ti (−0.76, −0.77, −0.78, −0.90), indicating a slightly decreased negative charge. The dissociated positive and negative charge densities of Ti substrate are (+0.13~+0.05)/(−0.23~−0.04) for $LiClO_4$/Ti and (+0.62~+0.08)/(−0.05~−0.01) for $LiClO_4$-PPy/Ti, respectively. Therefore, the electrostatic interaction is formed between negative-charged perchlorate O and positive-charged Ti. Additionally, the negative charge densities of nitrogen atom are (−0.58~−0.47). Obviously, the strong electrostatic interaction is further formed between the negative-charged pyrrole N and the positive-charged Ti in $LiClO_4$-PPy/Ti. Herein, $LiClO_4$ exhibits the surface adsorption state of Ti in $LiClO_4$/Ti and the doping state of PPy in $LiClO_4$-PPy/Ti. The Mulliken charge population analysis proves the shorter intermolecular interaction distance in $LiClO_4$-PPy/Ti in comparison with $LiClO_4$/Ti. Accordingly, $LiClO_4$-PPy/Ti achieves higher electroactivity than $LiClO_4$/Ti. The DFT simulation calculation results agree with the experimental measurement results in confirming the superior capacitance of $LiClO_4$-PPy/Ti.

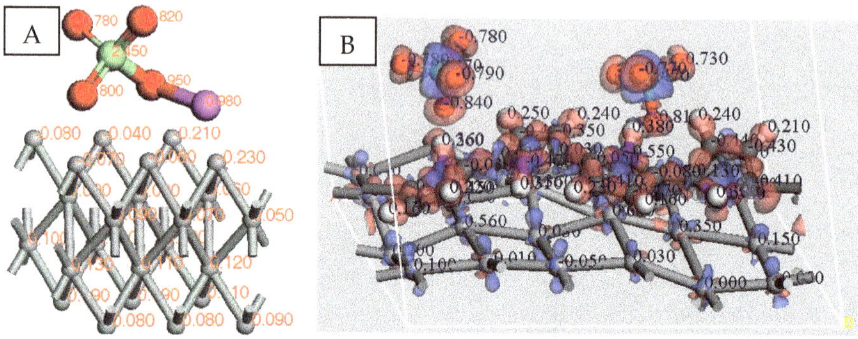

Figure 7. Mulliken charge population of (**A**) $LiClO_4$/Ti and (**B**) $LiClO_4$-PPy/Ti.

Table 2. Mulliken charge population of C, O, N, Cl, and Ti atoms in $LiClO_4$/Ti and $LiClO_4$-PPy/Ti.

Element	$LiClO_4$-Ti	$LiClO_4$-PPy-Ti
O1	−0.78	−0.76
O2	−0.80	−0.77
O3	−0.82	−0.78
O4	−0.95	−0.90
Cl	+2.45	+2.47
C1	/	+0.03
C2	/	+0.04
N	/	(−0.58~−0.47)
Ti	(−0.23~−0.04) (+0.13~+0.05)	(−0.05~−0.01) (+0.62~+0.08)

3. Experimental Section

3.1. Preparation of $LiClO_4$-PPy/Ti

The titanium sheet (Ti, purity 99.5%, size 10 × 50 mm, thickness 0.2 mm.) was washed using acetone, ethanol, and then deionized water to remove surface stains. The Ti sheet then conducted the chemical polishing treatment in HF-HNO_3 solution to dissolve the surface-covered oxide layer. The chemical cleaned Ti sheet was obtained. The Ti sheet conducted the electrostatic adsorption in lithium perchlorate solution form $LiClO_4$/Ti. A normal pulse voltammetry process was employed to synthetize $LiClO_4$-PPy/Ti using a three-electrode reaction system, which included the fresh Ti sheet working electrode, Pt foil counter electrode, the Hg/Hg_2Cl_2 reference electrode, and acetonitrile solvent solution of 0.1 M $LiClO_4$ electrolyte and 0.15 M pyrrole. The reaction parameters included a pulse potential range of 0.7~1.1 V and pulse potential increment of 0.001 V.

3.2. Characterization and Measurement

The surface microstructure characterization of $LiClO_4$-PPy/Ti and $LiClO_4$/Ti was conducted using a scanning electron microscope (SEM, ZEISS Ultra Plus, Germany). A Raman spectrum measurement of $LiClO_4$-PPy/Ti was conducted using a Raman spectroscope (Thermo Fisher DXR3, He-Ne laser, Emitting wavelength of 785 nm). A crystal phase structure characterization of titanium substrate was conducted using an X-ray diffraction (XRD, Bruker D8-Discover, Wavelength of X-ray source, 0.154 nm) apparatus. Electrochemical capacitive measurements of $LiClO_4$-PPy/Ti and $LiClO_4$/Ti were performed using an electrochemical workstation (CHI 760D, CH Instruments Co., Ltd., China), which includes galvanostatic charge/discharge (GCD) and cyclic voltammetry (CV). The electrochemical impedance spectrum (EIS) measurement was carried out at the frequency range of 10^{-2}~10^5 Hz. The 1.0 M $LiClO_4$ was used as the doping ion source and the electrolyte solution, contributing to maintaining the electrochemical stability of the titanium substrate in $LiClO_4$-PPy/Ti. electrode. The total surface area was used to evaluate electrochemical capacitance.

3.3. Theoretical Simulation Calculation

Density functional theory (DFT) simulation calculation was carried out to investigate electronic properties and atomic charge distribution of $LiClO_4$ and $LiClO_4$-PPy interaction at the interface of Ti substrate [38]. The total density of states (DOS), Mulliken charge population, interfacial binding energy, highest occupied molecular orbital (HOMO), lowest unoccupied molecular orbital (LUMO), and electrostatic potential were determined and discussed for better understanding of the electrostatic interfacial interaction of $LiClO_4$-PPy/Ti and $LiClO_4$/Ti. The DMol3 package was applied to calculate HOMO-LUMO energy gaps, and the CASTEP package was applied to calculate electronic energy gap and charge density difference and DOS value.

4. Conclusions

The PPy/Ti has been prepared through an electro-polymerization deposition process which is applied as supercapacitor electrode for electrochemical energy storage. Theoretical and experimental investigations are adopted to disclose the conductivity and interfacial interaction-dependent capacitance of $LiClO_4$-PPy/Ti. $LiClO_4$/Ti reveals ideal electrical double-layer capacitance due to the reversible adsorption/desorption of $LiClO_4$ electrolyte on Ti substrate in the potential range of 0~0.8 V. $LiClO_4$-PPy/Ti reveals Faradaic capacitance due to reversible doping/dedoping perchlorate ion in PPy. $LiClO_4$-PPy/Ti reveals lower charge transfer resistance, lower ohmic resistance, and lower Warburg resistance to present more active interface, higher electronic conductivity, and more feasible ion diffusion properties than $LiClO_4$/Ti. $LiClO_4$/Ti and $LiClO_4$-PPy/Ti achieve the mean response current of 0.13~2.47 $\mu A\ cm^{-2}$ and 2.28~9.011 $\mu A\ cm^{-2}$ at 5~200 mV s^{-1}. The corresponding specific capacitances are 0.010~0.0095 mF cm^{-2} and 0.123~0.0122 mF cm^{-2} at 0.01~0.10 mA cm^{-2}. $LiClO_4$-PPy/Ti keeps much higher electrochemical conductivity and higher specific capacitance than $LiClO_4$/Ti. Theoretical calculation results prove that interfacial electrostatic adsorption of $LiClO_4$/Ti and ClO_4^- anion doping of $LiClO_4$-PPy/Ti could change atomic charge density distribution. $LiClO_4$-PPy/Ti exhibits higher electronic conductivity and lower interface energy than $LiClO_4$/Ti. $LiClO_4$-PPy/Ti can act as a superior supercapacitor electrode to exhibit electroactivity and capacitance for the promising energy storage application.

Author Contributions: Formal analysis, Y.X. and J.X.; data curation, Y.X., L.L. and C.X.; writing—original draft preparation, Y.X.; writing—review and editing, Y.X. All authors have read and agreed to the published version of the manuscript.

Funding: The Science and Technology Program of Suzhou City, China (SYG202342).

Data Availability Statement: Data are contained within the article.

Acknowledgments: This research work is supported by Science and Technology Program of Suzhou City, China (SYG202342), and Big Data Computing Center of Southeast University, China.

Conflicts of Interest: The authors declare no conflicts of interest.

References

1. Chen, G.Z. Understanding supercapacitors based on nano-hybrid materials with interfacial conjugation. *Prog. Natl. Sci.* **2013**, *23*, 245–255. [CrossRef]
2. Adalati, R.; Sharma, M.; Sharma, S.; Kumar, A.; Malik, G.; Boukherroub, R.; Chandra, R. Metal nitrides as efficient electrode material for supercapacitors: A review. *J. Energy Stor.* **2022**, *56*, 105912. [CrossRef]
3. Yadav, M.S. Metal oxides nanostructure-based electrode materials for supercapacitor application. *J. Nanopart. Res.* **2020**, *22*, 367. [CrossRef]
4. Parveen, N.; Ansari, S.A.; Ansari, M.Z.; Ansari, M.O. Manganese oxide as an effective electrode material for energy storage: A review. *Environ. Chem. Lett.* **2022**, *20*, 283–309. [CrossRef]

5. Ansari, M.Z.; Nandi, D.K.; Janicek, P.; Ansari, S.A.; Ramesh, R.; Cheon, T.; Shong, B.; Kim, S.-H. Low-Temperature Atomic Layer Deposition of Highly Conformal Tin Nitride Thin Films for Energy Storage Devices. *Acs Appl. Mater. Interface* **2019**, *11*, 43608–43621. [CrossRef]
6. Parveen, N.; Hilal, M.; Han, J.I. Newly Design Porous/Sponge Red Phosphorus@Graphene and Highly Conductive Ni_2P Electrode for Asymmetric Solid State Supercapacitive Device with Excellent Performance. *Nano-Micro Lett.* **2020**, *12*, 25. [CrossRef] [PubMed]
7. Karnan, M.; Hari Prakash, K.; Badhulika, S. Revealing the super capacitive performance of N-doped hierarchical porous activated carbon in aqueous, ionic liquid, and redox additive electrolytes. *J. Energy Stor.* **2022**, *53*, 105189. [CrossRef]
8. Xie, Y. Electrochemical Performance of Polyaniline Support on Electrochemical Activated Carbon Fiber. *J. Mater. Eng. Perform.* **2022**, *31*, 1949–1955. [CrossRef]
9. Wang, H.; Xie, Y. Hydrogen bond enforced polyaniline grown on activated carbon fibers substrate for wearable bracelet supercapacitor. *J. Energy Stor.* **2022**, *52*, 105042. [CrossRef]
10. Xie, Y. Electrochemical and hydrothermal activation of carbon fiber supercapacitor electrode. *Fiber Polym.* **2022**, *23*, 10–17. [CrossRef]
11. Xie, Y.; Wang, Y.; Wang, L.; Liang, J. Theoretical and Experimental Investigations of Oxygen Activation Effect of Carbon Nanofibers Interacting with Polypyrrole. *Fibers* **2024**, *12*, 4. [CrossRef]
12. Zhai, Z.; Zhang, L.; Du, T.; Ren, B.; Xu, Y.; Wang, S.; Miao, J.; Liu, Z. A review of carbon materials for supercapacitors. *Mater. Des.* **2022**, *221*, 111017. [CrossRef]
13. Jáquez-Muñoz, J.M.; Gaona-Tiburcio, C.; Chacón-Nava, J.; Cabral-Miramontes, J.; Nieves-Mendoza, D.; Maldonado-Bandala, E.; Delgado, A.D.; Flores-De los Rios, J.P.; Bocchetta, P.; Almeraya-Calderón, F. Electrochemical Corrosion of Titanium and Titanium Alloys Anodized in H_2SO_4 and H_3PO_4 Solutions. *Coatings* **2022**, *12*, 325. [CrossRef]
14. Ansari, S.A.; Khan, N.A.; Hasan, Z.; Shaikh, A.A.; Ferdousi, F.K.; Barai, H.R.; Lopa, N.S.; Rahman, M.M. Electrochemical synthesis of titanium nitride nanoparticles onto titanium foil for electrochemical supercapacitors with ultrafast charge/discharge. *Sustain. Energy Fuels* **2020**, *4*, 2480–2490. [CrossRef]
15. Liu, X.Y.; Chen, H.; Li, G.; Peng, J.H.; Zhang, Y.X. One-pot synthesis of pearl-chain-like manganese dioxide-decorated titanium grids as advanced binder-free supercapacitors electrodes. *Ceram. Int.* **2016**, *42*, 9227–9233. [CrossRef]
16. Wang, L.; Ma, Y.; Yang, M.; Qi, Y. Titanium plate supported MoS_2 nanosheet arrays for supercapacitor application. *Appl. Surf. Sci.* **2017**, *396*, 1466–1471. [CrossRef]
17. Wei, J.; Wei, S.; Wang, G.; He, X.; Gao, B.; Zhao, C. PPy modified titanium foam electrode with high performance for supercapacitor. *Eur. Polym. J.* **2013**, *49*, 3651–3656. [CrossRef]
18. Zhang, J.; Yu, X.; Zhao, Z.-y.; Zhang, Z.; Li, J. Influence of pore size of Ti substrate on structural and capacitive properties of Ti/boron doped diamond electrode. *J. Alloys Compd.* **2019**, *777*, 84–93. [CrossRef]
19. Chang, L.; Chen, B.; Qiao, H.; Huang, H.; Guo, Z.; He, Y.; Xu, R.; Xionghui, X. Study of the Effects of Pretreatment Processing on the Properties of Metal Oxide Coatings on Ti-Based Sheet. *J. Electrochem. Soc.* **2021**, *168*, 033501. [CrossRef]
20. Kishimoto, A.; Yamada, Y.; Funatsu, K.; Uda, T. Suitable Electrode Materials for Titanium Sheet Deposition. *Adv. Eng. Mater.* **2020**, *22*, 1900747. [CrossRef]
21. Lamberti, A. Flexible supercapacitor electrodes based on MoS_2-intercalated rGO membranes on Ti mesh. *Mater. Sci. Semicond. Process.* **2018**, *73*, 106–110. [CrossRef]
22. Seo, H.-S.; Bae, J.-U.; Kim, D.-H.; Park, Y.; Kim, C.-D.; Kang, I.B.; Chung, I.-J.; Choi, J.-H.; Myoung, J.-M. Reliable Bottom Gate Amorphous Indium-Gallium-Zinc Oxide Thin-Film Transistors with TiO_x Passivation Layer. *Electrochem. Solid State Lett.* **2009**, *12*, H348–H351. [CrossRef]
23. Reddy, P.C.H.; Amalraj, J.; Ranganatha, S.; Patil, S.S.; Chandrasekaran, S. A review on effect of conducting polymers on carbon-based electrode materials for electrochemical supercapacitors. *Synth. Met.* **2023**, *298*, 117447. [CrossRef]
24. Tsekova, D.S.; Karastoyanov, V.; Peychev, D.; Valova, I. Crystallization of ferritin on biocompatible Surfaces—Bare Ti and Ti covered by polypyrrole (PPy). *J. Cryst. Growth* **2024**, *631*, 127616. [CrossRef]
25. Chen, J.; He, Y.; Li, L. Real-time probing electrodeposition growth of polyaniline thin film via in-situ spectroscopic ellipsometry. *Thin Solid Films* **2022**, *762*, 139565. [CrossRef]
26. Kondratiev, V.V.; Holze, R. Intrinsically conducting polymers and their combinations with redox-active molecules for rechargeable battery electrodes: An update. *Chem. Pap.* **2021**, *75*, 4981–5007. [CrossRef]
27. Stejskal, J.; Sapurina, I.; Vilčáková, J.; Humpolíček, P.; Truong, T.H.; Shishov, M.A.; Trchová, M.; Kopecký, D.; Kolská, Z.; Prokeš, J.; et al. Conducting polypyrrole-coated macroporous melamine sponges: A simple toy or an advanced material? *Chem. Pap.* **2021**, *75*, 5035–5055. [CrossRef]
28. Ji, S.; Yang, J.; Cao, J.; Zhao, X.; Mohammed, M.A.; He, P.; Dryfe, R.A.W.; Kinloch, I.A. A Universal Electrolyte Formulation for the Electrodeposition of Pristine Carbon and Polypyrrole Composites for Supercapacitors. *Acs Appl. Mater. Interface* **2020**, *12*, 13386–13399. [CrossRef] [PubMed]
29. Zhu, H.; Xie, Y. Electrochemical performance of bridge molecule-reinforced activated carbon fiber-m-aminobenzenesulfonic acid-polyaniline for braidable-supercapacitor application. *Chem. Eng. J.* **2023**, *478*, 147416. [CrossRef]
30. Jakhar, P.; Shukla, M.; Singh, V. Influence of $LiClO_4$ Concentration on 1-D Polypyrrole Nanofibers for Enhanced Performance of Glucose Biosensor. *J. Electrochem. Soc.* **2018**, *165*, G80–G89. [CrossRef]

31. Santino, L.M.; Acharya, S.; D'Arcy, J.M. Low-temperature vapour phase polymerized polypyrrole nanobrushes for supercapacitors. *J. Mater. Chem. A* **2017**, *5*, 11772–11780. [CrossRef]
32. Sharifi-Viand, A.; Mahjani, M.G.; Moshrefi, R.; Jafarian, M. Diffusion through the self-affine surface of polypyrrole film. *Vacuum* **2015**, *114*, 17–20. [CrossRef]
33. Wysocki, B.; Maj, P.; Sitek, R.; Buhagiar, J.; Kurzydłowski, K.J.; Święszkowski, W. Laser and Electron Beam Additive Manufacturing Methods of Fabricating Titanium Bone Implants. *Appl. Sci.* **2017**, *7*, 657. [CrossRef]
34. Hasoon, S. Electrochemical polymerization and Raman study of polypyrrole and polyaniline thin films HS Abdullah. *Int. J. Phys. Sci.* **2012**, *7*, 5468–5476.
35. Fernández Romero, A.J.; López Cascales, J.J.; Fernández Otero, T. Perchlorate Interchange during the Redox Process of PPy/PVS Films in an Acetonitrile Medium. A Voltammetric and EDX Study. *J. Phys. Chem. B* **2005**, *109*, 907–914. [CrossRef] [PubMed]
36. Wang, J.; Wu, C.; Wu, P.; Li, X.; Zhang, M.; Zhu, J. Polypyrrole capacitance characteristics with different doping ions and thicknesses. *Phys. Chem. Chem. Phys.* **2017**, *19*, 21165–21173. [CrossRef] [PubMed]
37. Chen, S.; Liu, H.; Wang, Y.; Xu, S.; Liu, W.; He, D.; Liu, X.; Liu, J.; Hu, C. Electrochemical Capacitance of Spherical Nanoparticles Formed by Electrodeposition of Intrinsic Polypyrrole onto Au Electrode. *Electrochim. Acta* **2017**, *232*, 72–79. [CrossRef]
38. Xie, Y.; Mu, Y. Interface Mo-N coordination bonding MoSxNy@Polyaniline for stable structured supercapacitor electrode. *Electrochim. Acta* **2021**, *391*, 138953. [CrossRef]

inorganics

MDPI

Article

An Electrochemically Prepared Mixed Phase of Cobalt Hydroxide/Oxyhydroxide as a Cathode for Aqueous Zinc Ion Batteries

Fuwei Li [1,2], Yunbo Zhu [1,2], Hiroshi Ueno [3,4] and Ting Deng [1,2,*]

1 Key Laboratory of Automobile Materials of MOE, School of Materials Science and Engineering, Jilin University, Changchun 130012, China
2 Jilin Provincial International Cooperation Key Laboratory of High-Efficiency Clean Energy Materials, Jilin University, Changchun 130012, China
3 Creative Interdisciplinary Research Division, The Frontier Research Institute for Interdisciplinary Sciences (FRIS), Tohoku University, Sendai 980-8578, Japan; hiroshi.ueno.d5@tohoku.ac.jp
4 Department of Chemistry, Graduate School of Science, Tohoku University, Sendai 980-8578, Japan
* Correspondence: tdeng@jlu.edu.cn

Abstract: Cobalt hydroxide is a widely studied electrode material for supercapacitor and alkaline zinc ion batteries. The large interlayer spacing of $Co(OH)_2$ is also attractive to store Zn ions. However, $Co(OH)_2$ is quite unstable in the acidic $ZnSO_4$ electrolyte due to its amphoteric nature. Herein, we synthesized a mixed phase of $Co(OH)_2$/CoOOH via a two-step electrochemical preparation. As the cathode material for an aqueous zinc ion battery (AZIB), $Co(OH)_2$/CoOOH delivered a maximum capacity of 164 mAh g^{-1} at 0.05 A g^{-1} and a high energy density of 275 Wh kg^{-1}. Benefiting from the low charge-transfer resistance, a capacity of 87 mAh g^{-1} was maintained at 1.6 A g^{-1}, showing a good rate performance of the mixed phase. Various spectroscopy analyses and simulations based on the density functional theory (DFT) suggested a higher thermal stability of the mixed phase than pure $Co(OH)_2$, due to its less local structural disorder. The reduced Co-Co and Co-O shells increased the mechanical strength of the mixed phase to accommodate Zn^{2+} ions and endure the electrostatic repulsion, resulting in an enhanced cycling stability. The mixed phased also delivered a good stability at the current density of 0.05 A g^{-1}. After 200 cycles, a capacity retention of 78% was retained, with high Coulombic efficiencies. These results provide a new route to synthesize high-performance LDH for aqueous zinc ion batteries.

Keywords: aqueous zinc battery; cobalt hydroxide; CoOOH; mixed phase

1. Introduction

The utilization of renewable energy sources, such as wind, solar, and tidal energies, is expected to alleviate the environmental pollution resulted from the combustion of fossil fuels. Due to the seasonal and regional factors, the storage of these energies with high efficiency, therefore, is of great significance to the wide application of these renewable energy sources. Among various energy storage systems, lithium ion batteries, due to their high energy density, are dominant in the markets of portable electronic devices and electric automobiles [1,2]. The safety issue and high cost of lithium ion batteries have driven us to search for alternative energy storage devices (ESDs) with comparable performance, but with a higher safety and lower cost. Aqueous ESDs, such as supercapacitor and nickel/metal hydride batteries, have been extensively investigated due to their excellent power densities and long lifespans. Unfortunately, the further application of these aqueous ESDs is restricted by the low energy density. Recently, aqueous zinc ion batteries (AZIBs) have drawn an exceptional amount of attention due to their environmental friendliness, high theoretical capacity of the Zn anode (820 mAh g^{-1}), and their abundancy in the Earth's crust [3,4]. However, the application of AZIBs is restricted by the unsatisfying electrochemical performance of cathode materials. Currently, Mn- and V-based oxides are the most investigated cathode materials for AZIBs [5,6]. In addition to the low redox potential of V-based oxides, along with the dissolution problem of Mn-based oxides, their performance is also affected by the strong interaction between zinc ions and the crystal structure of oxides [7]. Therefore, seeking new cathode materials is still of great significance to develop AZIBs with high energy densities and long lifespans.

Layered double hydroxides (LDHs), with tunable chemical compositions and 2D pore structures, exhibit excellent pseudocapacitive activities in battery/supercapacitor hybrid ESDs [8–10]. Facilitated via the H^+ intercalation/deintercalation mechanism and the structural similarity, cobalt hydroxide ($Co(OH)_2$) has demonstrated an extraordinary performance in supercapacitors [11]. Its atoms are very exposed on the surface of its unique 2D morphology, and the large interlayer spacings facilitates the charge transfer between the electrolyte and the active material. However, the interlayer spacings of $Co(OH)_2$ are usually occupied by the anion residues (NO_3^-, CO_3^{2-}, organic anions, etc.) from the preparation to neutralize the positive charge of the Co-O layers [12]. As a result, Zn ion storage is prohibited by either the interlayer species or the electrostatic repulsion from the positively charged Co-O layers. Our previous study has proven that the excellent pseudocapacitive behavior originates from the rapid transformation between $Co(OH)_2$ and CoOOH based on their structural similarity [11]. Compared to $Co(OH)_2$, CoOOH not only preserves the morphology of $Co(OH)_2$, but also shows a higher level of thermal stability than $Co(OH)_2$, since each H atom is shared by two adjacent Co-O layers, and the positive charges are subsequently reduced, which presents CoOOH as a promising cathode material for AZIBs. Nevertheless, it remains uncertain to fabricate a high-performance CoOOH cathode for AZIBs, since no relevant study has been reported.

Herein, a mixed phase of $Co(OH)_2$/CoOOH was synthesized via a two-step electrochemical preparation. As an AZIB cathode material, $Co(OH)_2$/CoOOH showed a high average potential of 1.7 V vs. Zn^{2+}/Zn, which is higher than most of the reported cathode materials for AZIBs. Correspondingly, the mixed phase of $Co(OH)_2$/CoOOH can deliver a maximum capacity of 164 mAh g^{-1} at 0.05 A g^{-1} in 1 M $ZnSO_4$ electrolyte. The $Co(OH)_2$/CoOOH cathode also exhibited an excellent rate performance; a high capacity of 87 mAh g^{-1} was obtained at 1.6 A g^{-1}. X-ray adsorption spectroscopy (XAS), along with theoretical simulations, indicated the high crystallinity and thermal stability of the mixed phase. The even distribution of CoOOH in the mixed phase suppresses the structure's degradation. The reduced Co-Co and Co-O shells increased the mechanical strength of the mixed phase to accommodate Zn^{2+} ions and endure the electrostatic repulsion, resulting in an enhanced cycling stability. The mixed phase of $Co(OH)_2$/CoOOH showed a stable cycle life of 200 cycles at 0.05 A g^{-1}, with a capacity retention of 78%. The excellent performance of $Co(OH)_2$/CoOOH has demonstrated the promising future of LDHs as cathode materials for AZIBs.

2. Results

The mixed phase of $Co(OH)_2$/CoOOH was synthesized via a two-step electrochemical preparation, as illustrated in Figure 1A. Firstly, $Co(OH)_2$ was electrodeposited on the carbon fiber paper (CFP) as the substrate in a three-electrode system, in which a CFP, a platinum plate, a saturated calomel electrode, and a 1 M $Co(NO_3)_2$ solution served as the working electrode, counter electrode, reference electrode, and electrolyte, respectively. As the supercapacitor electrode, $Co(OH)_2$ loses one H atom and is then oxidized to CoOOH to store charges in alkaline media. Accordingly, the as-prepared $Co(OH)_2$ electrode was placed into a 1 M KOH solution for a cyclic voltammetry (CV) treatment to prepare the mixed phase of $Co(OH)_2$/CoOOH. The CV potential was set in the range of −0.1~0.45 V. To acquire the structural information, X-ray diffraction (XRD) was conducted on each step of the preparation. Figure 1B depicts the XRD patterns of the CFP after the electrodeposition, alkaline media contact, and the CV treatment, in which α, β, and C stand for α-$Co(OH)_2$, β-$Co(OH)_2$, and CoOOH, respectively. Two peaks were observed at 33° and 59°, which can be indexed to the (100) and (110) planes of α-$Co(OH)_2$ (PDF card No.460605). Compared to the pure CFP (Figure S1), the feature peak of the carbon materials (002) at 26° was far more reduced, indicating the well coating of α-$Co(OH)_2$ on the CFP. The α-$Co(OH)_2$ is known for its large interlayer spacing due to either anion or molecule insertion from the electrolyte. Our previous work has manifested that the electrochemically deposited α-$Co(OH)_2$ was unstable, which could easily transform into β-$Co(OH)_2$ in alkaline media by removing the NO_3^- species from the structure. The same phenomenon was also observed in this work. After being soaked in a 1 M KOH solution, the (001), (100), (101), (102), (110), and (111) planes of β-$Co(OH)_2$, with a high level of crystallinity, were detected in the red line (PDF card No.300443), while no signs of α-$Co(OH)_2$ were detected in the XRD patterns, indicating the transformation from α-$Co(OH)_2$ into β-$Co(OH)_2$. After the CV treatment, three new peaks were detected at 20°, 38°, and 65°, which can be indexed to the (001), (012), and (018) planes of CoOOH (PDF card No.070159), respectively, indicating the co-existence of β-$Co(OH)_2$ and CoOOH.

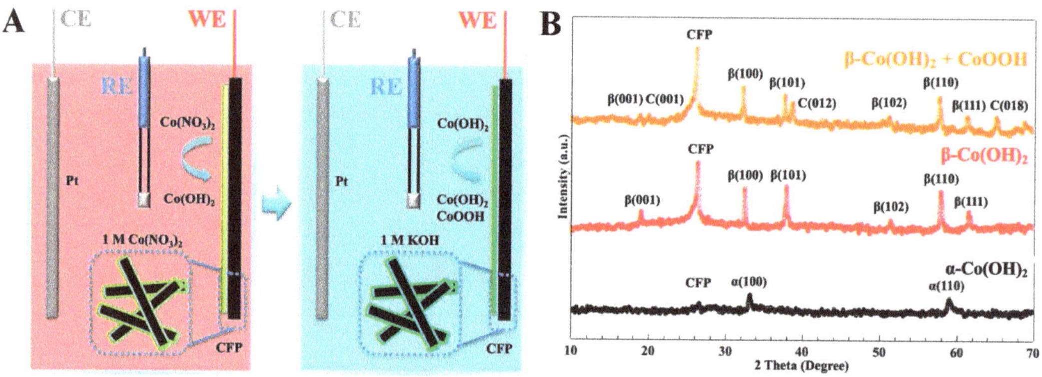

Figure 1. (**A**) The illustration of the synthesis of the $Co(OH)_2$/CoOOH mixed phase on the CFP. (**B**) The XRD patterns of α-$Co(OH)_2$, β-$Co(OH)_2$, and the $Co(OH)_2$/CoOOH mixed phase on the CFP. Black line: α-$Co(OH)_2$; red line: β-$Co(OH)_2$; orange line: $Co(OH)_2$/CoOOH mixed phase.

The morphology evolution of $Co(OH)_2$ was analyzed via scanning electron spectroscopy (SEM) and transmission electron spectroscopy (TEM). As shown in Figure 2A, the as-prepared α-$Co(OH)_2$ displayed a petal-like morphology, with an average thickness of 200 nm. With a closer scrutiny, each petal was loosely stacked, consisting of multiple $Co(OH)_2$ layers with large interlayer spacings (Figure 2B). After being soaked in a 1 M KOH solution, the morphology of $Co(OH)_2$ changed dramatically from the loosely stacked petals to the hexagonal platelets, corresponding to the change in the XRD pattern (Figure 2C). In the hexagonal platelets, no obvious void was observed, suggesting a decreased interlayer spacing. In addition, TEM images also confirmed the morphology evolution from α- to β-$Co(OH)_2$. The porous structure was observed in α-$Co(OH)_2$ (Figure 2D), which disappeared and changed to the compact stack of $Co(OH)_2$ layers (Figure 2E), when the electrode was put into alkaline media. No obvious pores were identified. These results indicate that the actual reactant was β-$Co(OH)_2$. In addition, the space group of hexagonal β-$Co(OH)_2$ was R3m. The (010) and (100) crystal planes exhibited an acute angle of 60° (Figure 2F), also suggesting the formation of β-$Co(OH)_2$. After the CV treatment, the morphology of the mixed phase of β-$Co(OH)_2$ and CoOOH was quite similar to β-$Co(OH)_2$ (Figures S2 and S3), showing a morphology of well-shaped hexagonal platelets, which resulted from the similar structural parameters of β-$Co(OH)_2$ and CoOOH.

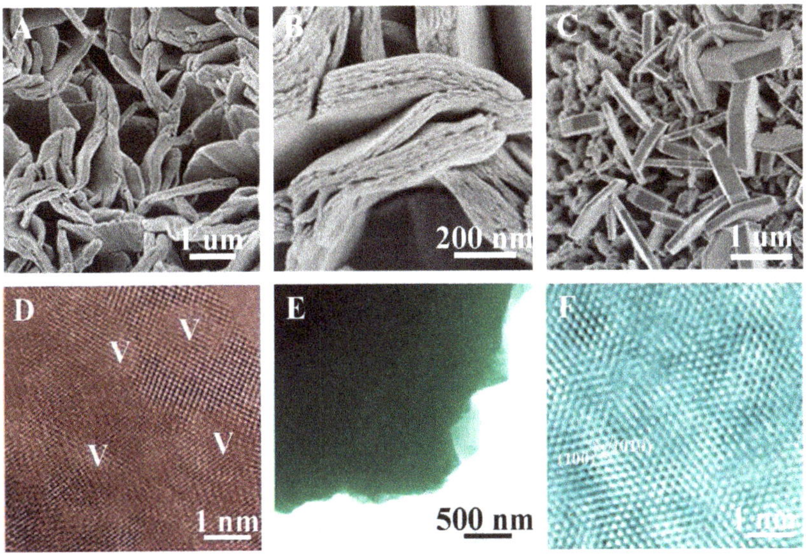

Figure 2. (**A**,**B**) The SEM images of α-$Co(OH)_2$ with different magnitudes. (**C**) The SEM image of β-$Co(OH)_2$. (**D**) TEM image of α-$Co(OH)_2$. (**E**,**F**) TEM and HRTEM images of β-$Co(OH)_2$.

Fourier transform infrared spectroscopy (FTIR) and X-ray photoemission spectroscopy (XPS) analyses were conducted to probe the surface state of the $Co(OH)_2/CoOOH$ mixed phase. Figure 3A shows the FTIR spectra of the as-prepared $Co(OH)_2$ and the $Co(OH)_2$ after their immersion in the alkaline media. There was a sharp absorption band observed at 1383 cm^{-1} in the spectrum of α-$Co(OH)_2$, representing the N-O vibration from the inserted NO_3^- species [13]. The intensity of the N-O vibration was substantially reduced after diving the as-prepared $Co(OH)_2$ in the KOH solution, indicating the removal of NO_3^-. In addition, an intense band appeared at 3572 cm^{-1} in the spectrum of β-$Co(OH)_2$, which is the typical feature of OH groups. The transformation from α- to β-$Co(OH)_2$ could be explained by the NO_3^- replacement of OH^- with a smaller size, leading to the compact arrangement of the Co-O layers. According to the FTIR analysis, α-$Co(OH)_2$ is quite unstable in the alkaline media, which can rapidly transform into β-$Co(OH)_2$. So, the actual reactant is β-$Co(OH)_2$, rather than α-$Co(OH)_2$, to store Zn^{2+} ions. Figure 3B depicts the XPS spectrum of C 1s of α-$Co(OH)_2$ on the CFP. In addition to the typical signals of the sp^2 carbon and some surface function groups, such as C-O, C=O, and COOH, strong Co-O-C interactions were generated between the CFP and $Co(OH)_2$ at 294 eV and 298 eV, which can provide a robust bonding between the active material and the substrate to prevent their pulverization [14]. X-ray adsorption spectroscopy (XAS) measurements were also conducted to evaluate the structural evolution from $Co(OH)_2$ to CoOOH. X-ray absorption near-edge structure (XANES) spectra of β-$Co(OH)_2$ and the mixed phase of $Co(OH)_2/CoOOH$ are shown in Figure 3C. The intensity of the $Co(OH)_2/CoOOH$ mixed phase was much higher than that of β-$Co(OH)_2$, indicating a decreased disorder in the local environment and stronger Co-Co shells [15,16]. The K-edge of the $Co(OH)_2/CoOOH$ mixed phase shows a positive shift compared to β-$Co(OH)_2$ in the inset of Figure 3C, which demonstrates an increased Co valence in the mixed phase. In addition, the overall oscillation patterns of β-$Co(OH)_2$ and the mixed phase were quite similar, suggesting similar coordination environments of the Co atom in β-$Co(OH)_2$ and the mixed phase. The extended X-ray absorption fine structure (EXAFS) spectra of β-$Co(OH)_2$ and the mixed phase are displayed in Figure 3D. The similar patterns also suggest similar coordination environments in β-$Co(OH)_2$ and the mixed phase. Moreover, two peaks were also observed, standing for the Co-O and Co-Co shells. The increased intensities of both peaks in the mixed phase were detected, which indicate a decreased structural disorder in the mixed phase [17,18]. Furthermore, negative shifts of the Co-O and Co-Co shells were also observed in the mixed phase, which suggests the reduced Co-O and Co-Co bond lengths due to the existence of CoOOH. Compared to $Co(OH)_2$, each H atom is shared by the two adjacent Co-O layers in CoOOH. The density functional theory (DFT) calculation simulated the transformation from $Co(OH)_2$ into CoOOH, and its computed energy file is provided in Figure 4. CoOOH (a = b = 3.036 Å; c = 8.862 Å) was obtained by losing one H atom per formula unit from $Co(OH)_2$ (a = b = 3.176 Å; c = 9.358 Å). We found that the energy difference between CoOOH and $Co(OH)_2$ was 0.6 eV, and that the activation energy from CoOOH to $Co(OH)_2$ was higher than for the opposite process, indicating that the mixed phase is more stable than pure $Co(OH)_2$.

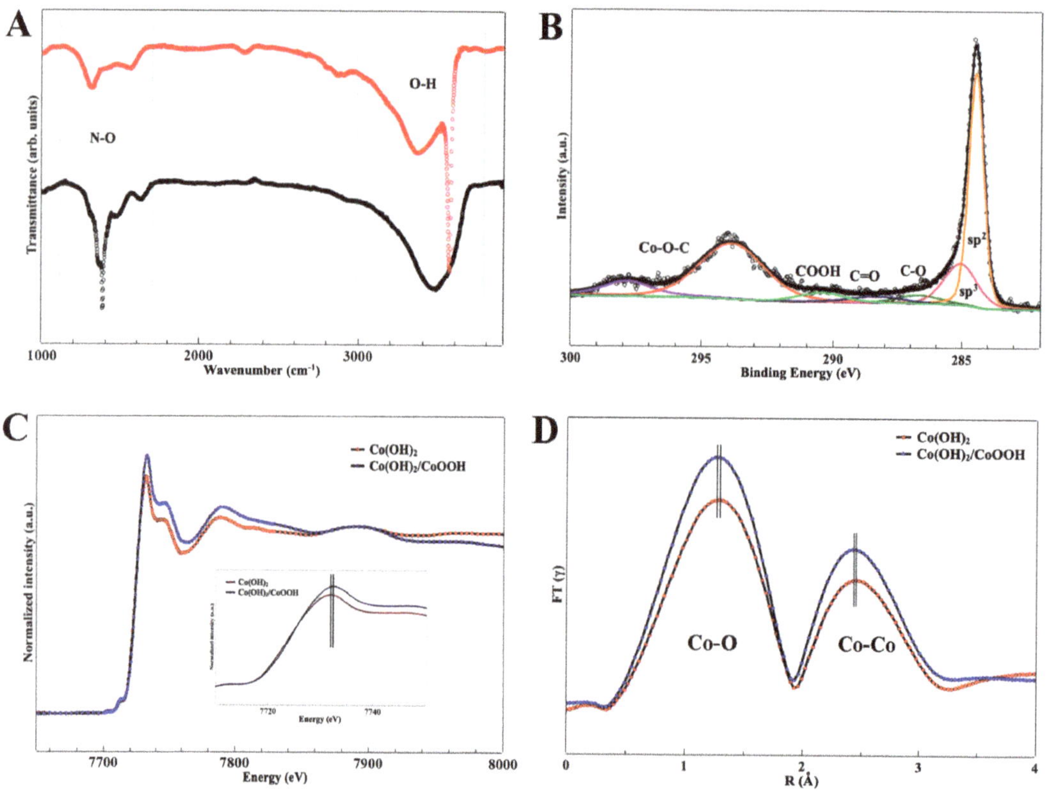

Figure 3. (**A**) FTIR analysis of the as-prepared $Co(OH)_2$ and β-$Co(OH)_2$. Red line: α-$Co(OH)_2$; black line: β-$Co(OH)_2$. (**B**) XPS spectrum of the C 1s of α-$Co(OH)_2$/CFP. (**C**) XANES spectra of β-$Co(OH)_2$ and the mixed phase $Co(OH)_2$ and CoOOH. (**D**) EXAFS spectra of β-$Co(OH)_2$ and the mixed phase $Co(OH)_2$ and CoOOH. Co-O and Co-Co stand for the nearest Co-O and Co-Co shell to the Co atom.

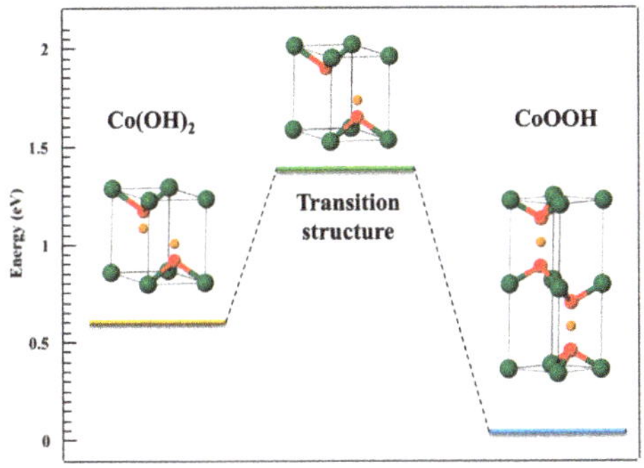

Figure 4. Phase transformation energy profile from $Co(OH)_2$ to CoOOH. CoOOH is considered as the ground state. Energy values for all the minima and transition states are provided in Table S1. The green, red, and orange balls stand for Co, O, and H atoms, respectively.

The electrochemical performance of the $Co(OH)_2$/CoOOH mixed phase as the cathode for AZIBs was investigated in a coin cell, in which the $Co(OH)_2$/CoOOH mixed phase, a Zn foil, and 1 M $ZnSO_4$ solution were used as the cathode, anode, and electrolyte, respectively (Figure 5A). The mixed phase of $Co(OH)_2$/CoOOH exhibited a maximum capacity of 164 mAh g^{-1} at 0.05 A g^{-1}, as well as a high energy density of 275 Wh kg^{-1} (Figure 5B). Moreover, no obvious decay in capacity was observed in the first five cycles, suggesting a good stability of the mixed phase of $Co(OH)_2$/CoOOH. The differential capacity (dQ/dV) curve is depicted in Figure 5C, and two discharge peaks located at 1.6 eV and 1.8 V vs. Zn^{2+}/Zn were observed. The average discharge voltage of 1.7 V vs. Zn^{2+}/Zn was higher than for most of the reported cathode materials, such as V_2O_5 (0.6 V vs. Zn^{2+}/Zn) and MnO_2 (1.4 V vs. Zn^{2+}/Zn). Figure 5D shows the Nyquist plot of the mixed phase. The inset is the electrical equivalent circuit, in which R_{ct}, R_s, W, and CPE stand for the charge-transfer resistance, the equivalent series resistance, the Warburg impedance, and the constant phase element, respectively. According to the fitting results (Table S2), R_{ct} was calculated to be 2.1 Ω, and such a small charge-transfer resistance indicates an excellent electron transfer rate in the mixed phase of β-$Co(OH)_2$ and CoOOH. Figure 5E displays the CV curves of the mixed phase of $Co(OH)_2$/CoOOH at different scan rates, and the reaction kinetics were evaluated based on the scan rate and the corresponding current. The capacity contribution of the diffusion- and surface-controlled processes can be deconvoluted according to the following equation [19]:

$$i = k_1 v + k_2 v^{\frac{1}{2}}$$

in which $k_1 v$ and $k_2 v^{\frac{1}{2}}$ stand for the surface- and diffusion-controlled processes, respectively. As the scan rate increased from 0.1 mV s^{-1} to 5 mV s^{-1}, the contribution ratio of the diffusion-controlled capacity was stabilized at ~70%, indicating that both surface and diffusion synergically regulate the electrode reaction. As a result, the mixed phase of $Co(OH)_2$/CoOOH exhibited an excellent rate performance. Its capacity can reach 164, 124, 108, 103, 98, and 87 mAh g^{-1} at 0.05, 0.1, 0.2, 0.4, 0.8, and 1.6 A g^{-1}, respectively (Figure 5G,H). In contrast, β-$Co(OH)_2$ only showed a maximum capacity of 48.8 mAh g^{-1} at 0.1 A g^{-1} (Figure S4), along with an inferior rate performance (Figure S5). Compared to recently reported cathode materials for AZIBs, such as PBA (99.7 mAh g^{-1} at 0.05 A g^{-1}) [20], FeHCF (76 mAh g^{-1} at 1 A g^{-1}) [21], and $Na_3V_2(PO_4)_2F_3$ (65 mAh g^{-1} at 0.08 A g^{-1}) [22], the mixed phase of $Co(OH)_2$/CoOOH exhibited an increased capacity, an improved rate performance, and a higher discharge voltage of 1.7 V (Table 1). The mixed phased also delivered a good stability at the current density of 0.05 A g^{-1}. After 200 cycles, a capacity retention of 78% was retained, with high Coulombic efficiencies (Figure 5I). However, the capacity of β-$Co(OH)_2$ rapidly dropped to 38% after 50 cycles at the current density of 0.1 A g^{-1} (Figure S6). Both $Co(OH)_2$ and CoOOH are amphoteric materials, which could be dissolved in acidic media. The pH value of the 1 M $ZnSO_4$ electrolyte is about 4~4.5, meaning that the dissolution of the active material is responsible for the capacity loss. On the other hand, the interaction between the Zn^{2+} ions and the Co-O layer was enhanced after the removal of NO_3^- species, which neutralize the positive charge of the Co-O layers. However, the high thermal stability and the shortened Co-Co and Co-O shells, according our XAS analysis and DFT calculation, increased the mechanical strength of the mixed phase to accommodate Zn^{2+} ions and endure the electrostatic repulsion.

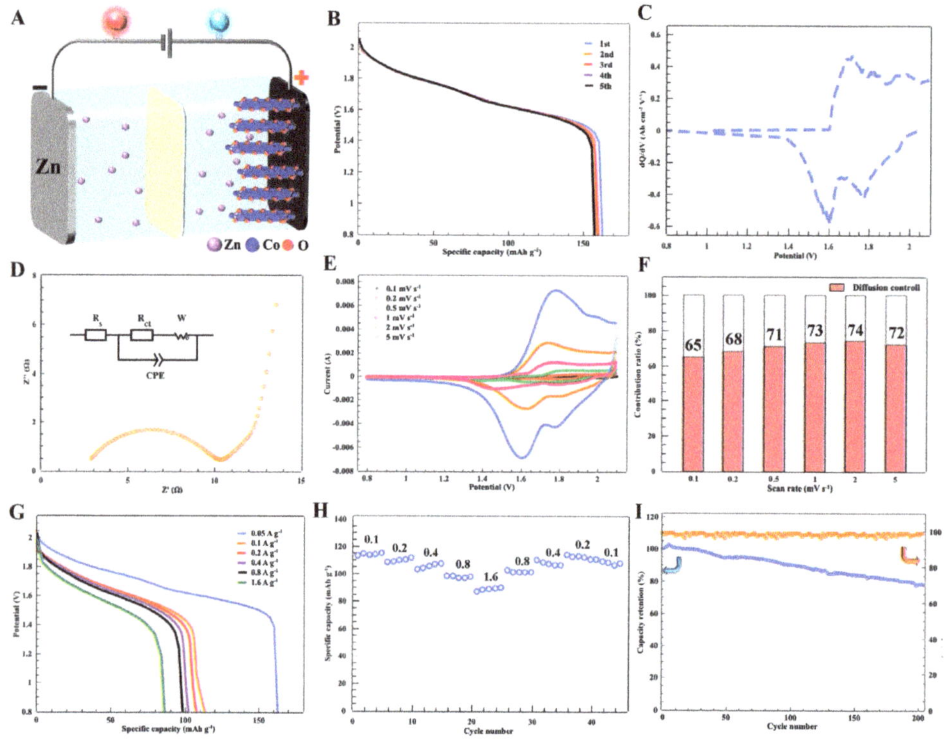

Figure 5. (**A**) Schematic illustration of a Zn-$Co(OH)_2$/CoOOH cell in a 1 M $ZnSO_4$ electrolyte. (**B**) Discharge profiles of the first 5 cycles of the mixed phase at 0.05 A g^{-1}. (**C**) The differential capacity curves of the mixed phase $Co(OH)_2$/CoOOH. (**D**) The Nyquist plot and the equivalent electrical circuit of the Zn-$Co(OH)_2$/CoOOH cell in a 1 M $ZnSO_4$ electrolyte. (**E**) The CV curves of the mixed phase of $Co(OH)_2$/CoOOH at different scan rates from 0.1 mV s^{-1} to 5 mV s^{-1}. (**F**) The contribution ratio of the diffusion-controlled capacity and the surface-controlled capacity. (**G**) Discharge profiles of the mixed phase of $Co(OH)_2$/CoOOH at different current densities from 0.05 A g^{-1} to 1.6 A g^{-1}. (**H**) Rate performance of the mixed phase of $Co(OH)_2$/CoOOH. (**I**) The stability (200 cycles at 0.05 A g^{-1}) and Coulombic efficiency of the mixed phase of $Co(OH)_2$/CoOOH.

Table 1. Electrochemical performance comparison of cathode materials for AZIBs.

Cathode	Maximum Capacity	Rate Performance	Discharge Voltage	Refs.
PBA	99.7 mAh g^{-1} at 0.05 A g^{-1}	39.7 mAh g^{-1} at 2 A g^{-1}	1.4	[20]
FeHCF	76 mAh g^{-1} at 1 A g^{-1}	41 mAh g^{-1} at 8 A g^{-1}	1.5	[21]
$Na_3V_2(PO_4)_2F_3$	65 mAh g^{-1} at 0.08 A g^{-1}	33 mAh g^{-1} at 8 A g^{-1}	1.6	[22]
$Co(OH)_2$/CoOOH	164 mAh g^{-1} at 0.05 A g^{-1}	87 mAh g^{-1} at 1.6 A g^{-1}	1.7	This work

3. Discussion

Recently, LDHs with abundant H vacancies were obtained via electrochemical methods [23,24], and these LDHs have exhibited promising performance as the cathode material for AZIBs in mild acidic media. Coincidently, the transition metal oxyhydroxide contains one less H atom than its hydroxide, thus indicating that the excellent performance may originate from oxyhydroxides or the synergic effect of hydroxide and oxyhydroxide. Unfortunately, the effect of oxyhydroxide was overlooked. CoOOH and $Co(OH)_2$ are quite alike in structure, but the oxyhydroxide is more stable, which could stabilize the whole structure against the strong interaction with Zn^{2+} ions. Therefore, according to our work, the transition metal oxyhydroxide is supposed to be a promising cathode material for AZIBs.

4. Materials and Methods

Synthesis of α-$Co(OH)_2$. The α-$Co(OH)_2$ was electrochemically deposited on the carbon fiber paper (CFP) as the substrate. A 1 M $Co(NO_3)_2$ solution (analytical reagent (AR), Xilong Chemical Co., Ltd., Guangzhou, China) was used as the electrolyte. Before the electrodeposition, the CFP was washed with acetone, ethanol, and distilled water for 30 min each, and the CFP was dried in a 60 degree oven overnight. Then, the α-$Co(OH)_2$ thin film was electrodeposited on the CFP in a conventional three-electrode electrolytic cell, in which the CFP, a platinum plate, and a saturated calomel electrode (SCE) served as the working, counter, and reference electrode, respectively. The electrodeposition was conducted using a CHI660E potentiostat, and the voltage was maintained at −1 V for 30 min. The formation of α-$Co(OH)_2$ can be expressed as follows: $NO_3^- + 7H_2O + 8e^- \rightarrow NH_4^+ + 10\,OH^-$, $Co^{2+} + 2OH^- \rightarrow Co(OH)_2$. Following the electrodeposition, the prepared electrode was washed with distilled water for several times and dried in a 60 degree vacuum oven overnight.

Synthesis of the mixed phase of $Co(OH)_2$/CoOOH. The as-prepared α-$Co(OH)_2$ was utilized as the precursor to obtain the mixed phase of $Co(OH)_2$ and CoOOH via an electrochemical method in a conventional three-electrode electrolytic cell for a CV treatment, in which the as-prepared α-$Co(OH)_2$, a platinum plate, a saturated calomel electrode (SCE), and a 1 M KOH solution served as the working electrode, counter electrode, reference electrode, and the electrolyte, respectively. The CV treatment was conducted using a CHI660E potentiostat. The potential range of the CV treatment was set from −0.1 V to 0.45 V, and the scan rate was 25 mV s^{-1} for 50 cycles. The electrochemical process was expressed as follows: $Co(OH)_2 + OH^- \rightarrow CoOOH + H_2O + e^-$. After the CV treatment, the electrode was washed using the distilled water for several times to wash off the alkaline residues and dried in a 60 degree vacuum oven overnight.

Physical characterization. The morphologies of α-$Co(OH)_2$, β-$Co(OH)_2$, and the mixed phase of $Co(OH)_2$/CoOOH were characterized via scanning electron microscopy (SEM, Hitach SU8000 scanning electron microscope, Hitachi High-Tech, Tokoy, Japan) and transmission electron microscopy (TEM, JEM ARM 1300S for high-resolution TEM images, Thermo Fisher, Shanghai, China). The structural evolution from α-$Co(OH)_2$ to the mixed phase of $Co(OH)_2$/CoOOH was probed via X-ray diffraction (XRD, RIGAKU D/MAX2500, Rigaku, Tokoy, Japan). Fourier transform infrared spectra were conducted using a JW-BK132F from the JWGB SCI. & TECH instrument (Beijing, China), and X-ray photoelectron spectroscopy (XPS) spectra were obtained using the ESCALAB-250 instrument with a monochromatic Al Kα radiation source and a hemisphere detector with an energy resolution of 0.1 eV. The X-ray absorption (XAS) spectra were obtained on the beamline 1W1B at the Beijing Synchrotron Facility (BSRF), with an electron of 2.2 GeV and a beam current of 250 mA.

Electrochemical evaluation. The electrochemical performances of β-$Co(OH)_2$ and the mixed phase were carried out in a coin cell, in which Zn foil and 1 M $ZnSO_4$ served as the anode and the electrolyte, respectively. The cathode and anode were separated via a glass microfiber filter (CAT No. 1823-125). The galvanostatic profiles, dQ/dV plot, and cycling performance were conducted using a Neware battery testing system. The CV curves and Nyquist plot were obtained using a CHI660E potentiostat.

DFT simulations. The Perdew–Burke–Ernzerhof (PBE) functional and supercell approach were utilized as implemented in the Vienna ab initio simulation package (VASP) to perform the spin-polarized DFT calculations. Co (3s, 3p, 3d, and 4s) and O (2s and 2p) electrons were treated as valence states, with a cut-off energy of 520 eV in plane waves, and PBE-based projector-augmented wave potentials were used to replace the other electrons. All energies were computed using the DFT + U method. $Co_2O_4H_2$ (two formula units) was used as a unit cell, and the equilibrium lattice constants were simulated with cell shape, a lattice vector, and atomic position using a residual force of 0.02 eV $Å^{-1}$. These conditions ensured a convergence in the equilibrium distance. As for the transition state, the considered phase transformation was located using the elastic band algorithm.

5. Conclusions

To sum up, we synthesized a mixed phase of $Co(OH)_2$/CoOOH via a two-step electrochemical preparation. The inserted NO_3^- species was removed in the alkaline media, resulting in the transformation from α- into β-$Co(OH)_2$, and CoOOH was obtained based on the energy storage mechanism of $Co(OH)_2$ in supercapacitors. As the cathode material for AZIBs, the electrochemical performance of the mixed phase of $Co(OH)_2$/CoOOH was much improved than that of β-$Co(OH)_2$. The mixed phase delivered a maximum capacity of 164 mAh g^{-1} at 0.05 A g^{-1}, a high average discharge plateaus of 1.7 V, as well as a high energy density of 275 Wh kg^{-1}. Benefiting from the low charge-transfer resistance, $Co(OH)_2$/CoOOH showed a good rate performance, maintaining a capacity of 87 mAh g^{-1} at 1.6 A g^{-1}. DFT calculations and various spectroscopy analyses suggested the high thermal stability of the mixed phase of $Co(OH)_2$/CoOOH than pure $Co(OH)_2$ because of less local structural disorder. The reduced Co-Co and Co-O shells increased the mechanical strength of the mixed phase to accommodate Zn^{2+} ions and endure the electrostatic repulsion, resulting in an enhanced cycling stability. This work provides a new perspective to fabricating LDH materials for high-performance AZIBs.

Supplementary Materials: The following supporting information can be downloaded at: https://www.mdpi.com/article/10.3390/inorganics11100400/s1, Figure S1: the XRD pattern of CFP; Figure S2: the SEM image after the CV treatment; Figure S3: the TEM image after the CV treatment; Figure S4: discharge profiles of β-$Co(OH)_2$ at 0.1 A g^{-1}; Figure S5: the rate performance comparison of β-$Co(OH)_2$ and the mixed phase; Figure S6: the stability test of β-$Co(OH)_2$ at 0.1 A g^{-1}; Table S1: total energy values (in eV) for the minima and the transition state shown in Figure 4; Table S2: the fitting results of the EIS spectra of the mixed phase of $Co(OH)_2$ and CoOOH.

Author Contributions: Conceptualization, T.D.; Funding acquisition, T.D.; Investigation, F.L. and Y.Z.; Resources, T.D.; Supervision, T.D.; Validation, H.U.; Visualization, F.L.; Writing—original draft, F.L. and Y.Z.; Writing—review and editing, H.U. and T.D. All authors have read and agreed to the published version of the manuscript.

Funding: This research was funded by the International Collaboration Program of Jilin Provincial Department of Science and Technology (No. 20230402051GH), the Open Project Program of Key Laboratory of Preparation and Application of Environmental Friendly Materials (Jilin Normal University) of the Ministry of China (No. 2021006), the Fundamental Research Funds for the Central Universities JLU, and "Double-First Class" Discipline for Materials Science and Engineering.

Data Availability Statement: The data presented in this study are available on request from the corresponding author. The data are not publicly available due to the data protection.

Acknowledgments: We acknowledge Beijing Synchrotron Radiation Facility (BSRF) for their XAS measurements on the beamline 1W1B.

Conflicts of Interest: The authors declare no conflict of interest. The funders had no role in the design of the study.

References

1. Xu, J.; Cai, X.; Cai, S.; Shao, Y.; Hu, C.; Lu, S.; Ding, S. High-Energy Lithium-Ion Batteries: Recent Progress and a Promising Future in Applications. *Energy Environ. Mater.* **2023**, *6*, e123450. [CrossRef]
2. Zhu, Z.; Jiang, T.; Ali, M.; Meng, Y.; Jin, Y.; Cui, Y.; Chen, W. Rechargeable Batteries for Grid Scale Energy Storage. *Chem. Rev.* **2022**, *122*, 16610–16751. [CrossRef]
3. Zheng, S.; Zhao, W.; Chen, J.; Zhao, X.; Pan, Z.; Yang, X. 2D Materials Boost Advanced Zn Anodes: Principles, Advances, and Challenges. *Nano-Micro Lett.* **2023**, *15*, 46. [CrossRef]
4. Zeng, Y.; Luan, D.; Lou, X.W. Recent advances in electrode engineering strategies for aqueous Zn-based batteries. *Chem* **2023**, *9*, 1118–1146. [CrossRef]
5. Zhang, N.; Wang, J.-C.; Guo, Y.-F.; Wang, P.-F.; Zhu, Y.-R.; Yi, T.-F. Insights on rational design and energy storage mechanism of Mn-based cathode materials towards high performance aqueous zinc-ion batteries. *Coordin. Chem. Rev.* **2023**, *479*, 215009. [CrossRef]
6. Liu, N.; Li, B.; He, Z.; Dai, L.; Wang, H.; Wang, L. Recent advances and perspectives on vanadium- and manganese-based cathode materials for aqueous zinc ion batteries. *J. Energy Chem.* **2021**, *59*, 134–159. [CrossRef]
7. Liu, H.; Zhou, Q.; Xia, Q.; Lei, Y.; Long Huang, X.; Tebyetekerwa, M.; Song Zhao, X. Interface challenges and optimization strategies for aqueous zinc-ion batteries. *J. Energy Chem.* **2023**, *77*, 642–659. [CrossRef]
8. Ban, J.; Wen, X.; Lei, H.; Cao, G.; Liu, X.; Niu, C.; Shao, G.; Hu, J. In-plane grain boundary induced defect state in hierarchical NiCo-LDH and effect on battery-type charge storage. *Nano Res.* **2022**, *16*, 4908–4916. [CrossRef]
9. Zhang, R.; Dong, J.; Zhang, W.; Ma, L.; Jiang, Z.; Wang, J.; Huang, Y. Synergistically coupling of 3D FeNi-LDH arrays with Ti3C2Tx-MXene nanosheets toward superior symmetric supercapacitor. *Nano Energy* **2022**, *91*, 106633. [CrossRef]
10. Han, X.; Li, J.; Lu, J.; Luo, S.; Wan, J.; Li, B.; Hu, C.; Cheng, X. High mass-loading NiCo-LDH nanosheet arrays grown on carbon cloth by electrodeposition for excellent electrochemical energy storage. *Nano Energy* **2021**, *86*, 106079. [CrossRef]
11. Deng, T.; Zhang, W.; Arcelus, O.; Kim, J.G.; Carrasco, J.; Yoo, S.J.; Zheng, W.; Wang, J.; Tian, H.; Zhang, H.; et al. Atomic-level energy storage mechanism of cobalt hydroxide electrode for pseudocapacitors. *Nat. Commun.* **2017**, *8*, 15194. [CrossRef] [PubMed]
12. Wang, L.; Dong, Z.H.; Wang, Z.G.; Zhang, F.X.; Jin, J. Layered α-Co(OH)$_2$Nanocones as Electrode Materials for Pseudocapacitors: Understanding the Effect of Interlayer Space on Electrochemical Activity. *Adv. Funct. Mater.* **2013**, *23*, 2758–2764. [CrossRef]
13. Li, T.; Han, S.; Wang, Y.; Zhou, J.; Zhang, B.; Yu, Y. A Spectroscopic Study on Nitrogen Electrooxidation to Nitrate. *Angew. Chem. Int. Edit.* **2023**, *62*, e202217411. [CrossRef] [PubMed]
14. Wang, H.; Wang, D.; Deng, T.; Zhang, X.; Zhang, C.; Qin, T.; Cheng, D.; Zhao, Q.; Xie, Y.; Shao, L.; et al. Insight into graphene/hydroxide compositing mechanism for remarkably enhanced capacity. *J. Power Sources* **2018**, *399*, 238–245. [CrossRef]
15. Lee, S.-Y.; Oh, H.-J.; Kim, M.; Cho, H.-S.; Lee, Y.-K. Insights into enhanced activity and durability of hierarchical Fe-doped Ni(OH)2/Ni catalysts for alkaline oxygen evolution reaction: In situ XANES studies. *Appl. Catal. B Environ.* **2023**, *324*, 122269. [CrossRef]
16. Ren, Y.; Yang, Y.; Chen, L.; Wang, L.; Shi, Y.; Yin, P.; Wang, W.; Shao, M.; Zhang, X.; Wei, M. Synergetic effect of Cu^0-Cu^+ derived from layered double hydroxides toward catalytic transfer hydrogenation reaction. *Appl. Catal. B Environ.* **2022**, *314*, 121515. [CrossRef]
17. Lin, Y.; Wang, H.; Peng, C.K.; Bu, L.; Chiang, C.L.; Tian, K.; Zhao, Y.; Zhao, J.; Lin, Y.G.; Lee, J.M.; et al. Co-Induced Electronic Optimization of Hierarchical NiFe LDH for Oxygen Evolution. *Small* **2020**, *16*, e2002426. [CrossRef] [PubMed]
18. Chen, J.; Zheng, F.; Zhang, S.-J.; Fisher, A.; Zhou, Y.; Wang, Z.; Li, Y.; Xu, B.-B.; Li, J.-T.; Sun, S.-G. Interfacial Interaction between FeOOH and Ni–Fe LDH to Modulate the Local Electronic Structure for Enhanced OER Electrocatalysis. *ACS Catal.* **2018**, *8*, 11342–11351. [CrossRef]
19. Huang, Z.; Li, X.; Chen, Z.; Li, P.; Ji, X.; Zhi, C. Anion chemistry in energy storage devices. *Nat. Rev. Chem.* **2023**, *7*, 616–631. [CrossRef]
20. Zeng, Y.; Lu, X.F.; Zhang, S.L.; Luan, D.; Li, S.; Lou, X.W.D. Construction of Co-Mn Prussian Blue Analog Hollow Spheres for Efficient Aqueous Zn-ion Batteries. *Angew. Chem. Int. Edit.* **2021**, *60*, 22189–22194. [CrossRef] [PubMed]
21. Yang, Q.; Mo, F.; Liu, Z.; Ma, L.; Li, X.; Fang, D.; Chen, S.; Zhang, S.; Zhi, C. Activating C-Coordinated Iron of Iron Hexacyanoferrate for Zn Hybrid-Ion Batteries with 10,000-Cycle Lifespan and Superior Rate Capability. *Adv. Mater.* **2019**, *31*, e1901521. [CrossRef]
22. Li, G.; Yang, Z.; Jiang, Y.; Jin, C.; Huang, W.; Ding, X.; Huang, Y. Towards polyvalent ion batteries: A zinc-ion battery based on NASICON structured $Na_3V_2(PO_4)_3$. *Nano Energy* **2016**, *25*, 211–217. [CrossRef]
23. Meng, J.; Song, Y.; Qin, Z.; Wang, Z.; Mu, X.; Wang, J.; Liu, X.X. Cobalt–Nickel Double Hydroxide toward Mild Aqueous Zinc-Ion Batteries. *Adv. Funct. Mater.* **2022**, *32*, 2204026. [CrossRef]
24. Zhao, Y.; Zhang, P.; Liang, J.; Xia, X.; Ren, L.; Song, L.; Liu, W.; Sun, X. Unlocking Layered Double Hydroxide as a High-Performance Cathode Material for Aqueous Zinc-Ion Batteries. *Adv. Mater.* **2022**, *34*, 2204320. [CrossRef]

inorganics MDPI

Article

Two Birds with One Stone: Ammonium-Induced Carbon Nanotube Structure and Low-Crystalline Cobalt Nanoparticles Enabling High Performance of Lithium-Sulfur Batteries

Qi Tan [1], Hongliang Liu [2], Guozhu Liang [2], Kaigui Jiang [2], Hangxuan Xie [3], Weijie Si [4], Jiajv Lin [4,*] and Xiongwu Kang [4,*]

1 Guangdong Electric Power Development Co., Ltd., Guangzhou 510080, China
2 Guangdong Zhuhai Jinwan Power Co., Ltd., Zhuhai 519000, China
3 China Southern Power Grid Technology Co., Ltd., Guangzhou 510080, China
4 New Energy Research Institute, School of Environment and Energy, South China University of Technology, Guangzhou 510006, China
* Correspondence: linjiaju@dg.gd.csg.cn (J.L.); esxkang@scut.edu.cn (X.K.)

Abstract: The electrochemical performance of lithium–sulfur batteries (LiSBs) has been hampered by the slow redox kinetics and shuttle effect of lithium polysulfides (LiPSs), which require the rational design and synthesis of highly active electrocatalysts towards this reaction. Herein, worm-like N-doped porous carbon nanotube-supported low-crystalline Co nanoparticles (a-Co-NC@C) were derived from binary Zn–Co ZIF via a two-step thermal annealing method. Initial thermal annealing 950 °C in Ar + H_2 atmosphere results in the carbonization of binary Zn–Co ZIF and the formation of high crystalline Co nanoparticles. Thermal annealing in ammonia atmosphere at 350 °C not only results in the reduced crystallinity of cobalt nanoparticles; it also promotes the growth of highly graphitized and heavily N-doped intertwined carbon nanotubes. The enlarged porous carbon nanotube structure offers accommodation for sulfur content, while the doped carbon and Co nanoparticles with reduced crystallinity facilitate the redox kinetics of LiPSs, improving the cycling stability, rate performance and capacity of LiSBs batteries. As a result, the a-Co-NC@C cathode displays a specific capacity of 559 mAh g^{-1} after 500 cycles at 1 C, and a specific capacity of 572 mAh g^{-1} at 3 C. It delivers a specific capacity of 579 mAh g^{-1} at high sulfur loading of a 2.55 mg cm^{-2} at 1 C after 400 cycles. This work highlights the importance of phase engineering of carbon matrix and transition metal nanoparticles in electrochemical performance of Li-S batteries.

Keywords: porous carbon nanotube; low crystalline; cobalt nanoparticles; promoted redox kinetics; Li–S batteries

1. Introduction

As the demand for energy storage continues to grow, commercially mature and widely used lithium-ion batteries, which are close to their theoretical energy density, are no longer able to meet society's energy storage needs [1,2]. In particular, the emergence of electric vehicles requires batteries with high energy density, small size and high safety performance [3–6]. Among the many secondary batteries, lithium–sulfur (Li–S) batteries have become a hot research topic in recent decades due to their high theoretical specific capacity and energy density, as well as their low environmental contamination [7–9]. However, the practical application of Li–S batteries is constrained by their own disadvantages [10–12], such as the non-conductive nature of the sulfur itself, the dissolution and "shuttle effect" of lithium polysulfide (Li_2PSs) and the slow redox kinetics of polysulfide [13–15].

Extensive efforts have been made to improve the charge capacity, cycling and rate performance of Li–S batteries. Porous carbon materials have been used as the sulfur host, which can not only improve the electrical conductivity of the cathode but also suppress the dissolution of LiPSs. For instance, the coral-like N-doped multilayer porous carbon [16], when utilized as the cathode of Li–S batteries, delivers an initial discharge capacity of 1617 mAh g^{-1} at 0.1 C and a capacity decay rate of only 0.05% per cycle up to 500 cycles at 0.5 C. N-doped porous carbon microspheres (NPCSM), derived from microalgae [17], greatly inhibit the diffusion of PSs by strong physical and chemical adsorption and exhibit a specific capacity of 1030.7 mAh g^{-1} at a current density of 0.1 A g^{-1} as a cathode of Li–S batteries and a capacity of 692.3 mAh g^{-1} at a high current density of 5 A g^{-1}.

Due to the weak interaction between the non-polar carbon material and PSs, the electrochemical performance of Li–S batteries is still unsatisfactory. Therefore, metal contents are introduced into the porous carbon materials, which behave as electrocatalyst towards redox reaction of PSs. Metal organic frameworks (MOF) have been widely used to prepare porous carbon as the sulfur host in Li–S batteries due to their tunable structural composition, rich void structure and large specific surface area [18]. In particular, heteroatom-doped carbon materials based on the zeolitic imidazole framework (ZIF-67) have been extensively investigated in Li–S batteries. The N-doping carbon materials provide high adsorption capacity to LiPSs, and the cobalt nanoparticles promote the redox kinetics of PSs and improve the utilization of the active sulfur content [19]. In another study, the Co–Nx active sites derived from carbonization of Zn/Co-ZIF [20], when used as a separator interlayer, demonstrate good electrocatalytic activity towards the redox reaction of Li_2PSs and a 31% increase in capacity. Searching for excellent sulfur-hosting carbon materials and electrocatalysts towards the redox reaction of Li_2PSs is central to the community of Li–S batteries.

Additionally, engineering of the morphology and phase of the carbon and metal nanomaterials has been demonstrated to be effective in regulating electrochemical performance. For instance, the amorphous transition metal nanomaterials, which are featured via short-range ordering and anisotropy, have been demonstrated as superior to their crystalline counterparts' electrocatalytic water splitting, etc. [21,22]. The reactions involved in morphology, phase and surface and interface engineering are very complicated, and how the carbon materials and metal species interact with each other remains elusive and challenging to understand.

Herein, a composite composed of worm-like N-doped intertwined porous carbon tubes and Co nanoparticles with low crystallinity (a-Co-NC@C) was prepared by a sequential thermal annealing method, as shown in Scheme 1. Firstly, Zn–Co-ZIF was wrapped by a PAN through a phase inversion method, which was further carbonized in Ar/H_2 atmosphere at 950 °C, resulting in the formation of high-crystalline Co nanoparticles supported on carbon materials. Further annealing in ammonia atmosphere at 350 °C results in the growth of intertwined porous carbon nanotubes with large surface area, possibly benefiting from the catalytic effect of Co nanoparticles, and also their reduced crystallinity. Due to the synergy of the uniformly distributed worm-like porous carbon nanotubes and the heavily N-doped low-crystalline cobalt nanoparticles, a-Co-NC@C delivers a superior performance, with a capability of 572 mAh g^{-1} at a current of 3 C and a capacity of 559 mAh g^{-1} after 500 cycles at a current density of 1 C. Even at a high sulfur loading of 3.95 mg cm^{-2}, a specific capacity of 564.1 mAh g^{-1} remains after 200 cycles at 0.1 C.

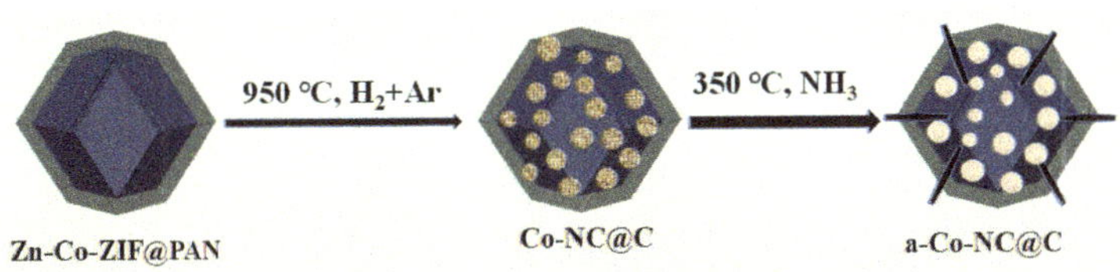

Scheme 1. Illustration of the synthetic process of low-crystalline N-doped porous carbon supported Co nanoparticles.

2. Results

The crystalline structure and polyhedron morphology of Zn–Co-ZIF was examined by XRD (Figure S1) and SEM (Figure S2). The XRD patterns of Zn–Co-ZIF are consistent with those of ZIF-8 and ZIF-67, indicating the successful preparation of Zn–Co-ZIF. The clear sharp peaks in XRD indicates the high crystallinity of Zn–Co-ZIF. The morphology of Co-NC@C and a-Co-NC@C nanocomposites is characterized by SEM and TEM. In Figure 1a,d, Co-NC@C displays a smoother surface and denser structure, and a-Co-NC@C exhibits a rougher surface and porous structure. In the further magnified image in Figure 1b,e, a more detailed surface morphology can be observed. It can be seen that Co-NC@C was covered by finer nanoparticles, which are densely packed together. In contrast, a-Co-NC@C shows a worm-like structure, which consists of carbon nanotubes grown during thermal annealing in ammonia atmosphere under the catalysis of Co nanoparticles.

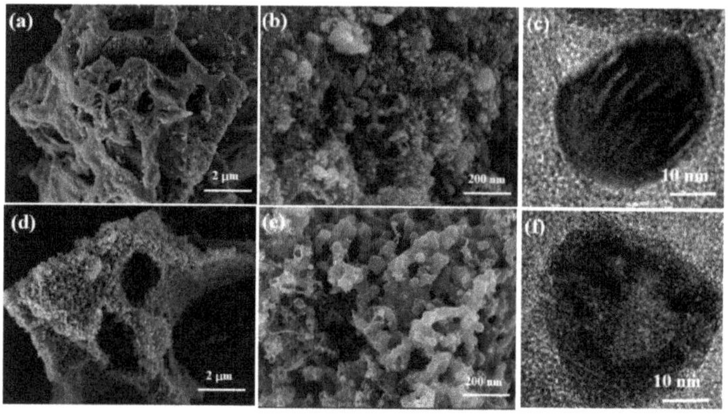

Figure 1. SEM images of at low magnification (**a**); high magnification (**b**) and high-resolution TEM (**c**) for Co-NC@C; SEM images of at low magnification (**d**); high magnification (**e**) and high-resolution TEM (**f**) for a-Co-NC@C.

It can be seen that thermal annealing at 950 °C causes the carbonization of Zn–Co-ZIF@PAN and the formation of a porous carbon-supported Co nanoparticles composite (Co-NC@C). The coating of PAN on Zn–Co-ZIF results in the thorough dispersion and reduced size of the Co nanoparticles on the carbon matrix [23]. Further annealing of Co-NC@C in ammonia atmosphere induces the formation of worm-like heavily N-doped intertwined porous carbon nanotubes in the Co-NC@C composite [24,25]. Such porous carbon nanotubes show high electronic conductivity and a low-weight network for sulfur accommodation [26], and they behave as an excellent electrocatalyst towards the redox reaction of polysulfides [27]. The size of the Co nanoparticles was determined to be 30.6 and 33.3 nm in dimeter for Co-NC@C and a-Co-NC@C, respectively, as shown in Figure 1c,f, which apparently does not change. Usually, such large Co nanoparticles will not show catalytic activity [24,25] towards the growth of carbon nanotubes during thermal annealing in Ar or Ar/H_2 atmosphere. It is speculated that such a strong interaction between Co nanoparticles and ammonia may result in reduced Co nanoparticles and further dissolution of Co species into the adjacent carbon support, which may show good catalytic activity towards the carbon nanotube growth.

The crystallinity of Co nanoparticles after thermal annealing in ammonia atmosphere was monitored by XRD, as shown in Figure 2a. For the Co-NC@C sample, the very sharp diffraction peaks are indexed to Co, indicating the high crystallinity of Co nanoparticles. On the contrary, only very weak and broad diffraction peaks are observed for a-Co-NC@C, indicating the much-reduced crystallinity of Co nanoparticles due to thermal annealing in ammonia atmosphere. This process might be caused by the insertion and subsequent coordination interaction of NH_3 molecules with Co, which disrupts the crystalline structure of Co nanoparticles, spreading the Co atoms into a larger space and thus reducing the crystallinity of Co nanoparticles. Thermal annealing in ammonia atmosphere may further change the graphitization degree, porosity and N-doping level. As shown in Figure 2b, the D (defect) and G (graphite) bands are observed at 1340 and 1590 cm^{-1}, respectively [28]. The intensity ratio of D to G (I_D/I_G) is similar for Co-NC@C and a-Co-NC@C. However, much high noise was observed for a-Co-NC@C, possibly induced by the high doping content of nitrogen into carbon in Co-NC@C [29].

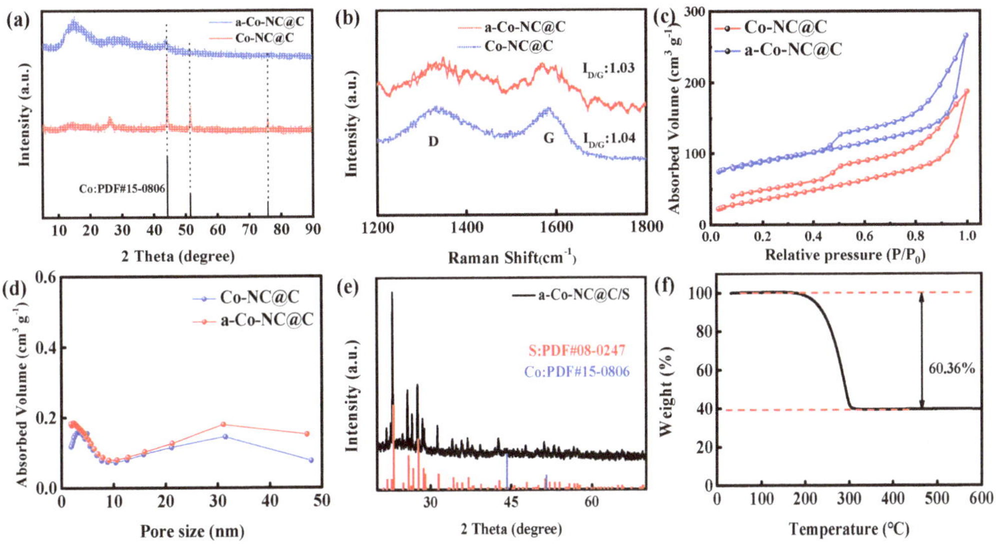

Figure 2. (**a**) XRD, (**b**) Raman spectra, (**c**) nitrogen adsorption/desorption isotherm and (**d**) pore size distributions of Co-NC@C and a-Co-NC@C, respectively. (**e**) XRD pattern and (**f**) TGA of a-Co-NC@C/S.

The porosity of Co-NC@C and a-Co-NC@C is characterized by nitrogen adsorption/desorption isotherms, as shown in Figure 2c,d. The a-Co-NC@C composite shows a surface area of 291.77 m^2 g^{-1} and a pore volume of 0.411 cm^3 g^{-1}, which is much higher than those of the Co-NC@C composite (131.45 m^2 g^{-1} and 0.289 cm^3 g^{-1}), indicating remarkably enhanced surface area and pore volume by thermal annealing in ammonia atmosphere, which offers a large surface area for electrochemical reaction and is also beneficial to the penetration of the electrolyte and mass transfer [30]. Moreover, the pore size of a-Co-NC@C and Co-NC@C is distributed around 1–7 nm and 30 nm, indicating a mesoporous structure of the carbon matrix. When a-Co-NC@C is loaded with sulfur, a-Co-NC@C/S exhibits clear diffraction peaks of sulfur in XRD spectra (Figure 2e), with a sulfur loading of approximately 60 wt%, as revealed by TGA (Figure 2f).

The chemical content of carbon, nitrogen and Co in Co-NC@C and a-Co-NC@C is explored by X-ray photoelectron spectroscopy (XPS). As shown in Figure S3, the fitted C 1*s* spectrum can be deconvoluted into three peaks which, at 284.8 eV, 285.8 eV and 287.6 eV, respectively, can be ascribed to sp^2 carbon, C=N bonds and C–N species. The surface sites of the N-doped carbon can be used as a conductive Lewis base catalyst to improve the adsorption of PSs [31]. Figure 3a shows the high-resolution spectrum of N 1*s*, which can be deconvoluted into a pyridinic N at 398.5 eV and a graphitic N peak at 401.14 ev [32]. It is observed that the content of graphitic N decreases and the content of pyridinic N increases for a-Co-NC@C. In addition, the N content has been increased from 3.62 at% to 4.57 at%. The ability of pyridinic N to introduce strong elements near the Fermi energy level to give electronic states has been reported to change the reaction site and increase the adsorption on PSs [33]. Therefore, increasing the content of the pyridinic N of a-Co-NC@C can improve the anchoring and catalytic capacity of the LiPSs by increasing the immobilization and catalytic sites. From the high-resolution XPS spectrum of Co 2*p*3/2 in Figure 3b, it is observed that the amount of metallic Co at 778.49 eV in Co-NC@C is higher than that in a-Co-NC@C, while only the Co–N bonding component located at 780.06 eV [34] was observed for a-Co-NC@C, indicating heavily doped N content in Co nanoparticles via thermal annealing in ammonia atmosphere. Such positively charged and low-crystalline N-doped Co nanoparticles may provide enhanced interactions with the PSs through both Co–S and Li–N bonds concurrently [35], thus suppressing the shuttle effect [36].

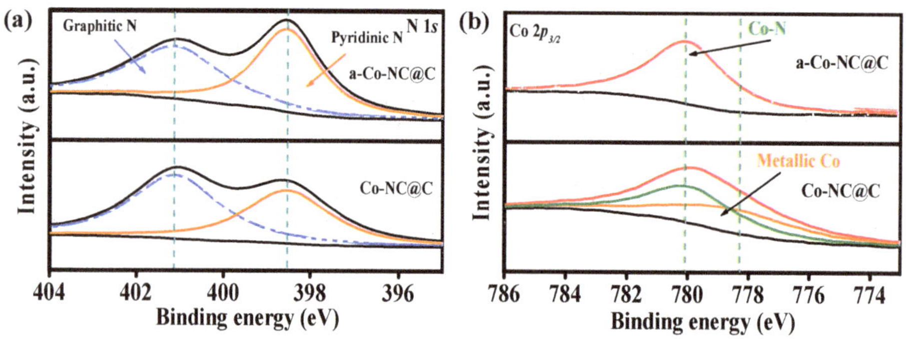

Figure 3. High-resolution XPS spectrum of (**a**) N 1s and (**b**) Co $2p_{3/2}$ for Co-NC@C and a-Co-NC@C.

The electrochemical performances of Co-NC@C/S and Co-NC@C-N/S as cathodes of Li–S batteries are shown in Figure 4. The cyclic voltammetry (CV) profiles are shown in Figure 4a. The reduction peaks A and B correspond to the conversion of solid S_8 to long-chain PSs and of short-chain PSs to Li_2S/Li_2S_2, respectively. The oxidation peaks C and D correspond to the conversion of Li_2S/Li_2S_2 to short-chain PSs and, further, to long-chain PSs as well as S_8, respectively [37–39]. It is observed that the current density on the a-Co-NC@C composite is 1.5~2 times that of Co-NC@C, and the polarization voltage between peak A and C is lower than that of Co-NC@C (0.270 vs. 0.354 V) (Table S1), which indicates the evidently enhanced redox kinetics of PSs on a-Co-NC@C composite. Figure 4b shows the rate performance, where a-Co-NC@C displays a capacity of 651 and 572 mAh g^{-1} 2 C and 3 C, much higher than that of Co-NC@C (514 and 342 mAh g^{-1}) [40,41]. The polarization voltage at different current densities of a-Co-NC@C from rate curves (Figure S4) is smaller than that of Co-NC@C, as shown in Figure 4c. This indicates that a-Co-NC@C/S exhibits higher redox kinetics of PSs than Co-NC@C [42]. The high redox kinetics and holding capacity of a-Co-NC@C compared to Co-NC@C may induce an improved cycling performance. As shown in Figure 4d, a-Co-NC@C, as a cathode of Li–S batteries, delivers a specific capacity of 559 mAh g^{-1} after 500 cycles at 1 C and a decay rate of 0.066% per cycle, which is superior to those of the Co-NC@C composite (a capacity of 371 mAh g^{-1} after 500 cycles at 1 C and decay rate of 0.11%). The long cycling performance of a-Co-NC@C/S as a cathode of the Li–S battery was tested at 2 C and 3 C, respectively (as shown in Figure S5), which still exhibits a specific capacity of 509 mAh g^{-1} 500 after cycles at 2 C, and a capacity of 393 mAh g^{-1} after 800 cycles at 2 C.

GITT tests were further performed for Li–S batteries in the first charge and discharge cycle, as shown in Figure 5a for Co-NC@C and Figure 5b for a-Co-NC@C electrodes, respectively, from which the reaction resistance was derived (Figure 5c,d). It is observed from the reaction that the resistance of a-Co-NC@C during discharging is apparently smaller than that of the Co-NC@C electrode, which indicates faster redox kinetics and lithium-ion diffusion on the former than on the latter [39].

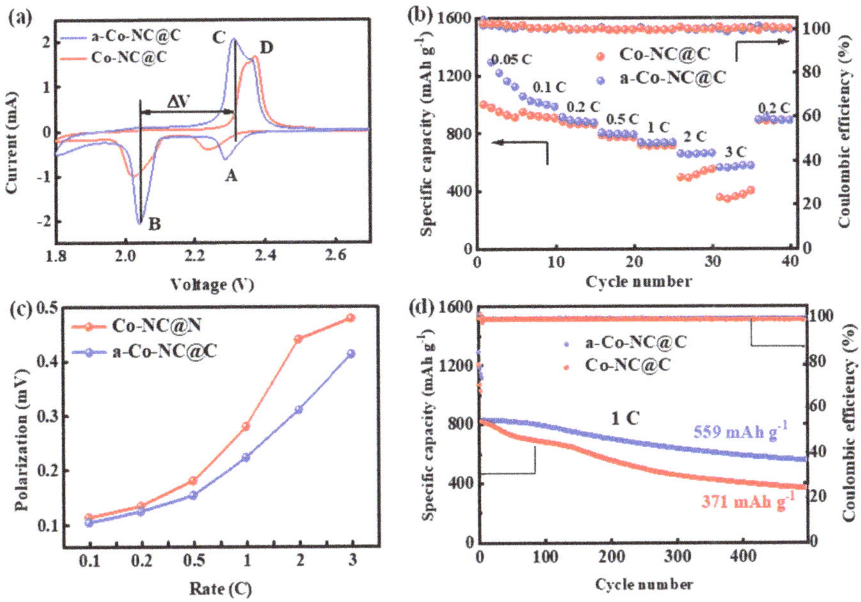

Figure 4. (**a**) CV curves; (**b**) rate performance; (**c**) polarization voltages at different current densities; (**d**) cycling performance at 1 C of Li–S batteries with Co-NC@C/S and a-Co-NC@C/S as cathodes, respectively. The arrows indicate the related axis for the plot in the panels.

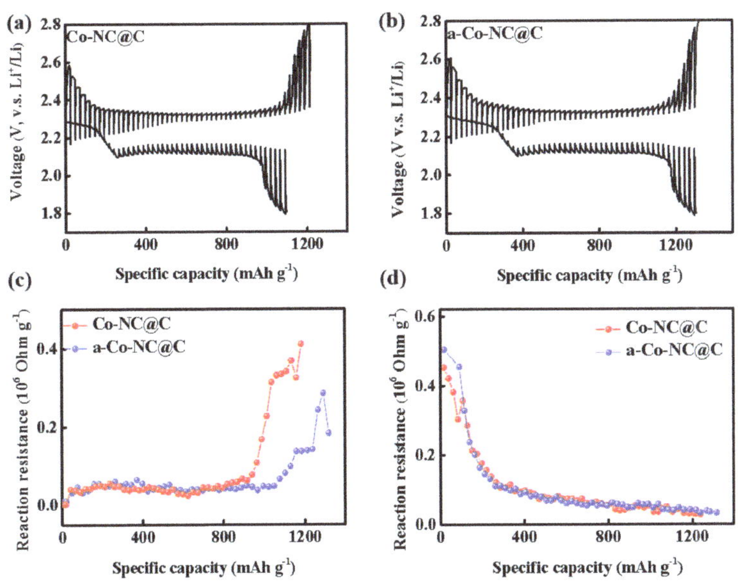

Figure 5. GITT voltage profiles of (**a**) Co-NC@C, (**b**) a-Co-NC@C in the 1st cycle; the reaction resistance of Co-NC@C and a-Co-NC@C in the (**c**) discharging process and (**d**) charging process derived from (**a**,**b**).

The commercialization of Li–S batteries requires high sulfur loading for high capacity and energy density [39,40]. As shown in Figure 6a, at a sulfur loading of 3.95 mg cm^{-2}, an initial capacity of 1071 mAh g^{-1} at the first cycle and a starting discharge capacity of 737.7 mAh g^{-1} after activation at 0.05 C was achieved. The capacity remains at 564.1 mAh g^{-1} after 200 cycles at 0.1 C. At a lower sulfur loading of 2.29 mg cm^{-2}, the starting discharge capacity after activation at 0.05 C was 973.5 mAh g^{-1} and a capacity of 642.8 mAh g^{-1} was achieved after 200 cycles at a current of 0.2 C (Figure 6b). The rate performance of a-Co-NC@C composite at 2 mg cm^{-2} was shown in Figure 6c, which display a capacity of 813, 669, 583, 534, 478 and 428 mAh g^{-1} at 0.1, 0.2, 0.5, 1, 2 and 3 C, respectively, indicating good rate performance for the a-Co-NC@C composite at a high sulfur loading. The a-Co-NC@C composite also showed superior long-cycling performance at high sulfur loading. As shown in Figure 6d, at a load of 2.55 mg cm^{-2}, a capacity of 945 mAh g^{-1} after activation and a capacity of 579 mAh g^{-1} after 400 cycles at 1 C were achieved, corresponding to a capacity decay rate of 0.097% per cycle.

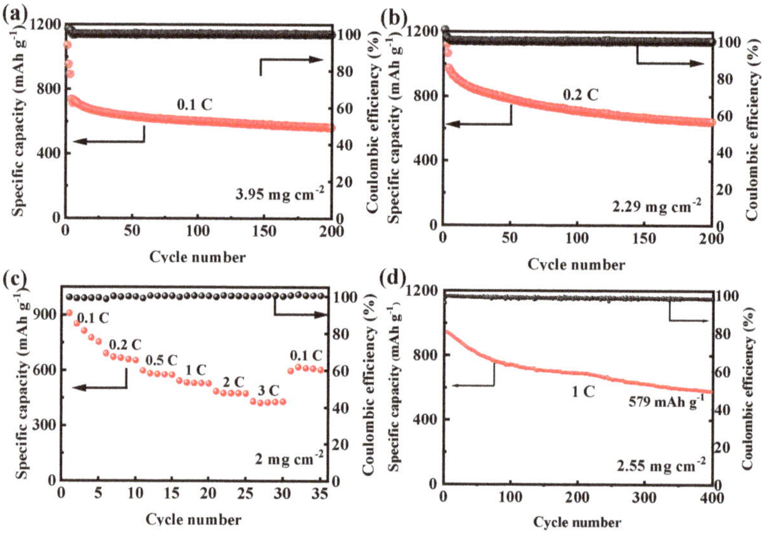

Figure 6. The performance of a-Co-NC@C cathode at high sulfur loading: (**a**) cycling performance at 0.1 C for a sulfur loading of 3.95 mg cm^{-2}; (**b**) cycling performance at 0.2 C at a sulfur loading of 2.29 mg cm^{-2}; (**c**) rate performance at a sulfur loading of 2 mg cm^{-2}; (**d**) cycling performance at 1 C at a sulfur loading of 2.55 mg cm^{-2}. The arrows indicate the related axis for the plot in the panels.

3. Materials and Methods

3.1. Materials

$Co(NO_3)_2{\cdot}6H_2O$, $Zn(NO_3)_2{\cdot}6H_2O$, 2-methylimidazole (2-MeIM), methanol, N,N-Dimethylformamide (DMF) and polyacrylonitrile (PAN, MW = 150,000) was obtained from Shanghai Macklin Biochemical Co., Ltd., Shanghai, China. All materials have not been further purified before use.

3.1.1. Synthesis of Zn–Co-ZIF and Zn–Co-ZIF@PAN

Typically, 20 mmol of $Co(NO_3)_2{\cdot}6H_2O$, 20 mmol of $Zn(NO_3)_2{\cdot}6H_2O$ and 80 mmol 2-methylimidazole (molar ratio of Co^{2+} to Zn^{2+} is 1:1) were added in a beaker containing 500 mL anhydrous methanol and stirred for 1 h. The solution was centrifuged at a rate of 8000× g r min^{-1} for 24 h, and the precipitate was collected, washed three times with methanol and desiccated in an oven at 60 °C for 12 h. The collected purple product was named Zn–Co-ZIF.

A total of 0.5 g Zn–Co-ZIF powder was added into 15 mL DMF and dispersed by ultrasonication for 1 h, into which 0.5 g PAN was then added into and stirred for 4 h. The resulting mixture was dripped into 500 mL of deionized water using a syringe to obtain a flake product, which was collected by filtration, washed several times with deionized water and desiccated overnight in an oven at 60 °C. The obtained sample is denoted as Zn–Co-ZIF@PAN.

3.1.2. Preparation of Co-NC@C and a-Co-NC@C

Co-NC@C nanocomposites were prepared by thermal annealing Zn–Co-ZIF@PAN powder at 950 °C for 3 h under Ar/H_2(10%). Co-NC@C was further annealed at 350 °C for 3 h in an ammonia atmosphere to obtain a-Co-NC@C.

3.1.3. Sulfur Loading to Co-NC@C and a-Co-NC@C

Co-NC@C (or a-Co-NC@C) and sulfur powder were ground at a mass ratio of 3:7 for 30 min to achieve a homogeneous mixture, which was then placed in a tube furnace under an Ar atmosphere and annealed at 155 °C for 12 h and then heated to 200 °C for 2 h to remove the excess sulfur. The resulting samples are named Co-NC@C/S and a-Co-NC@C/S, respectively.

3.2. Electrochemical Measurements

Cyclic voltammetry (CV) tests were carried out on an electrochemical workstation CHI 750E with a scan rate of 0.1 mV s^{-1} and an operating voltage window of 1.8–2.7 V. The electrochemical impedance spectroscopy (EIS) was carried out on the CHI 750E in the frequency range 0.01 to 100,000 Hz. Electrochemical performance tests such as long cycling performance, rate performance and charge/discharge curves were carried out on the LAND electric test system.

3.3. The Assembly of Batteries

To load sulfur into Co-CN (or a-Co-CN), Co-CN (or a-Co-CN) was mixed with sulfur powder by grinding for 30 min, transferred into an ampoule and thermally annealed at 155 °C for 12 h with a ramp rate of 2 °C min^{-1} and then 200 °C for another 2 h to obtain Co-NC@C/S (a-Co-NC@C/S). To prepare an active electrode, a slurry was prepared in advance by dispersing Co-CN (or a-Co-CN), super P and PVDF in a mass ratio of 8:1:1 into NMP in a beaker, which was then coated on an Al foil and dried in an oven at 60 °C for 24 h. Battery assembly was carried out in the glove box. The button cell was type CR2016. The cathode material was the Co-NC@C/S and a-Co-NC@C/S obtained above, the anode electrode was lithium metal flake, the separator was a Celgard 2400 and the main component of the electrolyte was LiTFSI dissolved in a 1:1 volume ratio DOL/DME (1, 3-dioxolane = DOL; dimethyl ether = DME) mixture, to which 0.2 M $LiNO_3$ was added. The sulfur loading of the cathode was about 1.2–1.5 mg cm^{-2}, and the ratio of the electrolyte to the sulfur content on the cathode was 25 mL g^{-1}.

4. Conclusions

In this work, we reported the preparation of heavily N-doped worm-like porous carbon nanotube structure mixed low-crystalline Co nanoparticles by sequential thermal annealing of ZnCo-ZIF in $Ar+H_2$ and ammonia atmosphere. The porous carbon was heavily doped with pyridinic N content, and the Co species were completely converted to N-bonded. The formation of porous carbon nanotube structure was ascribed to the catalytic effect of Co nanoparticles endowed by N-doping and reduced crystallinity, which facilitates the adsorption and redox kinetics of LiPSs. Therefore, the a-Co-NC@C/S composite, as sulfur host of the cathode, delivers a specific capacity of 559 mAh g^{-1} after 500 cycles at 1 C and a specific capacity of 572 mAh g^{-1} at 3 C. At a high sulfur loading of 2.55 mg cm^{-2}, it achieves a specific capacity of 579 mAh g^{-1} after 400 cycles at 1 C. Under 2 C and 3 C, it is able to cycle up to 500 and 800 cycles, respectively, and still releases a capacity of 509 and 391 mAh g^{-1}, respectively. This work highlights the importance of the low crystallinity of Co nanoparticles in improving the redox kinetics of LiPSs in LiSB.

Supplementary Materials: The following supporting information can be downloaded at https://www.mdpi.com/article/10.3390/inorganics11070305/s1: Figure S1: XRD pattern of Zn–Co-ZIF; Figure S2: SEM of Zn–Co-ZIF; Figure S3: High resolution XPS spectra of C 1*s* of Co-NC, Co-NC@C and a-Co-NC@C; Figure S4: Charge/discharge profiles at various rates of the cell with cathode of Co-NC@C and a-Co-NC@C; Figure S5: Long term cycling performance at 1 C, 2 C of a-Co-NC@C/S electrode. Table S1: Current density values at different redox peaks and polarization voltages of composite cathodes of Co-NC@C and a-Co-NC@C.

Author Contributions: Conceptualization, J.L. and X.K.; methodology, J.L. and X.K.; validation, H.L., G.L., W.S., K.J. and H.X.; formal analysis, Q.T., K.J. and J.L.; investigation, Q.T. and J.L.; resources, Q.T.; data curation, Q.T. and J.L.; writing—original draft preparation, Q.T. and J.L.; writing—review and editing, Q.T. and X.K.; supervision, X.K.; funding acquisition, X.K. All authors have read and agreed to the published version of the manuscript.

Funding: This research was funded by National Nature Science Foundation of China, grant number 2032151.

Data Availability Statement: The data presented in this study are available on request from the corresponding author.

Conflicts of Interest: The authors declare no conflict of interest. The funders had no role in the design of the study.

References

1. Wu, Q.; Zhou, X.; Xu, J.; Cao, F.; Li, C. Carbon-Based Derivatives from Metal-Organic Frameworks as Cathode Hosts for Li–S batteries. *J. Energy Chem.* **2019**, *38*, 94–113. [CrossRef]
2. Yu, S.; Cai, W.; Chen, L.; Song, L.; Song, Y. Recent Advances of Metal Phosphides for Li–S Chemistry. *J. Energy Chem.* **2021**, *55*, 533–548. [CrossRef]
3. Xu, R.; Lu, J.; Amine, K. Progress in Mechanistic Understanding and Characterization Techniques of Li-S Batteries. *Adv. Energy Mater.* **2015**, *5*, 1500408. [CrossRef]
4. Jamesh, M.-I. Recent advances on flexible electrodes for Na-ion batteries and Li–S batteries. *J. Energy Chem.* **2019**, *32*, 15–44. [CrossRef]
5. Tong, Z.; Huang, L.; Lei, W.; Zhang, H.; Zhang, S. Carbon-containing Electrospun Nanofibers for Lithium–Sulfur Battery: Current status and future directions. *J. Energy Chem.* **2021**, *54*, 254–273. [CrossRef]
6. Chen, J.; Dong, Q. Research Progress of Key Components in Lithium-Sulfur Batteries. *J. Electrochem.* **2020**, *26*, 648–662. [CrossRef]
7. Seh, Z.W.; Sun, Y.; Zhang, Q.; Cui, Y. Designing High-energy Lithium-Sulfur Batteries. *Chem. Soc. Rev.* **2016**, *45*, 5605–5634. [CrossRef] [PubMed]
8. Zhao, G.; Ahmed, W.H.Z.; Zhu, F. Nitrogen-Sulfur Co-Doped Porous Carbon Preparation and Its Application in Lithium-Sulfur Batteries. *J. Electrochem.* **2021**, *27*, 614–623. [CrossRef]
9. Kai, W. Preparation and Process Optimization of Cathode Materials for Lithium-Sulfur Batteries. *J. Electrochem.* **2020**, *26*, 825–833. [CrossRef]
10. Manthiram, A.; Fu, Y.; Chung, S.H.; Zu, C.; Su, Y.S. Rechargeable lithium-sulfur batteries. *Chem. Rev.* **2014**, *114*, 11751–11787. [CrossRef]
11. Ali, T.; Yan, C. 2D Materials for Inhibiting the Shuttle Effect in Advanced Lithium-Sulfur Batteries. *ChemSusChem* **2020**, *13*, 1447–1479. [CrossRef] [PubMed]
12. Wei, Z.; Zhang, N.; Wu, F.; Chen, R. Progress and Prospects on Multifunctional Coating Separators for Lithium-Sulfur Battery. *J. Electrochem.* **2020**, *26*, 716–730. [CrossRef]
13. Chen, Y.; Wang, T.; Tian, H.; Su, D.; Zhang, Q.; Wang, G. Advances in Lithium-Sulfur Batteries: From Academic Research to Commercial Viability. *Adv. Mater.* **2021**, *33*, e2003666. [CrossRef]
14. Chung, S.H.; Manthiram, A. Current Status and Future Prospects of Metal-Sulfur Batteries. *Adv. Mater.* **2019**, *31*, e1901125. [CrossRef]
15. Zhang, S.S. Liquid electrolyte lithium/sulfur battery: Fundamental Chemistry, Problems, and Solutions. *J. Power Sources* **2013**, *231*, 153–162. [CrossRef]
16. Ji, S.; Imtiaz, S.; Sun, D.; Xin, Y.; Li, Q.; Huang, T.; Zhang, Z.; Huang, Y. Coralline-Like N-Doped Hierarchically Porous Carbon Derived from Enteromorpha as a Host Matrix for Lithium-Sulfur Battery. *Chem.—A Eur. J.* **2017**, *23*, 18208–18215. [CrossRef]
17. Xia, Y.; Fang, R.; Xiao, Z.; Huang, H.; Gan, Y.; Yan, R.; Lu, X.; Liang, C.; Zhang, J.; Tao, X.; et al. Confining Sulfur in N-Doped Porous Carbon Microspheres Derived from Microalgaes for Advanced Lithium-Sulfur Batteries. *ACS Appl. Mater. Interfaces* **2017**, *9*, 23782–23791. [CrossRef]
18. Yu, L.; Yu, X.Y.; Lou, X.W.D. The Design and Synthesis of Hollow Micro-/Nanostructures: Present and Future Trends. *Adv. Mater.* **2018**, *30*, e1800939. [CrossRef] [PubMed]

19. Li, Y.-J.; Fan, J.-M.; Zheng, M.-S.; Dong, Q.-F. A Novel Synergistic Composite with Multi-functional Effects for High-Performance Li–S Batteries. *Energy Environ. Sci.* **2016**, *9*, 1998–2004. [CrossRef]
20. Kong, Z.; Liu, Q.; Liu, X.; Wang, Y.; Shen, C.; Zhan, L. Co-Nx bonds as Bifunctional Electrocatalytic Sites to Drive The Reversible Conversion of Lithium Polysulfides for Long Life Lithium Sulfur Batteries. *Appl. Surf. Sci.* **2021**, *546*, 148914. [CrossRef]
21. Zhou, Y.; Fan, H.J. Progress and Challenge of Amorphous Catalysts for Electrochemical Water Splitting. *ACS Mater. Lett.* **2021**, *3*, 136–147. [CrossRef]
22. Anantharaj, S.; Noda, S. Amorphous Catalysts and Electrochemical Water Splitting: An Untold Story of Harmony. *Small* **2020**, *16*, 1905779. [CrossRef]
23. Gao, H.; Ning, S.; Zhou, Y.; Men, S.; Kang, X. Polyacrylonitrile-induced Formation of Core-Shell Carbon Nanocages: Enhanced Redox Kinetics towards Polysulfides by Confined Catalysis in Li-S Batteries. *Chem. Eng. J.* **2021**, *408*, 127323. [CrossRef]
24. Meng, J.; Niu, C.; Xu, L.; Li, J.; Liu, X.; Wang, X.; Wu, Y.; Xu, X.; Chen, W.; Li, Q.; et al. General Oriented Formation of Carbon Nanotubes from Metal–Organic Frameworks. *J. Am. Chem. Soc.* **2017**, *139*, 8212–8221. [CrossRef]
25. Li, R.; Rao, D.; Zhou, J.; Wu, G.; Wang, G.; Zhu, Z.; Han, X.; Sun, R.; Li, H.; Wang, C.; et al. Amorphization-induced Surface Electronic States Modulation of Cobaltous Oxide Nanosheets for Lithium-sulfur Batteries. *Nat. Commun.* **2021**, *12*, 3102. [CrossRef] [PubMed]
26. Zhang, S.; Zeng, X.T.; Xie, H.; Hing, P. A Phenomenological Approach for The Id/Ig Ratio and sp3 Fraction of Magnetron Sputtered a-C Films. *Surf. Coat. Technol.* **2000**, *123*, 256–260. [CrossRef]
27. Li, C.; Sui, X.-L.; Wang, Z.-B.; Wang, Q.; Gu, D.-M. 3D N-doped Graphene Nanomesh Foam for Long Cycle Life Lithium-Sulfur Battery. *Chem. Eng. J.* **2017**, *326*, 265–272. [CrossRef]
28. Wang, J.; Luo, D.; Li, J.; Zhang, Y.; Zhao, Y.; Zhou, G.; Shui, L.; Chen, Z.; Wang, X. "Soft on Rigid" Nanohybrid as The Self-supporting Multifunctional Cathode Electrocatalyst for High-performance Lithium-Polysulfide Batteries. *Nano Energy* **2020**, *78*, 105293. [CrossRef]
29. Chen, J.-J.; Yuan, R.-M.; Feng, J.-M.; Zhang, Q.; Huang, J.-X.; Fu, G.; Zheng, M.-S.; Ren, B.; Dong, Q.-F. Conductive Lewis Base Matrix to Recover the Missing Link of Li2S8 during the Sulfur Redox Cycle in Li–S Battery. *Chem. Mater.* **2015**, *27*, 2048–2055. [CrossRef]
30. Cho, Y.J.; Kim, H.S.; Im, H.; Myung, Y.; Jung, G.B.; Lee, C.W.; Park, J.; Park, M.-H.; Cho, J.; Kang, H.S. Nitrogen-Doped Graphitic Layers Deposited on Silicon Nanowires for Efficient Lithium-Ion Battery Anodes. *J. Phys. Chem. C* **2011**, *115*, 9451–9457. [CrossRef]
31. Yuan, H.; Zhang, W.; Wang, J.-g.; Zhou, G.; Zhuang, Z.; Luo, J.; Huang, H.; Gan, Y.; Liang, C.; Xia, Y.; et al. Facilitation of Sulfur Evolution Reaction by Pyridinic Nitrogen Doped Carbon Nanoflakes for Highly-Stable Lithium-Sulfur Batteries. *Energy Storage Mater.* **2018**, *10*, 1–9. [CrossRef]
32. Luo, M.; Cai, J.; Zou, J.; Jiang, Z.; Wang, G.; Kang, X. Promoted Alkaline Hydrogen Evolution by An N-Doped Pt–Ru Single Atom Alloy. *J. Mater. Chem. A* **2021**, *9*, 14941–14947. [CrossRef]
33. Yao, S.; Xue, S.; Peng, S.; Jing, M.; Qian, X.; Shen, X.; Li, T.; Wang, Y. Synthesis of Graphitic Carbon Nitride at Different Thermal-Pyrolysis Temperature of Urea and Its Application in Lithium–Sulfur Batteries. *J. Mater. Sci.-Mater. Electron.* **2018**, *29*, 17921–17930. [CrossRef]
34. Wang, N.; Xu, Z.; Xu, X.; Liao, T.; Tang, B.; Bai, Z.; Dou, S. Synergistically Enhanced Interfacial Interaction to Polysulfide via N,O Dual-Doped Highly Porous Carbon Microrods for Advanced Lithium–Sulfur Batteries. *ACS Appl. Mater. Interfaces* **2018**, *10*, 13573–13580. [CrossRef]
35. Zhang, B.; Liu, X.; Li, D. Electrochemical Properties of Sulfur in Different Carbon Support Materials. *J. Electrochem.* **2019**, *25*, 749–756. [CrossRef]
36. Wang, J.; Peng, Q.; Guan, Y.; Wang, W.; Wang, A.; Liu, Q.; Huang, Y. Efficient Interface Enabled by Nano-Hydroxyapatite@Porous Carbon for Lithium-Sulfur Batteries. *J. Electrochem.* **2022**, *28*, 2219008. [CrossRef]
37. Ji, X.; Wang, J.Y.; Wang, A.B.; Wang, W.K.; Yao, M.; Huang, Y.Q. Preparation of Highly-Cyclized Sulfurized Polyacrylonitrile for Lithium-Sulfur Batteries. *J. Electrochem.* **2022**, *28*, 2219010. [CrossRef]
38. Li, X.Y.; Zhao, C.X.; Li, B.Q.; Huang, J.Q.; Zhang, Q. Advances on Composite Cathodes for Lithium-Sulfur Batteries. *J. Electrochem.* **2022**, *28*, 2219013. [CrossRef]
39. Gao, H.C.; Ning, S.L.; Lin, J.J.; Kang, X.W. Molecular Perturbation of 2D Organic Modifiers on Porous Carbon Interlayer: Promoted Redox Kinetics of Polysulfides in Lithium-Sulfur Batteries. *Energy Storage Mater.* **2021**, *40*, 312–319. [CrossRef]
40. Gao, H.; Ning, S.; Zou, J.; Men, S.; Zhou, Y.; Wang, X.; Kang, X. The Electrocatalytic Activity of $BaTiO_3$ Nanoparticles towards Polysulfides Enables High-Performance Lithium–Sulfur Batteries. *J. Energy Chem.* **2020**, *48*, 208–216. [CrossRef]
41. Rana, M.; Ahad, S.A.; Li, M.; Luo, B.; Wang, L.; Gentle, I.; Knibbe, R. Review on Areal Capacities and Long-Term Cycling Performances of Lithium Sulfur Battery at High Sulfur Loading. *Energy Storage Mater.* **2019**, *18*, 289–310. [CrossRef]
42. Peng, J.; Huang, J.; Cheng, X.; Zhang, Q. Review on High-Loading and High-Energy Lithium–Sulfur Batteries. *Adv. Energy Mater.* **2017**, *7*, 1700260. [CrossRef]

inorganics

MDPI

Review

Multi-Dimensional Inorganic Electrode Materials for High-Performance Lithium-Ion Batteries

Musab Hammas Khan *, Patrizia Lamberti and Vincenzo Tucci

Department of Information and Electrical Engineering and Applied Mathematics, University of Salerno, Via Giovanni Paolo II, 132, 84084 Fisciano, SA, Italy; plamberti@unisa.it (P.L.); vtucci@unisa.it (V.T.)
* Correspondence: muskhan@unisa.it

Abstract: Energy storage devices are essential for enhancing the effectiveness and sustainability of electrical energy. Lithium-ion batteries (LIBs) are one of the most efficient energy storage solutions available. The choice of electrode materials plays a vital role in defining the performance of an energy storage device. A range of electrode materials have been developed utilizing both organic and inorganic substances. Due to their notable electrochemical characteristics, strong chemical stability, and well-established technological approaches, inorganic materials have been extensively studied to achieve high-performance devices. This review paper aims to provide a thorough and analytical review of different materials ranging from zero- to three-dimensional (3D), like quantum dots, nanotubes, and nanosheets that have been proposed for high-performance LIBs. This study also includes challenges and future pathways to address the issues with inorganic materials utilized as electrode materials for high-performance energy storage LIBs.

Keywords: batteries; inorganic electrode materials; lithium-ion batteries; 0D, 1D, 2D, and 3D nano-materials

1. Introduction

Devices for energy storage are more and more diffusely applied across various domains, including electric power distribution, electronics, automotive, aerospace, etc. [1–3], for tackling the global challenges related to the finite availability of fossil fuels and to air pollution reduction.

Many different technologies have been proposed to fulfill the constraints of the required power and energy in such diversified domains. Fuel cells demonstrate significant energy density, yet they possess a relatively low power density. Supercapacitors have higher power but less energy densities and longer cycle lifetimes due to surface charge storage processes [4–6] whereas rechargeable batteries provide high energy densities as a result of the faradaic charge storage mechanism [7–10]. Among these last electrochemical systems, throughout the last three decades, lithium-ion batteries (LIBs) have been used as the power source for a variety of products, including electric vehicles, power tools, laptops, mobile phones, etc. As well as being lightweight, LIBs have the highest energy and power densities compared to the other battery technologies. They also have a long lifecycle and a low self-discharge rate, which makes them reliable and easy to maintain.

The three primary components of a LIB (as of any other battery) are an electrolyte and two electrodes (anode and cathode). The cathode stores ions when charging whereas the anode stores lithium ions during discharging. The presence of electrolytes in batteries enables the movement of ions between the anode and cathode, thereby ensuring efficient energy storage and release. The most commonly used liquid electrolytes employed in LIBs are Lithium salts, like $LiPF_6$. Solid electrolytes, including those composed of ceramic or polymer, as well as gel electrolytes, represent two additional types of electrolytes. Each of these categories offers a distinct array of benefits, encompassing aspects such as safety and stability.

The electrodes significantly influence a battery's energy storage capacity, efficiency, and overall lifespan. During the charging and discharging processes, the anode and cathode are crucial for ion mobility, directly affecting the electric battery's overall performance. The choice of material influences ion storage capacity, conductivity, and stability during several charge-discharge cycles. Efforts are now underway to produce and use the most appropriate materials for electrode components in batteries.

Even though there have been improvements over the years, the best electrode materials for LIBs that use the intercalation mechanism have reached their peak performance. This is when lithium ions are reversibly incorporated into the structure of the host material. This implies that investigating novel materials or ion storage processes other than conventional intercalation may be necessary to achieve additional notable increases in energy capacity or efficiency. Because of this, a significant amount of attention has been focused on high-capacity conversion reaction-type cathodes, such as those containing sulfur (lithium-sulfur batteries) and oxygen (lithium-air batteries). Furthermore, safe and affordable rechargeable batteries hold great potential as workable alternatives for large electrical systems [11,12]. Promising potential for enhancing battery performance, such as cyclic stability, specific capacity, specific energy, and charge/discharge rate, has been shown for a variety of electrodes based on nanomaterials [13–16].

Nanomaterials with a significantly enhanced surface area markedly enhance ion storage and reaction speeds, increasing battery capacity and power output. Their nanoscale dimensions facilitate accelerated ion diffusion and more compact electron transport pathways, leading to enhanced efficiency. Furthermore, nanomaterials possess the potential to accommodate volumetric changes during cycling, thereby enhancing the longevity and stability of batteries. Nanoparticles can be categorized based on their size, shape, and structural morphology.

Researchers have made a lot of progress in the past few years in studying zero-dimensional (0D), one-dimensional (1D), two-dimensional (2D), and three-dimensional (3D) nanostructures like quantum dots, nanotubes, and nanosheets as high-performance electrode materials for energy storage devices. Even though they have a lot of interesting properties, problems like material aggregation, limited electrolyte diffusion, and slow reaction kinetics limit their applicability [17,18].

More than 15,000 publications in 2024 that seek the best electrode material for high-performance LIBs are on the Google Scholar website. We conducted this search in December 2024 using the keywords anode, cathode, and electrode materials for lithium-ion batteries. Based on the literature review, this study aims to analyze the characteristics of 0D, 1D, 2D, and 3D inorganic materials applied for LIB electrodes in an organized way. It focuses on their electrochemical properties, energy and power densities, and important performance metrics like charge transfer efficiency, capacity retention, and cycle stability. The presented results offer a wide-ranging overview of the pros, cons, and trade-offs of each dimensionality by showing how structural morphology affects energy storage mechanisms. This may provide a suitable reference for researchers engaged in designing long-lasting, high-performance energy storage systems.

2. Electrode Materials for Use in High-Performance ESD

The investigation of high-performance electrodes, electrocatalysts, and photoelectrodes stands as a pivotal element in the commercialization of these renewable energy technologies. Considering that a battery functions as a dynamic chemical reactor, enhancing its performance through the use of high-performance materials necessitates a profound understanding of the processes occurring at every level. Novel electrode materials for Lithium-ion batteries, which are renowned for their exceptional cycle stability and appealing specific capacity, have drawn a lot of interest in recent years [19]. More than 15,000 publications in 2024 that seek the best electrode material for high-performance LIBs are on the Google Scholar website. We conducted this search in December 2024 using the keywords anode, cathode, and electrode materials for lithium-ion batteries.

Typical materials for LIBs include carbon-based materials, metal oxides, conducting polymers, and different combinations of these elements as electrode components. In recent studies, anode materials for LIBs have encompassed carbon and carbon-based materials [20–23], metal oxides [24–26], nitrides [27,28], and carbides [29,30], along with their composites [31,32]. $LiCoO_2$ [33], $LiMnO_2$ [34], and transition-metal phosphates, especially olivine $LiFePO_4$ [35], exemplify layered compounds featuring an anion close-packed or very-close-packed lattice. They have garnered significant attention in both academic inquiry and commercial applications concerning cathode materials. Other materials that are widely utilized include conducting polymers. Electrodes adopted for high-performance energy storage devices include a variety of compounds for applications, such as carbonaceous compounds like graphene and carbon nanotubes [36,37], conducting polymers like PANI [38], metal oxides like MnO_2 [39], carbon nanocomposites [40,41], and metal oxide composites [42], among others. Inorganic materials hold considerable importance because of their ability to store lithium through a conversion mechanism [43]. Metals, in combination with carbon, like SnO_2 [44], Mo_2N [45], FeN [46], VN [47], Ni_3N [48], and NbN [49], have also been proposed as electrode materials due to the elevated specific energy and power densities, superior electrical conductivity, thermal stability, and electrochemical behavior.

2.1. Organic Electrode Materials

The swift advancement of OAMs necessitates collaborative efforts across various fields, including computer science, chemistry, and materials science. Organic electrode active materials provide a number of advantages, including their low cost, low impact on the environment, sustainability, adjustable design, and exceptional electrical storage activity. We can source these materials from biomass or produce them through straightforward synthesis methods [50,51]. A wide variety of batteries are available, such as lithium-ion, sodium-ion, aluminium-ion, magnesium-ion, and proton batteries, with considerable emphasis on the investigation of organic active materials (OAMs) [52]. In contrast to the rigidity of inorganic materials, the flexibility of organic materials renders them more suitable for bendable energy storage devices. However, achieving large-scale commercialization requires addressing several challenges [53,54]. At the moment, the main problems that OAMs face in the research field are low conductivity, the ability to dissolve in electrolytes, limited working current, low specific capacity, synthesis technology, and safety.

In particular, it is crucial to consider the safety of organic electrode materials. They exhibit significantly different thermal stability, making them more susceptible to inactivation at elevated temperatures [53].

2.2. Inorganic Electrode Materials

Although inorganic materials are the subject of the majority of the present research, interest in their wider potential in energy storage is growing due to their significant benefits, which include increased surface areas, superior electrical conductivity, enhanced thermal stability, greater number of active sites, and higher capacities compared to organic compounds [54]. A substantial array of inorganic materials serve as anode or cathode materials for LIBs, with numerous reports available in the literature. The mechanism of their response enables them to be classified into three separate categories: alloying reaction (e.g., Si and Sn), intercalation reaction (e.g., $Li_4Ti_5O_{12}$), and conversion reaction (e.g., MnO and Fe_2O_3) [55,56].

Silicon has a very high specific capacity of 4200 mA h g^{-1} due to a 4.4-electron alloying reaction. It also has a very low operational voltage of about 0.4 V compared to Li/Li^+ [56]. Phosphorus (P) is an interesting anode material because it has a theoretical specific capacity of 2595 mA h g^{-1}. This comes from the reaction of 3Li + P ↔ Li3 P. Metal oxides exhibit high experimental and theoretical specific capacities, as documented in numerous studies [57,58]. Over the past 10 years, according to data available on Google Scholar, more than 50,000 have been published on inorganic electrode materials for lithium-ion batteries. The cathode material serves as a fundamental element that directly influences the overall performance of batteries. Inorganic cathode materials have garnered significant interest, primarily due to their impressive discharge capacity. The metallic elements typically exhibit accessible oxidation states (e.g., $Mn^{2+}/Mn^{3+}/Mn^{4+}$, $V^{3+}/V^{4+}/V^{5+}$, and $Fe^{2+}/Fe^{3+}/Fe^{4+}$), resulting in multi-electron reactions throughout the charge and discharge cycles. The Li_2FeSiO_4 (LFSO) material is cheap and good for the environment. It can help with a two-electron reaction that has a theoretical specific capacity of 332 mA h g^{-1} [59]. Newly discovered materials, like P14AQ/CNT nanocomposite and Triplite $LiFeSO_4F$ [60], have a lot of potential as cathodes.

Many inorganic nano-materials have been thoroughly investigated as LIB electrode materials, including metal oxides, metal phosphides, and oxysalt nanoparticles. The tiny size of these nanoparticles enables a reduction in the collective electrode tension (mechanical stress/strain) resulting from Li insertion/removal and the diffusion channel between lithium ions. Nevertheless, a significant issue with these nanoparticles is their comparatively poor conductivity, particularly when they are in weak contact [61–63].

3. Nanostructured Inorganic Electrode Materials

In the sequel, the classification proposed by Pokropivny and Skorokhod for NSMs [64], encompassing 0D, 1D, 2D, and 3D NSMs, is adopted. The advancements observed in recent years indicate that NSMs possess significant potential for the innovation of new technologies aimed at meeting the enduring demand for energy. But there are still constraints on using different types of nanostructures with different properties.

Zero-dimensional (0D) nanostructures like quantum dots are strongly reactive with a quick ion exchange that is made possible by high surface-to-volume ratios, although charge transport may be hampered by their poor connectivity. Additionally, 1D materials are perfect for enhancing conductivity in energy devices because they provide continuous channels for the passage of electrons and ions. Two-dimensional materials in layered structures provide huge surface areas and high ion diffusion rates, which are essential for energy storage in batteries and supercapacitor electrodes. Additionally, 3D materials show a stability and energy storage capacity enhanced by porous structures that optimize surface area and facilitate effective molecular movement. Every dimensionality has a distinct contribution and customized morphologies are becoming more and more important for high-energy storage devices [64]. Figure 1 illustrates the diverse dimensions of categorized nanomaterials [65].

Table 1, Shows a summary of some relevant properties of nanomaterials with different sizes (0D, 1D, 2D, and 3D) used in LIBs. This comparison underscores the possible advantages of each solution in enhancing electrochemical performance and energy storage efficiency. Due to their unique structural properties, 0D and 3D materials often have high initial capacities in some cases. However, they often face problems like poor cycling stability, agglomeration (in 0D), and inefficient ion diffusion (in 3D). These problems arise because of shape restrictions, such as particle aggregation, volume expansion, and limited access to electrolytes. In contrast, 1D and 2D materials typically offer a better balance of capacity, cycle life, and rate performance due to their optimized morphology for ion transport and structural integrity. Therefore, a material's suitability for lithium-ion batteries depends not only on how much power it can hold but also on how well it can keep working repeatedly and in real-world situations.

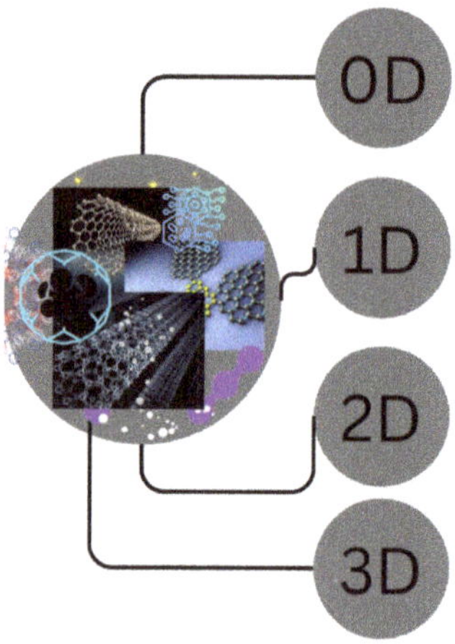

Figure 1. Different possible dimensions of the nanomaterials.

Table 1. Zero to Three Dimensional inorganic nano-materials materials with their characteristics for high-performance LIBs.

Material	Capacity ($mA\ h\ g^{-1}$)	Specific Current ($mA\ g^{-1}$)	No. of Cycles	Reference
		0D materials		
TiO_2@GNRs	320.8	100	500	[66]
NiO@Co_3O_4@GQDs	1158	100	250	[67]
SnO_2-GNS	793	400	150	[68]
$Ca_3Co_4O_9$	316.3	-	100	[69]
Si QDs	1232	200	100	[70]
SnO_2/Graphene	635	60	100	[71]
ZnO@ZnO QDs/C NRAs	699	500	100	[72]
SnO_2/graphene composite	634	100	50	[73]
MoS_2@CNT Fabric	700	100	50	[74]
SnO_2/GNS	570	50	30	[75]
GNS/Fe_3O_4	1026	35	30	[76]
		1D materials		
Ge-CNFs	281.6	5000	5000	[77]
Si	1200	2000	500	[78]
$ZnMnO_3$	950	500	500	[79]
$Li_{1.2}Ni_{2.5}B_2$	350	100	500	[80]
S@$Sr_8Ti_7S_{21}$	1315	260	400	[16]
Fe_2O_3	812	100	300	[81]
$ZnCo_2O_4$	1050	400	200	[82]
Co_3O_4	1800	300	100	[83]

Table 1. *Cont.*

Material	Capacity (mA h g^{-1})	Specific Current (mA g^{-1})	No. of Cycles	Reference
Cu-Si	1500	1400	100	[84]
SnO_2-CuO	600	500	100	[85]
Zn_4Sb_3	450	100	100	[86]
Nb_2Se_9	542.2	100	100	[87]
	2D materials			
Cu_3Si/Si	1675	200	100	[15]
NPCN/Si-2	1977	100	100	[13]
Graphene/MoS_2 composite	1351	100	200	[88]
Si nanosheets/graphene composite	1650	840	50	[89]
Co_3Sn_2/Co–N-doped graphene composite	1615	2501	100	[90]
MoO_2/carbon	1051	500	100	[91]
Si–reduced graphene oxide composite	1500	1350	100	[92]
SiA/NC anode	1224	20	100	[93]
Si/graphene composite	1390	2000	200	[94]
Graphene/$NiCo_2O_4$	1267	100	10	[95]
FeS-rGO composite	978	100	40	[96]
	3D materials			
Rutile TiO_2	140	75	1000	[97]
$Cs_2NaErCl_6$	120	300	500	[98]
$CsPbBr_3$@CNTs	470.2	100	200	[99]
$CsPbBr_3$	261	60	100	[100]
$CsPbI_3$	235	40	100	[101]
CoO QD/Graphene	1008	1000	50	[102]
Co_3O_4	866	178	50	[103]
CoO@3DOM	560.8	1000	50	[104]
$Cs_2NaBiCl_6$	775	75	25	[105]
$CsPbCl_3$	612.3	50	-	[106]

3.1. *Zero-Dimensional Inorganic Electrode Materials*

Zero-dimensional (0D) materials are defined as structures whose dimensions are less than 100 nm on all three sides. These include nanoparticles shaped like onions, spheres, and cages, as well as quantum dots. These structures may provide useful surfaces for electrochemical activity because of their specific dimensions [107]. Published literature extensively covers zero-dimensional electrode materials and the blending of 0D nanoparticles with other materials for use as LIB electrode materials. The electrochemical characteristics of zero-dimensional materials are highlighted, with particular attention paid to specific capacity, current density, and cyclic performance. Their usefulness and potential for use in cutting-edge energy storage systems are shown by this viewpoint.

Using co-precipitation and molten salt methods, Wang et al. created the zero-dimensional anode material $Ca_3Co_4O_9$, which works very well in LIBs. The 0D polyhedron anode had a high initial charge capacity of 391.9 mA h g^{-1}. It maintained 316.3 mA h g^{-1} with more than 95% Coulombic efficiency after 100 cycles. The material had a lot of surface area and ion channels that helped lithium ions (Li^+) move through it. This makes the capacity 33% higher at five rates. This material proved that micro/nanostructure engineering may enhance cycle stability and rate performance [69].

Choi et al. showed that the π-conjugated molecule had better electrochemical performance. They found that the silicon quantum dot (Si QD), a bridge structure, could discharge at the value of 1957 mA h g^{-1} initially and keep 63% of its capacity after 100 cycles at a rate of 200 mA g^{-1}. The π-conjugated molecules can make the electrochemical performance of the material better. They protect electrical conductivity and structural integrity from the mechanical stress that comes from the alloying process. The structure of the Si QD cluster improves stability and ion transport, which makes the cycle stability better than with Si QDs that are not clustered. By adding functional groups and altering organic structures, Si QD clusters' electrochemical properties may be further improved. This approach has significant promise for enhancing the anode materials in high-performing lithium-ion batteries [70]. According to Chen et al. [108], there are numerous dead zones internally present during the fabrication of a material that essentially remains inactive in terms of electrochemical activity. The diffusion depth of an electrolyte is limited to approximately 20 nanometres. It has been demonstrated that quantum dot (QD) structures can offer the necessary small size and a large specific surface area with a specific pore structure, both of which are crucial for improving performance and covering all relevant zones. They synthesized $NiCo_2S_4$ quantum dots and applied them onto nickel foam. After CV tests, it was clear that the material $NiCo_2S_4$ exhibits a larger CV area compared to the CoS_2 and NiS_2 in the KOH electrolyte. The GCD analysis revealed the specific charge capacity (calculate the precise charge that a material can deliver or store in relation to its mass) of $NiCo_2S_4$ was 987.2 C g^{-1}, which was greater than CoS_2 and NiS_2. They also constructed and tested a real two-electrode device that consists of nitrogen-doped reduced graphene oxide nanosheets used as the anode material and $NiCo_2S_4$ as the cathode material. The real device reached a value of 168.1 F g^{-1} specific capacitance at a low specific current of 1 A g^{-1}. The GCD curves have almost symmetric charge-discharge branches and a small potential drop. This suggests that the capacitance is very reliable. The cyclic examination of the device demonstrated a stability level of 91.9% after 15,000 cycles. This real device demonstrated a high energy density and power density of 67.5 W h kg^{-1} and 850 W kg^{-1}, respectively [108].

Wang et al. [109] created and tested a core-shell heterogeneous nanoparticle structure with a carbon shell around a $LiFePO_4$ core. This composite demonstrated a value of 169 mA h g^{-1} of specific capacity at a 0.6 C discharge rate, which was almost equal to the theoretically reported values for $LiFePO_4$. It maintained a reasonable specific capacity of 90 mA h g^{-1}, even when discharged at a high rate of 60 C. Furthermore, the material's cyclic ability was remarkable; after 1100 cycles at a discharge rate of 0.6, it had lost less than 5% of its specific capacity.

Oxide coatings, including MgO, Al_2O_3, and ZnO, have been added to LIB cathode materials to improve their performance. As a result, strain during phase transition may be reduced and the materials' structural stability can be enhanced. In their study, Ting and colleagues showed that a 10 nm layer of nanocrystalline ZnO may improve the $LiCoO_2$ cathode's electrochemical performance [110]. $LiCoO_2$, $LiNiO_2$, and $LiMn_2O_4$ are examples of electroactive cathode materials that may have oxide coatings applied to their surfaces to minimize phase transitions, enhance structural stability, and prevent electrolyte–cathode contact. It is evident in several additional instances that the coatings could be able to enhance conductivity [111].

Another kind of electrode material utilized in energy storage batteries is transition metal dichalcogenides. When 0D inorganic materials bind with organic ones, they may function effectively. Rana et al. synthesized CNTF/MoS_2, a zero-dimensional nanocomposite, for use as a battery electrode. Enhancing conductivity, diffusion, and cycle stability was the main goal. This technique uses MoS_2-coated carbon nanotube fibers (CNTFs) to make composite electrodes that do not need metallic collectors or polymer additives. The electrodes' excellent electrical conductivity and durability surpass those of conventional designs. They achieve a specific capacity of 0.7 Ah g^{-1} with retention rates of 108% after 50 cycles at a specific current of 0.1 Ag^{-1} and 89% after 250 cycles at a specific current of 1 A g^{-1}. The flexible composite fabrics provide remarkable mechanical strength and consistent performance at varying current densities. However, CNTFs may be expensive and complicated to produce and process, which could prevent them from being widely used in certain applications. Furthermore, even though carbon nanotubes show remarkable performance metrics, issues with their long-term stability and environmental effects still need to be resolved [74].

When combined with two-dimensional materials, zero-dimensional materials may achieve remarkable results. Because of its large surface area and excellent electrical properties, which are crucial for energy storage, graphene is a popular electrode material. A significant mass of published information concerns graphene pairings with zero-dimensional inorganic materials. This 2D/0D, graphene/inorganic material combination has led to the development of several novel composite electrode material architectures [112]. These include several anode materials, like Co_3O_4 and SnO_2, and cathode materials, including sulfur (S), $LiFePO_4$, V_2O_5 [113–115], MoO_2 [116], NiO [117], Si [118], Ge [119], and more.

Li et al. applied SnO_2 onto graphene in a structurally and morphologically controlled manner via atomic layer deposition [68]. They used the amorphous SnO_2 as an anode material for the real device testing. As a result, the recorded capacity was 793 mA h g^{-1} after 150 charge-discharge cycles. Over graphene nanosheets, Kim et al. decorated SnO_2 nanoparticles and created a SnO_2/graphene composite for enhancing the LIB performance [73]. In comparison to simply mixed samples, the produced composite showed better electrochemical performance. After 50 charge-discharge cycles, the material showed a specific capacity of 634 mA h g^{-1}, which was due to the uniformly distributed symmetry of nano particles. A one-step hydrothermal method was developed by Wang et al. to synthesize graphene–SnO_2 nanocomposites [71]. After 100 cycles, the generated nanocomposite showed remarkable rate capability and a 635 mA h g^{-1} specific capacity. SnO_2 nanoparticles were combined with graphene nanosheets by Paek et al., showing noticeably improved lithium storage performance [75]. After 30 cycles, the produced 2D/0D graphene–SnO_2 nanocomposite maintained a charge capacity of 570 mA h g^{-1}.

There are a lot of inorganic materials being used as an anode material of the LIBs. A straightforward two-step production of a graphene–Fe_2O_3 composite using the homogeneous precipitation of $FeCl_3$ was first reported by Zhu et al. [120]. They used a graphene composite for increasing the performance of LIBs, which showed an excellent cycle performance of 1355 and 982 mA h g^{-1} for discharge and charge capacities, respectively, depending on the mass of Fe_2O_3. There is also reported Mn_3O_4 material by Wang et al., who demonstrated that the material achieved a high specific capacity value of 810 mA h g^{-1} at the specific current value of 40 mA g^{-1}. The obtained capacity value was in the range of Mn_3O_4's theoretical capacity [76].

In storage applications, the morphology of the material is very important and may impart, unlike electrochemical behaviour. Figure 2 shows various 0D materials: SEM images highlight unique morphologies whereas cycling tests lead to different performances. The MoS_2 coating, with its nanoscale thickness (7–80 nm), exhibits 0D characteristics when analyzed independently but integrates into a 1D CNT-based hybrid system, forming a continuous conformal layer (Figure 2a). The large surface area of 0D MoS_2 helps ions move around better and the structural stability of 1D CNTs helps electrons move around better. This makes for a high capacity and great cycling performance (Figure 2b). Similarly, the amorphous SnO_2-GNS composite, with its 0D morphology of 3–5 nm nanoparticles, maximizes the electrolyte contact area, improving ion transport and capacity (Figure 2d). The isotropic structure of this battery responds well to changes in volume during cycling. This keeps the structure stable and gives it a high capacity of 793 mA h g^{-1} (Figure 2e). This makes it better than larger crystalline SnO_2 nanoparticles (30–40 nm), which have trouble adapting their structure (Figure 2c). The SnO_2/graphene composite has even more of the benefits of a 0D shape. The main SnO_2 nanoparticles are about 4 nm wide and spread out evenly on the graphene (Figure 2f). This makes it easier for ions to move through and for electrochemical reactions to happen. These small particles stick together to make bigger structures (about 14 nm), which keeps the capacity and stability in check (Figure 2g). After 50 cycles, they have a reversible capacity of 634 mA h g^{-1} and a 98% coulombic efficiency (Figure 2h). Finally, Si quantum dots (Si QDs) have a 0D shape that gives lithium ions a lot of surface area to interact with. The π-conjugated molecules around the material act as a buffer layer, reducing the mechanical stress caused by changes in volume and making it easier for electrons and Li^+ to move around (Figure 2i,j). This is what lets the material keep about 63% of its capacity (1232 mA h g^{-1}) after 100 cycles (Figure 2k). These studies collectively emphasize how 0D morphologies—whether as independent nanoparticles, integrated into hybrid systems or combined with secondary structures—play a pivotal role in maximizing electrochemical performance by enhancing surface area, ion transport, and structural stability. Due to their small size, 0D materials often generate a lot of aggregation and poor connectivity, which limits the routes that ions and electrons may take. To make matters worse, when 0D materials are often cycled (charge/discharge), they may lose their stability.

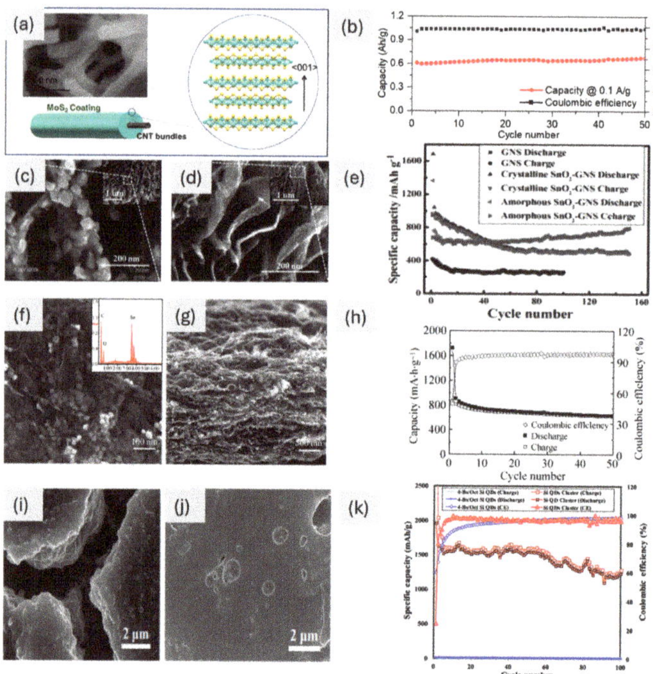

Figure 2. (**a**) Field emission SEM image of the CNTF/MoS_2 composite and a schematic picture of the MoS_2 coating around the CNTF bundles, with an inset showing a magnification; (**b**) cyclic performance of composite at 0.1 A g^{-1} [74]. Copyright © 2021 American Chemical Society. SEM images of (**c**) crystalline SnO_2-GNS, (**d**) amorphous SnO_2-GNS, (**e**) cycle performance at different rates but the same potential window of 0.01–3.00 V for GNS (100 mA g^{-1}) and for crystalline and amorphous SnO_2-GNS (400 mA g^{-1}) [68]. Copyright © 2012 Wiley. (**f**) SnO_2/graphene SEM image at 100 nm, with an inset of the energy spectrum, (**g**) SEM image (500 nm) of the composite, and (**h**) cyclic performance of SnO_2/graphene at a rate of 100 mA g^{-1} [73]. Copyright © 2010 Springer. SEM images of (**i**) 4-Bs/Oct Si QDs and (**j**) Si QD cluster and (**k**) galvanostatic cycling of Si QD cluster [70]. Copyright © 2020 American Chemical Society.

3.2. One-Dimensional Inorganic Electrode Materials

Unlike 0D nanostructures, 1D nanomaterials allow for effective charge transfer along their micro-scale axis by a direct conduit. Owing to their unique benefits, 1D nanostructures have been largely investigated for possible application in electrochemical energy storage systems [21,121–126]. With a lot of ongoing research, these kinds of materials are being employed commercially for high-performance energy storage systems. We have chosen to present 1D inorganic materials based on their electrochemical performance related to their energy outputs. This approach highlights their potential as high-performance electrode materials for energy storage applications.

Yan et al. successfully coated SnO_2 nanowires with amorphous MnO_2 through the straightforward method of immersing them in a hot $KMnO_4$ solution [127]. The material showed a high value of specific capacitance (637 F g^{-1}) at a scan rate of 2 mV s^{-1}. This material also performed well in terms of energy density and power density; the reported values are 35.4 Wh kg^{-1} and 25 kW kg^{-1}, respectively. A one-dimensional nanostructured composite of manganese dioxide and polypyrrole was synthesized by Li et al. [128]. Following 500 cycles (charging/discharging) in an electrolyte, the composite demonstrated a specific capacitance of 320 Fg^{-1} within the potential range of 0.4–0.6 V. In another published paper by Chen et al. [129], a capacity of 185 mA h g^{-1} is reported for nanotubes composed of $LiCoO_2$, which is greater than the typical value of 140 mA h g^{-1}; for such material at higher cycle rates, its capacity quickly decreases.

Plenty of polyanionic compounds have been studied on a large scale due to their high theoretical specific capacities and stability, i.e., $LiFePO_4$, $LiMnPO_4$, $Li_3V_2(PO_4)_3$, and others [130–132]. $LiFePO_4$, $LiMnPO_4$, and $Li_3V_2(PO_4)_3$ can be made to work better with other types of nanomaterials by studying their one-dimensional morphologies. This is a revolutionary way to make lithium-ion batteries work better, even in hybrid materials, as shown in Figure 3. These structures greatly lower internal resistance by making continuous paths for ionic and electronic conduction. They also make it easier for lithium ions to move around quickly, which is important for high-rate capability. This shape also provides mechanical integrity, which effectively reduces problems related to volume expansion and contraction during cycling. This makes sure that the battery lasts a long time and works reliably. Ultimately, the strategic balance of a high surface area with superior conductivity not only optimizes electrochemical performance but also paves the way for next-generation energy storage solutions that meet the demands of modern technology.

Figure 3. (**a**–**c**) SEM images of (**a**) $LiFePO_4$ and (**b**) $LiMn_{0.1}Fe_{0.9}PO_4$ at two different magnifications [133]. Copyright © 2021 scientific reports. (**c**) TEM image of the composite $Li_3V_2(PO_4)_3/(C + MWCNTs)$ at 50 nm [134]. Copyright © 2011 ScienceDirect.

A hierarchical-structure $Li_3V_2(PO_4)_3/C$ mesoporous nanowire was fabricated by Wei et al., which included uniformly implanted $Li_3V_2(PO_4)_3$ nanocrystals that maintained close contact with conductive carbon scaffolds [130]. Compared to regular LVP particles covered in carbon, these mesoporous nanowires showed better performance in terms of rate and cycle. The mesoporous nanowires retained 80% of their capacity even after 3000 cycles.

Nanofiber nanorod-structured Fe_2O_3-C was created by Kang and colleagues for use as electrode material in high-performance energy storage devices [81]. These nanofibers consist of nanosized hollow Fe_2O_3 spheres evenly distributed throughout the amorphous carbon nanofiber matrix. The improved rate and cycling performance were a consequence of the combination of the carbon matrix's continuous electron transport and the hollow nanospheres' ability to accept the volume shift that happens during cycling. This material shows a value of 812 mA h g^{-1} reversible capacity at a specific current of 1.0 A g^{-1} after 300 charge-discharge cycles.

Theoretically, silicon is another promising candidate for use in energy storage devices with high-capacity values of 4200 mA h g^{-1}. However, using Si material in lithium-ion batteries presents several challenges, such as the large volume variation (>300%) during lithiation/delithiation processes (removal and insertion of lithium-ions during charge-discharge), rapid capacity loss, and unstable SEI (Solid Electrolyte Interphase) layers. Furthermore, since Si is a semiconductor, its electronic conductivity is rather low. New research has found that mesoporous carbon nanofibers mixed with silicon nanoparticles or silicon nanofibers themselves work better for cycling and rate than silicon nanowires or particles in bulk. Hollow Si nanotubes may be used to achieve high performance [135]. Cui et al. presented the electrochemical characteristics of silicon nanowires [136]. These silicon nanowires have achieved a value of 1000 mA h g^{-1} and, even after 1000 charge-discharge cycles, they perform very well. Wu et al. [137] created a double-walled Si-SiOx nanotube structure. During lithiation, a stable outer SEI layer formed, enabling the expansion of the inner Si into the hollow region. Cycling performance was better with this SEI layer than with Si nanowires and nanotubes because it was thin and stable. Even after 6000 cycles and when tested at a high rate of 12, the capacity remained unchanged.

Wei et al. optimized $Li_{1.2}Ni_{2.5}B_2$ for use as an anode material for high-performance LIBs. This material shows outstanding stability during 500 charge-discharge cycles. Additionally, this anode material showed an excellent reversible capacity of 350 mA h g^{-1} at a current of 0.1 A g^{-1}. The results show that the 1D Ni/B structure is very promising as a commercially available LIB anode with the ability to charge quickly [80]. Dawei et al. presented the findings of a quasi-1D hexagonal chalcogenide perovskite material, $Sr_8Ti_7S_{21}$. The theoretical results of density functional theory showed that $Sr_8Ti_7S_{21}$ forms strong bonds with lithium polysulfides (LiPS). The effective sulfur host may circumvent the constraints. Consequently, S@$Sr_8Ti_7S_{21}$-based cathodes show a remarkable cycling stability of a 1315 mA h g^{-1} capacity over 400 cycles at 0.2 [16].

Woosung et al. reported an anode material (Nb_2Se_9) for Li-ion batteries (LIBs). This electrode exhibited a reversible specific capacity of 542.2 mA h g^{-1} when it was tested at a current rate of 0.1 A g^{-1} over one hundred cycles. Then, after the increase of specific current up to 3.2 A g^{-1}, further, it shows 272 mA h g^{-1} of capacity [87]. Tian et al. developed the Ge-CNF material specifically for use in high-performance energy storage devices. Nanotechnology and carbon coating modified this germanium/carbon nanofiber to enhance the LIBs' capacity and charging speed. This material was very durable and had excellent electrochemical properties. It could hold the capacity of 281.6 mA h g^{-1} after 5000 cycles (charging/discharging) at the specific current of 5 A g^{-1} and 795.9 mA h g^{-1} after 100 cycles at 0.1 A g^{-1}, with a coulombic efficiency of 99.76% [77].

The examination of TEM images, alongside cyclic performance data, reveals a compelling narrative about the intricate relationship between the morphologies of 1D materials and their electrochemical behaviors (Figure 4). The TEM image shows that $Sr_8Ti_7S_{21}$ perovskite has a flat structure (Figure 4a). The hexagonal Sr framework shows how well it is aligned, which improves lithium polysulphide binding and redox kinetics. This leads to a high capacity and great cycling stability, with only 0.08% decay over 400 cycles (Figure 4b).

Fe_2O_3 nanostructures, on the other hand, have a 1D bubble-nanorod shape, with hollow nanospheres surrounded by carbon nanofibers, as seen in TEM images (Figure 4c,d). This one-of-a-kind structure, which is held together by a 3.0 nm shell and carbon matrix, keeps the volume from changing too much during cycling. This lets it keep 84% of its capacity after 300 cycles (Figure 4e). The $Li_{1.2}Ni_{2.5}B_2$ has a 1D shape with Ni/B-based honeycomb channels and tunnel-type structures, which was confirmed by TEM (Figure 4f). The lattice spacing (0.42 nm) and even distribution of elements make it easy for Li^+ to intercalate and diffuse. This leads to good stability over 500 cycles (Figure 4g). Finally, Ge-CNFs-(700) nanofibers have a flat shape with even widths and smooth surfaces, as seen in TEM images (Figure 4h). These features make them good for moving Li^+ and electrons. This design, combined with a carbon coating, delivers ultra-high fast-charging performance, retaining 281.6 mA h g^{-1} after 5000 cycles (Figure 4i). These studies underscore how distinct 1D morphologies—ranging from hexagonal frameworks in $Sr_8Ti_7S_{21}$ to hollow nanospheres in Fe_2O_3, honeycomb channels in $Li_{1.2}Ni_{2.5}B_2$, and smooth nanofibers in Ge-CNFs—dictate their electrochemical behavior, emphasizing the importance of tailored structural design for optimizing energy storage performance.

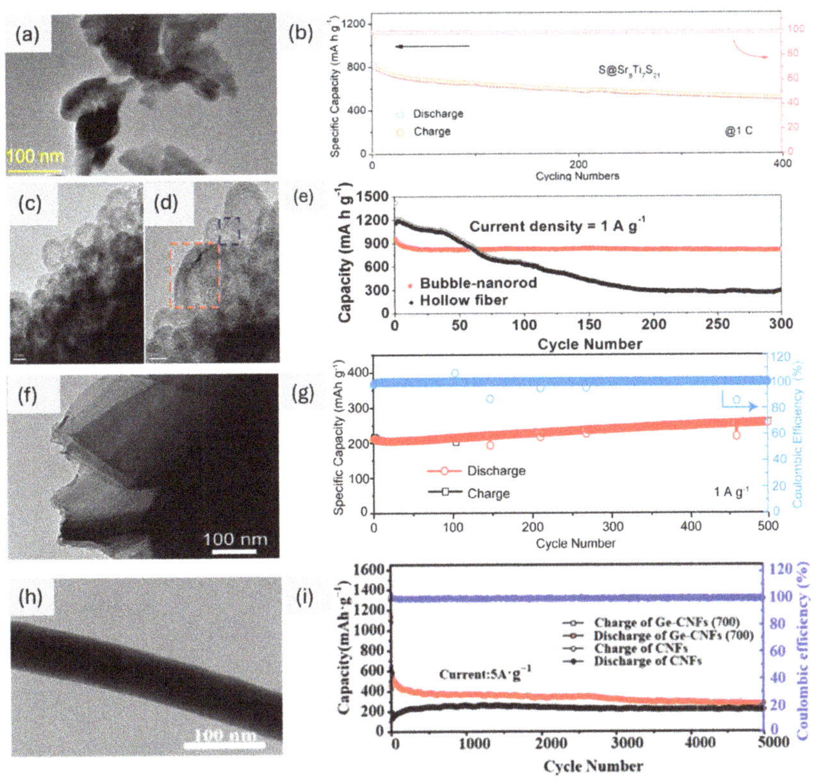

Figure 4. (**a**) TEM image of $Sr_8Ti_7S_{21}$ at 100 nm, (**b**) cycling performances of electrode $S@Sr_8Ti_7S_{21}$ [16]. Copyright © 2024 Wiley. (**c**,**d**) Fe_2O_3/C composite TEM images at 10 nm, (**e**) cyclic performance of different materials at a current density of 1 A g^{-1} [81]. Copyright © 2015, American Chemical Society. (**f**) TEM image of $Li_{1.2}Ni_{2.5}B_2$ at 100 nm, (**g**) cycling performance up to 500 cycles of $Li_{1.2}Ni_{2.5}B_2$ [80]. Copyright © 2024 Wiley. (**h**) TEM image (100 nm) of Ge-CNFs-(700), (**i**) cycling performance up to 5000 cycles of CNFs and Ge-CNFs-(700) [77]. Copyright © 2024, Elsevier.

3.3. Two-Dimensional Inorganic Electrode Materials

The remarkable physical and chemical properties of two-dimensional inorganic materials, such as exfoliated graphene, together with their high surface-to-mass ratio, have sparked significant interest in the field of lithium-ion batteries. Two-dimensional transition metal carbides and nitrides, or MXenes, due to their remarkable structural flexibility and intricate surface chemistry, have garnered significant attention in the field of energy storage.

Among all the 2D materials, several investigations have indicated the potential uses of MXenes in LIBs [138–140]. Many new studies have put into evidence interesting qualities, including strong mechanical properties, variable bandgap energies, large electrochemically active surfaces, and great electrical conductivity. Multilayer MXenes may have a range of inorganic chemicals added to their structure to improve their effective dispersion in aqueous solutions. They are intriguing options for electrode material for LIBs because their layered structure makes it easier to store ions. In fact, Li^+ can intercalate at low voltages (0.2 V to 0.6 V) with little diffusion barriers. It is necessary to assess the reversible potential of LIBs that include MXenes in light of the lowered diffusion barriers. Findings from theoretical studies indicate that MXenes with a smaller number of formula units have a higher reversible capacity and quicker diffusion rates than those with a larger number of formula units. On the other hand, the same relationship holds true in the inverse scenario [139].

For instance, despite having a low diffusion barrier of 0.42 eV, the material $V_4C_3O_2$ exhibits a relatively low capacity of 148 mA h g^{-1} compared to other materials because of faster ion movement but may lead to lower capacity if the material cannot host many ions. On the other side, V_2CO_2 has a far higher diffusion barrier of 0.82 eV than other materials, even though it has a substantially higher capacity value of 276 mA h g^{-1}. Also, the diffusion barrier of Ti_3C_2 is 0.07 eV, which is much lower than that of graphite, which is 0.3 eV. The previously mentioned benefits of MXenes are significant but we still need further study to fully understand their intercalated structures and the mechanisms associated with them [140]. Research on titanium carbide by M. Naguib et al. [141] revealed that Li^+ intercalates inside the material's layers. The study aimed to utilize titanium carbide as an anode material in lithium-ion batteries. Zhang et al. [142] reported highly conductive Ti_3C_2 MXene decorated with Sn^{4+} ions, emphasizing the material's promising application in lithium-ion anodes. Researchers have investigated several MXene materials, including Ti_2CT_x, Nb_2CT_x, $Ti_3C_2T_x$, and V_2CT_x, for their potential use as LIBs. The letter Tx corresponds to the functional surface terminal groups in these materials. Because of its remarkable conductivity and primarily electrochemical behavior, MXene exhibits significant volumetric capacitances. The materials show substantial improvements in gravimetric performance as compared to activated carbon, demonstrating capacitances of around 210 farads per gram at a rate of 10 volts per second [143]. Lukatskaya et al. found that Ti_3C_2 in a microporous configuration reached a value of 300 Fg^{-1} capacitance at a scan rate of 101 mV s^{-1} [143]. This result was exceptional for electrochemical behavior, highlighting the significant potential of MXenes in energy storage applications.

The inorganic 2D materials, aside from transition metal carbides/nitrides (MXenes), that have been stated for use in high-performance LIBs can be categorized into several groups: graphene and the derivatives of graphene, such as graphene oxides, silicene, borophene, stanine, phosphorene, and germanene (called elemental analogs of graphene); 2D structures like hBN, transition metal dichalcogenides (TMDs), and transition metal oxides (TMOs); and newly emerging 2D materials.

Among the reported 2D materials, graphene and its derivatives have undergone the most extensive study. Pure graphene exhibits an exceptionally large surface area of 2630 m^2 g^{-1}, alongside remarkable charge carrier mobility, a thermal conductivity of

5000 W mK^{-1}, and a Young's modulus of 1 TPa [144,145]. Graphite, being the most commonly utilized electrode material for high-energy storage devices, has a theoretical capacity that falls short of meeting the growing demand; it also lacks sufficient rate capability. Derived from graphite, graphene exhibits a larger surface area and increased interlayer spacing compared to graphite, prompting extensive exploration of its potential as an electrode material. The ability of both sides of graphene to adsorb ions suggests that its capacity could potentially be twice that of graphite when utilized as an electrode material. The experimental studies on graphene anodes [146] and graphene cathodes [147] have demonstrated the feasibility of the concept of graphene electrodes.

Shu et al. [148] developed a porous graphene paper as an anode material for LIBs. This material demonstrates a discharge capacity of 400 mA h g^{-1}, even at a high current density of 2000 mA g^{-1}. The electrochemical characteristics of a material also depend upon the synthesis process and techniques used for the testing. Cohn et al. [149] created a nano-manufactured material specifically designed for LIBs. It was a hybrid graphene-single-walled CNT. By testing this material, they calculated the capacity of 2640 mA h g^{-1} at the rate of 0.186 A g^{-1}. In another paper, Hassoun et al. [150] reported a capacity of 165 mA h g^{-1}, which was basically calculated by a graphene-based battery with an energy density of 190 W h kg^{-1}.

Researchers extensively examined graphene-based composites alongside other active materials that may achieve greater capacities. Layered transition metal oxides are very attractive options. They offer the potential for high capacity, exhibit excellent chemical stability, and are cost-effective. Ni et al. [91] developed a composite material of MoO_2 nanosheets with a carbon matrix. MoO_2 presents itself as a noteworthy electrode material, exhibiting moderate discharge and charge voltages. This composite, when utilized as the anode material in high-performance LIBs, has demonstrated a remarkable capacity of 1051 mA h g^{-1} over 100 cycles at a current of 0.5 A g^{-1}. Nb_2O_5 nanosheets have been synthesized through experimental methods and evaluated for their performance as the anode in Li-ion batteries. They achieved a lithium capacity of 184 mA h g^{-1} and demonstrated commendable rate capability at intermediate discharge/charge voltages between 1.0 and 2.5 V [151].

Additionally, 2D transition metal dichalcogenides (TMDs) represent an expanding category of two-dimensional materials. Recently, there has been a big rise in the discovery and production of new 2D TMDs with a wide range of electric properties that can be used as LIB electrode materials.

Xiao et al. [152] developed a composite (MoS_2/PEO), utilizing PEO to stabilize the disordered structure. The PEO/MoS_2 composite demonstrated an impressive capacity of 1000 mA h g^{-1}. Liu et al. [88] synthesized and reported the electrochemical characteristics of the graphene and MoS_2 composite. At the specific current value of 100 mAg^{-1}, the impressive capacity of 1351 mA h g^{-1} was shown by this material after 200 cycles (charging/discharging). Conversely, significant focus has been directed towards the fabrication of SnS_2 and graphene composites, with reported significant capacities up to 1000 mA h g^{-1}. SnS_2 nanosheets represent a notable option for creating composites with graphene [153–155]. Researchers are finding ways and techniques not only to increase the LIBs performance but also the combinations of lithium with other metals, for example, Qu et al. [153] reported on the development of composite material for high-performance lithium/sodium batteries. This SnS_2/rGO combined material shows a capacity of 544 mA h g^{-1} at a rate of 2A g^{-1}. Alongside phosphorene, recent studies have also explored other two-dimensional (2D) monoelemental materials that have a similar lattice structure to graphene, including silicene, germanene, borophene, and stanine [156–160].

Bo et al. created a 2D carbon/Si composite using the ice template technique supported by ultrasonic atomization. They used this composite (NPCN/Si-2) as an electrode material to evaluate its electrochemical characteristics. This material showed a capacity value of 1977 mA h g^{-1} at the rate of 100 mA g^{-1} after 100 cycles (charging/discharging). The material even tested for 1000 charge-discharge cycles and it shows a stable capacity of 889 mA h g^{-1} [13]. Do et al. stated a unique synthesis method for improving the performance of Si-based anode material using the spray drying technique. The material demonstrates the capacity of 1224 mA h g^{-1} and an outstanding retention rate of 82% after the 100 cycles at 20 mA g^{-1} current density; this special combination produces high capacities and robust retention behaviors. They constructed a LIB consisting of silicon-based nanocarbon material as an anode. After 200 cycles, the lithium-ion battery full cell shows 65% capacity retention with an energy density of 350 Wh kg^{-1} [93].

Abdul et al. reported a B_4C_3 monolayer with a storage capacity of 2770 mA hg^{-1}. They conducted theoretical studies using density functional theory to investigate the electrochemical viability of the material. Using the values of the diffusion coefficient and ionic conductivity, they explained the behavior of the material and predicted its potential for use as an anode in lithium-ion batteries [14]. Zaid et al. used a hydrothermal process to make titanium (TiO_2) and C-TiO_2 material, which improved the Li-battery's electrical performance. As an anode material, the composite was a well-established and very effective carbon-doped TiO_2 (C-TiO_2). The findings demonstrated 400 and 500 mA hg^{-1} charge capabilities at 100 mAg^{-1} [161]. Siwei et al. synthesized the Cu_3Si/Si nanowires that displayed an initial 89% coulombic efficiency and a specific capacity of 2630 mA h g^{-1}, after 100 cycles of charging and discharging, resulting in a capacity of 1675 mA h g^{-1} at a current density of 200 mA g^{-1} [15].

The three studies (Figure 5) highlight the significance of 2D morphology in enhancing the electrochemical performance of Si-based materials, as revealed by TEM analysis. The first study demonstrates a 2D crumpled nanosheet morphology in NPCN/Si-2, where Si nanoparticles are uniformly embedded, creating a 3D interconnected network. This structure, confirmed by TEM (Figure 5a,b), ensures efficient Li-ion accessibility, reduces aggregation, and provides structural stability, resulting in high capacity and cycling stability (Figure 5c). In the second study, the 2D structure of the rGO and SWCNT layers creates an interconnected network (Figure 5d,e). This restores the sp^2 carbon network, which makes the electrical conductivity and Li-ion transport better. This results in high capacity and retention (82%) after 100 cycles (Figure 5f). In the third study, Cu_3Si/Si nanowires have a shape that is similar to that of a flat surface. This is because Cu_3Si/Si nanoparticles are mixed in with Si nanowires, making a complex that can grow in size, as seen in the lattice fringes of TEM (Figure 5g). This morphology enhances cycling stability and supports high capacities (Figure 5l). In all cases, the inorganic materials (Si, Cu_3Si) play a critical role in maintaining structural integrity and electrochemical performance while TEM analysis provides detailed insights into the morphological features that directly influence the material's efficiency and stability (Figure 5).

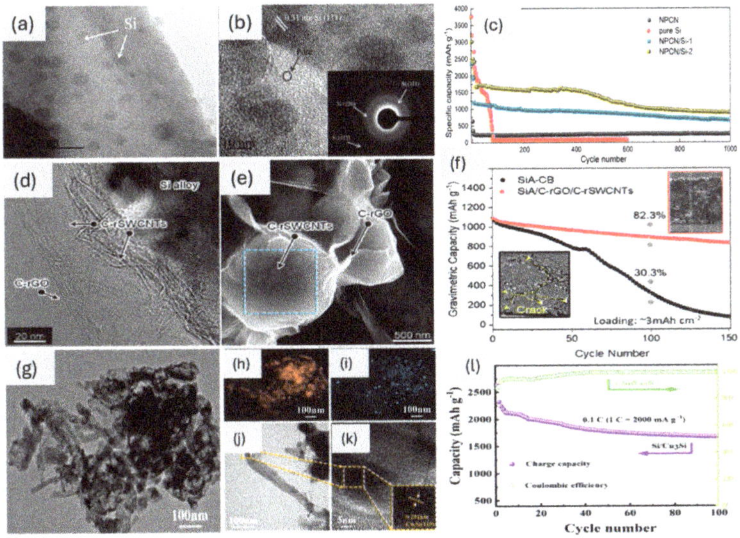

Figure 5. (a,b) High-resolution TEM images of NPCN/Si-2 with an inset of the SAED pattern and (c) electrochemical performance of pure Si, NPCN, and composites at the rate of 1 A g^{-1} [13]. Copyright © 2024 Elsevier. (d) TEM image (20 nm) of a SiA/NC, (e) SEM image (500 nm) of SiA/NC, (f) cyclic performance of materials with the inset showing cross-sectional images after 150 cycles of the materials, SiA and SiA/NC [93]. Copyright © 2023 Wiley. (g) TEM image (100 nm) of the produced nanowires, (h–k) Cu-Si 3:16 EDS map scanning, (l) cycle performance of Si/Cu_3Si for 100 cycles at 0.2 A g^{-1} [15]. Copyright © 2024 Elsevier.

3.4. Three-Dimensional Inorganic Electrode Materials

Additionally, 3D materials like SnO_2, Fe_2O_3, TiO_2, V_2O_5, and conducting polymers still face challenges because they have lower energy capacities than 1D and 2D materials [102,103,162,163]. A large-scale study is also being given to the electrochemical characteristics of 3D inorganic perovskite materials for LIBs. The three-dimensional shape of the highly linked crystal structure of the $CsPbCl_3$ perovskite makes it easier for electrons to move and for Li-ions to diffuse. For lithium-ion batteries to operate well over an extended period, this 3D design enhances structural stability and ionic conductivity. Furthermore, the strong $CsPbCl_3$ framework reduces volume expansion, improving cycle stability and capacity retention. $CsPbCl_3$ material with a capacity of 612.3 mA h g^{-1} at a rate of 50 mA g^{-1} was reported by Pal et al. [106]. After 100 cycles, Paul et al. found that the $CsPbBr_3$ material had a reduced capacity of 261 mA h g^{-1} at a current density of 60 mA g^{-1}. When we look at $CsPbBr_3$ with XPS, we can see that its chemical composition is stable (Figure 6c). This means that it will work well during cycling (charging/discharging) [100]. According to Kaisar et al., the $CsPbI_3$ material can sustain performance for up to 100 cycles with a capacity of 235 mA h g^{-1} at 40 mA g^{-1}. It was found that the crystal structures of the $CsPbBr_3$ and $CsPbI_3$ perovskites are orthorhombic by XRD. This is necessary for them to be stable in lithium-ion batteries from an electrochemical point of view. The Rietveld refinement of $CsPbI_3$ shows that the unit cell volume changes during lithiation, affecting how much lithium can be stored (Figure 6b) [101]. According to Liu et al., the $CsPbBr_3$@CNTs composite has an enhanced capacity of 470.2 mA h g^{-1} at 100 mA g^{-1} after 200 charge-discharge cycles. Improved conductivity and stability are shown by the $CsPbBr_3$@CNTs composite's ability to retain its structure (Figure 6a) [99]. The stability and lithium storage potential of 3D perovskites in energy storage applications are highlighted by these structural insights. Wu et al. found that the promising electrode material $Cs_2NaBiCl_6$

has the highest capacity at 775 mA h g^{-1} when the current density is 75 mA g^{-1} for 25 cycles [105]. Another material, $Cs_2NaErCl_6$, was described by Yang et al. in a different paper. It has a capacity of 120 mA hg^{-1} at a high current density of 300 mA g^{-1} after 500 cycles [98].

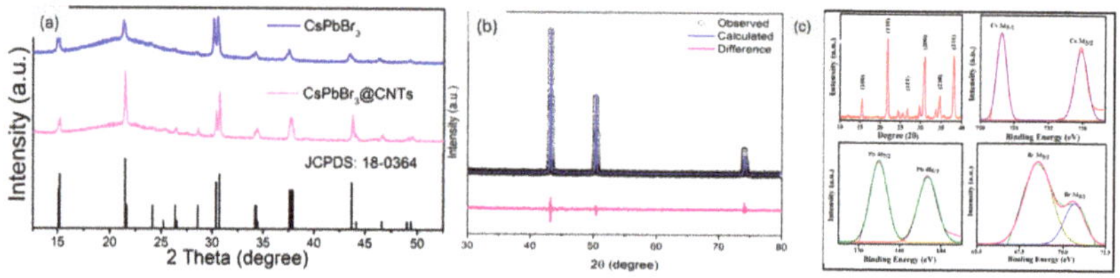

Figure 6. (**a**) XRD of $CsPbBr_3$ and composite [99], Copyright © 2021 Elsevier, (**b**) showing the XRD pattern of CsPbI3 after 100 charge-discharge cycles [101], Copyright © 2021 MDPI. (**c**) $CsPbBr_3$ XRD pattern and XPS spectra of Cs/Pb/Br_3 separately [100], Copyright © 2021, American Chemical Society.

At the moment, 3D-ordered porous (3DOP) electrode materials are being intensively recognized and investigated for their potential use in high-performance electrochemical energy storage systems. One of the advantages of 3DOP electrodes is that they have a periodic structure. This structure enables the connecting walls to reach into the unoccupied pores, which prevents the walls from being pulverized. These materials provide a high surface area along with interconnected pores, which enhance the efficiency of ion transport and electron pathways. The distinctive architecture boosts the charge and discharge rates as well as overall energy density. Furthermore, 3DOP materials demonstrate remarkable mechanical stability, accommodating volume fluctuations during cycling to enhance battery longevity [97].

Titanium dioxide (TiO_2) has undergone a significant examination for use in high-performance LIBs. TiO_2 is chemically stable, abundant in nature, very cost-effective, and environmentally friendly element. In most cases, it is used as an anode material for Li storage. It has been shown that the rutile TiO_2 electrode has an early discharge capacity of 608 mA h g^{-1}. This is a lot more than the theoretical capacity of TiO_2, which is 168 mA h g^{-1} for $Li_{0.5}TiO_2$ and 336 mA h g^{-1} for $LiTiO_2$. Furthermore, this 3DOP rutile TiO_2 electrode demonstrates remarkable performance, reaching up to 5000 cycles without the use of any binders [97]. Transition cobalt oxides are also important materials for use in high-performance LIBs with impressive theoretical capacities. In the case of Co_3O_4, the theoretical value of 890 mA h g^{-1} capacity is reported. And in the case of CoO, the theoretical value of 716 mA h g^{-1} capacity is reported [164–166]. It is not possible to use CoO in LIBs as an anode material in experiments because it loses a lot of its capacity after long cycles, which is caused by big changes in volume. In addressing this issue, three-dimensional ordered porous carbon inverse opals have been employed to encapsulate CoO nanoparticles [104]. After 1000 cycles, the CoO/C electrode exhibits a notable value of 674 mA h g^{-1} capacity, which is 94% of the theoretical capacity. For active nanoparticles, the three-dimensional ordered porous carbon structure serves both a dimensional limitation and an interpenetrated, continuous conducting network.

On the other hand, iron oxides stand out among the numerous transition metal oxides as a widely examined anode material due to their numerous advantages. These include impressive theoretical capacities as well as their abundant availability and environmentally friendly properties. For example, Fe_3O_4 has a reported capacity value of 926 mA h g^{-1} whereas Fe_2O_3 has a reported capacity value of 1007 mA h g^{-1}. In another publication, the capacities of 3DOP α-Fe_2O_3 are reported as 1883 (discharge) and 1139 mA h g^{-1} (charge), respectively [167].

Among all the many types of anode materials, investigations that compare Si anodes have shown that they have the greatest specific capacity of any of the other possibilities. As the need for energy storage solutions that work well and last a long time grows, studying bulk-silicon- and nano-silicon-based anodes in lithium-ion secondary cells looks like a good idea. That is the reason they have been studied on a large scale. It is thought that both bulk-silicon and nano-silicon (Si)-based anodes for lithium-ion secondary cells could be used to store energy in the future and they have a high theoretical capacity (about 4200 mA h g^{-1}) [168]. One of the most notable drawbacks, however, is the large volume expansion of this material (about 400%) that takes place during the electrochemical processes with Li ions. This significantly restricts the practical use of these materials [169,170]. There are several Si anode composites with Ni (Si/Ni) that were manufactured using the method of electrodeposition in order to solve the pulverization problem [171,172].

Additional electrochemically active anode materials, including graphene [173], $TiNb_2O_7$ [174], MoS_2 [175], and Ge [176,177], were also developed as 3DOP electrodes for Li-ion storage.

SEM and TEM confirm the 3D-ordered macroporous (3DOM) structure of $TiNbO_7$, which displays well-organized "honeycomb" pores with linked macropores (~100 nm diameter) and inorganic walls (~50 nm thick) (Figure 7). It makes it easier for lithium ions to move around and keeps the structure stable during cycling. This very porous three-dimensional framework is better at storing lithium ions. Because it is linked in three dimensions, this 3DOM-$TiNbO_7$ is a good material for the anode of high-performance lithium-ion batteries [174]. Also, tin dioxide is one of the examined anode materials for LIBs, with an irreversible conversion reaction alongside a reversible alloying reaction, (SnO_2/Sn + Li_2O) and (Sn/$Li_{4.4}Sn$), respectively [178–180]. The investigation of commercial inorganic cathode nanomaterials for LIBs primarily centers on $LiFePO_4$. The olivine-structured $LiFePO_4$ is a highly favorable alternative to $LiCoO_2$ due to its environmentally friendly nature. The primary drawback lies in its inadequate conductivity and rate capability [181]. To solve the problem of low conductivity, some conductive material can be used as a mixture with $LiFePO_4$ electrodes. $LiMnO_2$ has garnered significant attention thanks to its impressive theoretical capacity near 300 mA h g^{-1} along with good experimental capacities [182]. The $LiMnO_2$ crystal has both a layered monoclinic structure and an orthorhombic structure. The zigzag arrangement of Li and Mn ions highlights the orthorhombic structure.

Figure 7. (**a–c**) SEM images of $TiNb_2O_7$ at different magnifications and (**d–f**) TEM images of $TiNb_2O_7$ at different magnifications [174]. Copyright © 2017 Elsevier.

Although 3D materials have several benefits, poor cycle life; scaling issues; and expensive, complicated production restrict their use in high-performance LIBs. They are less appropriate for traditional usage since they often increase power density while decreasing volumetric energy density. Efficiency and longevity may be impacted by side reactions and unequal electrolyte penetration brought on by large surface areas. It is challenging to compete with cost-effective, optimized materials like graphite. Three-dimensional materials still have challenges with cost, performance, and scalability, despite continuous research aimed at enhancing their capabilities.

4. Conclusions and Future Outlook

Extensive efforts have been invested to develop novel and efficient electrode materials for high-performance energy storage devices based on inorganic oxides. However, the devices made up of these electrodes are characterized by lower energy and power density [183,184]. This study reveals that the electrode material properties like surface areas, charge storage sites, chemical reactions, unique bonding, and interactions with the electrolyte play a crucial role in defining the electrochemical performances of electrodes used in lithium-ion batteries.

LIBs can utilize zero-dimensional nanoparticles, such as quantum dots, and three-dimensional structures like graphene multilayers; however, these nanostructures are less prevalent compared to one- and two-dimensional alternatives [185–190]. One-dimensional nanoparticles, such as nanowires and nanotubes, show high conductivity and ability to improve charge transfer. Graphene and other layered materials exhibit exceptional electrical conductivity and a large surface area, both of which are essential for effective energy storage. Each type of nanostructure addresses specific performance metrics for LIBs. The most promising silicon-based architectures for high energy density are 0D and 1D structures. Additionally, 1D CNTs and 2D graphene provide a better rate performance for rapid charge/discharge. MXenes and 3D hierarchical structures are particularly noteworthy for long-term cycle stability.

This study allows us to highlight that optimizing the morphology of nanomaterials significantly enhances the electrochemical performance of LIB electrodes. To obtain high energy density, fast charge/discharge rates, and long-term stability, future research should focus on scalable ways to make hybrid structures that combine 0D, 1D, 2D, and 3D materials. A hybrid method that combines many nanostructures (for example, 3D frameworks with built-in 0D or 2D materials) could improve battery performance. This would bridge the gap between lab-based innovation and commercial scalability. Furthermore, investigating innovative surface alterations and electrolyte interactions may enhance economic viability even further.

Author Contributions: Conceptualization, M.H.K. and V.T.; methodology M.H.K. and V.T.: investigation, M.H.K. resources, P.L. and V.T.; data curation, M.H.K.; writing—original draft preparation, M.H.K.; writing—review and editing, M.H.K., P.L. and V.T.; supervision, V.T.; funding acquisition, P.L. and V.T. All authors have read and agreed to the published version of the manuscript.

Funding: The development of this work is framed in the Environmentally Friendly Aviation for All Classes of Aircraft (EFACA) project. This project is funded by the European Union Horizon Europe research and innovation programme (HORIZON-CL5-2021-D5-01-05) under grant agreement no. 101056866.

Data Availability Statement: No new data were created or analyzed in this study. Data sharing is not applicable to this article.

Acknowledgments: MHK acknowledges the support by the Dottorato di Ricerca Nazionale in Photovoltaics funded by Italian University Ministry.

Conflicts of Interest: The authors declare no conflicts of interest. The funders had no role in the design of this study; in the collection, analyses, or interpretation of data; in the writing of the manuscript; or in the decision to publish the results.

References

1. Zhang, K.; Han, X.; Hu, Z.; Zhang, X.; Tao, Z.; Chen, J. Nanostructured Mn-based oxides for electrochemical energy storage and conversion. *Chem. Soc. Rev.* **2015**, *44*, 699–728. [CrossRef] [PubMed]
2. Goodenough, J.B. Evolution of strategies for modern rechargeable batteries. *ACC Chem. Res.* **2013**, *46*, 1053–1061. [CrossRef] [PubMed]
3. Dunn, B.; Kamath, H.; Tarascon, J.-M. Electrical Energy Storage for the Grid: A Battery of Choices System power ratings, module size. *Science* **2011**, *334*, 928–935. [CrossRef] [PubMed]
4. Chodankar, N.R.; Dubal, D.P.; Kwon, Y.; Kim, D.H. Direct growth of $FeCo_2O_4$ nanowire arrays on flexible stainless steel mesh for high-performance asymmetric supercapacitor. *NPG Asia Mater.* **2017**, *9*, e419. [CrossRef]
5. Wang, F.; Wu, X.; Li, C.; Zhu, Y.; Fu, L.; Wu, Y.; Liu, X. Nanostructured positive electrode materials for post-lithium ion batteries. *Energy Environ. Sci.* **2016**, *9*, 3570–3611. [CrossRef]
6. Wang, F.; Wang, X.; Chang, Z.; Zhu, Y.; Fu, L.; Liu, X.; Wu, Y. Electrode materials with tailored facets for electrochemical energy storage. *Nanoscale Horiz.* **2016**, *1*, 272–289. [CrossRef]
7. Wang, F.; Wu, X.; Yuan, X.; Liu, Z.; Zhang, Y.; Fu, L.; Zhu, Y.; Zhou, Q.; Wu, Y.; Huang, W. Latest advances in supercapacitors: From new electrode materials to novel device designs. *Chem. Soc. Rev.* **2017**, *46*, 6816–6854. [CrossRef]
8. Liu, J.; Li, N.; Goodman, M.D.; Zhang, H.G.; Epstein, E.S.; Huang, B.; Pan, Z.; Kim, J.; Choi, J.H.; Huang, X. Mechanically and chemically robust sandwich-structured C@Si@C nanotube array Li-ion battery anodes. *ACS Nano* **2015**, *9*, 1985–1994. [CrossRef] [PubMed]
9. Wang, C.; Jiang, C.; Xu, Y.; Liang, L.; Zhou, M.; Jiang, J.; Singh, S.; Zhao, H.; Schober, A.; Lei, Y. Article type: Communication A selectively permeable membrane for enhancing cyclability of organic sodium-ion batteries. *Adv. Mater.* **2016**, *28*, 9182–9187. [CrossRef] [PubMed]
10. Sun, H.; Mei, L.; Liang, J.; Zhao, Z.; Lee, C.; Fei, H.; Ding, M.; Lau, J.; Li, M.; Wang, C.; et al. Three-dimensional holey-graphene/niobia composite architectures for ultrahigh-rate energy storage. *Science* **2017**, *356*, 599–604. [CrossRef]
11. Sun, Y.; Liu, N.; Cui, Y. *Promises and Challenges of Nanomaterials for Lithium-Based Rechargeable Batteries*; Nature Publishing Group: London, UK, 2016. [CrossRef]
12. Bruce, P.G.; Freunberger, S.A.; Hardwick, L.J.; Tarascon, J.M. *$LigO_2$ and LigS Batteries with High Energy Storage*; Nature Publishing Group: London, UK, 2012. [CrossRef]
13. Liang, B.; Tan, W.; Chen, M.; Yi, M.; Hu, J.; Zeng, K.; Wang, Y.; Li, Y.; Yang, G. Facile synthesis of two-dimensional carbon/Si composite assembled by ultrasonic atomization-assisted-ice template technology as electrode for lithium-ion battery. *J. Alloys Compd.* **2024**, *976*, 173030. [CrossRef]
14. Majid, A.; Najam, U.; Ahmad, S.; Alkhedher, M. On the prospects of using B_4C_3 as a potential electrode material for lithium-ion batteries. *Mater. Sci. Semicond. Process* **2024**, *176*, 108320. [CrossRef]
15. Jiang, S.; Cheng, J.; Nayaka, G.P.; Dong, P.; Zhang, Y.; Xing, Y.; Zhang, X.; Du, N.; Zhou, Z. Efficient electrochemical synthesis of Cu_3Si/Si hybrids as negative electrode material for lithium-ion battery. *J. Alloys Compd.* **2024**, *998*, 174996. [CrossRef]
16. Yang, D.; Han, Y.; Li, M.; Li, C.; Bi, W.; Gong, Q.; Zhang, J.; Zhang, J.; Zhou, Y.; Gao, H.; et al. Highly Conductive Quasi-1D Hexagonal Chalcogenide Perovskite Sr_8 Ti_7 S_{21} with Efficient Polysulfide Regulation in Lithium-Sulfur Batteries. *Adv. Funct. Mater.* **2024**, *34*, 2401577. [CrossRef]
17. Wei, Q.; Xiong, F.; Tan, S.; Huang, L.; Lan, E.H.; Dunn, B.; Mai, L. *Porous One-Dimensional Nanomaterials: Design, Fabrication and Applications in Electrochemical Energy Storage*; Wiley-VCH Verlag: Weinheim, Germany, 2017. [CrossRef]
18. Liu, Z.; Yuan, X.; Zhang, S.; Wang, J.; Huang, Q.; Yu, N.; Zhu, Y.; Fu, L.; Wang, F.; Chen, Y.; et al. *Three-Dimensional Ordered Porous Electrode Materials for Electrochemical Energy Storage*; Nature Publishing Group: London, UK, 2019. [CrossRef]
19. Cui, G.; Gu, L.; Thomas, A.; Fu, L.; van Aken, P.A.; Antonietti, M.; Maier, J. A Carbon/Titanium Vanadium Nitride Composite for Lithium Storage. *ChemPhysChem* **2010**, *11*, 3219–3223. [CrossRef]
20. Lee, K.T.; Lytle, J.C.; Ergang, N.S.; Oh, S.M.; Stein, A. Synthesis and rate performance of monolithic macroporous carbon electrodes for lithium-ion secondary batteries. *Adv. Funct. Mater.* **2005**, *15*, 547–556. [CrossRef]
21. Che, G.; Lakshmi, B.B.; Fisher, E.R.; Martin, C.R. Carbon nanotubule membranes for electrochemical energy storage and production. *Nature* **1998**, *393*, 346–349. [CrossRef]
22. Guo, B.; Wang, X.; Fulvio, P.F.; Chi, M.; Mahurin, S.M.; Sun, X.-G.; Dai, S. Soft-templated mesoporous carbon-carbon nanotube composites for high performance lithium-ion batteries. *Adv. Mater.* **2011**, *23*, 4661–4666. [CrossRef]

23. Li, S.; Luo, Y.; Lv, W.; Yu, W.; Wu, S.; Hou, P.; Yang, Q.; Meng, Q.; Liu, C.; Cheng, H.-M. Vertically aligned carbon nanotubes grown on graphene paper as electrodes in lithium-ion batteries and dye-sensitized solar cells. *Adv. Energy Mater.* **2011**, *1*, 486–490. [CrossRef]
24. Chen, J.; Xu, L.; Li, W.; Gou, X. α-Fe_2O_3 nanotubes in gas sensor and lithium-ion battery applications. *Adv. Mater.* **2005**, *17*, 582–586. [CrossRef]
25. Taberna, P.L.; Mitra, S.; Poizot, P.; Simon, P.; Tarascon, J.M. *High Rate Capabilities Fe_3O_4-Based Cu Nano-Architectured Electrodes for Lithium-Ion Battery Applications*; Nature Publishing Group: London, UK, 2006. [CrossRef]
26. Kim, W.T.; Jeong, Y.U.; Lee, Y.J.; Kim, Y.J.; Song, J.H. Synthesis and lithium intercalation properties of Li_3VO_4 as a new anode material for secondary lithium batteries. *J. Power Sources* **2013**, *244*, 557–560. [CrossRef]
27. Dong, S.; Chen, X.; Zhang, X.; Cui, G. Nanostructured transition metal nitrides for energy storage and fuel cells. *Coord. Chem. Rev.* **2013**, *257*, 1946–1956. [CrossRef]
28. Serhan, M.; Jackemeyer, D.; Long, M.; Sprowls, M.; Diez Perez, I.; Maret, W.; Chen, F.; Tao, N.; Forzani, E. Total iron measurement in human serum with a smartphone. In *AIChE Annual Meeting, Conference Proceedings*; American Institute of Chemical Engineers: New York, NY, USA, 2019.
29. Lee, J.T.; Zhao, Y.; Thieme, S.; Kim, H.; Oschatz, M.; Borchardt, L.; Magasinski, A.; Cho, W.-I.; Kaskel, S.; Yushin, G. Sulfur-infiltrated micro-and mesoporous silicon carbide-derived carbon cathode for high-performance lithium sulfur batteries. *Adv. Mater.* **2013**, *25*, 4573–4579. [CrossRef]
30. Naguib, M.; Halim, J.; Lu, J.; Cook, K.M.; Hultman, L.; Gogotsi, Y.; Barsoum, M.W. New two-dimensional niobium and vanadium carbides as promising materials for li-ion batteries. *J. Am. Chem. Soc.* **2013**, *135*, 15966–15969. [CrossRef] [PubMed]
31. Seng, K.H.; Park, M.H.; Guo, Z.P.; Liu, H.K.; Cho, J. Catalytic role of ge in highly reversible GeO_2/Ge/C nanocomposite anode material for lithium batteries. *Nano Lett.* **2013**, *13*, 1230–1236. [CrossRef] [PubMed]
32. Ji, L.; Lin, Z.; Alcoutlabi, M.; Zhang, X. Recent developments in nanostructured anode materials for rechargeable lithium-ion batteries. *Energy Environ. Sci.* **2011**, *4*, 2682–2699. [CrossRef]
33. Luo, S.; Wang, K.; Wang, J.; Jiang, K.; Li, Q.; Fan, S. Binder-free $LiCoO_2$/carbon nanotube cathodes for high-performance lithium ion batteries. *Adv. Mater.* **2012**, *24*, 2294–2298. [CrossRef] [PubMed]
34. Chen, S.; Cao, F.; Liu, F.; Xiang, Q.; Feng, X.; Liu, L.; Qiu, G. Facile hydrothermal synthesis and electrochemical properties of orthorhombic $LiMnO_2$ cathode materials for rechargeable lithium batteries. *RSC Adv.* **2014**, *4*, 13693–13703. [CrossRef]
35. Kobayashi, G.; Yamada, A.; Nishimura, S.-I.; Kanno, R.; Kobayashi, Y.; Seki, S.; Ohno, Y.; Miyashiro, H. Shift of redox potential and kinetics in Lix(MnyFe1-y)PO_4. *J. Power Sources* **2009**, *189*, 397–401. [CrossRef]
36. Hecht, D.S.; Hu, L.; Irvin, G. Emerging transparent electrodes based on thin films of carbon nanotubes, graphene, and metallic nanostructures. *Adv. Mater.* **2011**, *23*, 1482–1513. [CrossRef] [PubMed]
37. Kim, Y.K.; Min, D.H. Durable large-area thin films of graphene/carbon nanotube double layers as a transparent electrode. *Langmuir* **2009**, *25*, 11302–11306. [CrossRef] [PubMed]
38. Yu, M.; Zeng, Y.; Zhang, C.; Lu, X.; Zeng, C.; Yao, C.; Yang, Y.; Tong, Y. Titanium dioxide@polypyrrole core-shell nanowires for all solid-state flexible supercapacitors. *Nanoscale* **2013**, *5*, 10806–10810. [CrossRef]
39. Yu, M.; Zhai, T.; Lu, X.; Chen, X.; Xie, S.; Li, W.; Liang, C.; Zhao, W.; Zhang, L.; Tong, Y. Manganese dioxide nanorod arrays on carbon fabric for flexible solid-state supercapacitors. *J. Power Sources* **2013**, *239*, 64–71. [CrossRef]
40. Tang, Z.; Tang, C.H.; Gong, H. A high energy density asymmetric supercapacitor from nano-architectured $Ni(OH)_2$/Carbon nanotube electrodes. *Adv. Funct. Mater.* **2012**, *22*, 1272–1278. [CrossRef]
41. Yuan, L.; Lu, X.-H.; Xiao, X.; Zhai, T.; Dai, J.; Zhang, F.; Hu, B.; Wang, X.; Gong, L.; Chen, J.; et al. Flexible solid-state supercapacitors based on carbon nanoparticles/MnO_2 nanorods hybrid structure. *ACS Nano* **2012**, *6*, 656–661. [CrossRef]
42. Lu, X.; Zhai, T.; Zhang, X.; Shen, Y.; Yuan, L.; Hu, B.; Gong, L.; Chen, J.; Gao, Y.; Zhou, J.; et al. WO_3-x@Au@MnO_2 core-shell nanowires on carbon fabric for high-performance flexible supercapacitors. *Adv. Mater.* **2012**, *24*, 938–944. [CrossRef]
43. Balogun, M.-S.; Qiu, W.; Wang, W.; Fang, P.; Lu, X.; Tong, Y. Recent Advances in Metal Nitrides as High-Performance Electrode Materials for Energy Storage Devices. *J. Mater. Chem. A* **2015**, *3*, 1364–1387. [CrossRef]
44. Li, Y.; Levine, A.M.; Zhang, J.; Lee, R.J.; Naskar, A.K.; Dai, S.; Paranthaman, M.P. Carbon/tin oxide composite electrodes for improved lithium-ion batteries. *J. Appl. Electrochem.* **2018**, *48*, 811–817. [CrossRef]
45. Yi, H.; Huang, Y.; Sha, Z.; Zhu, X.; Xia, Q.; Xia, H. Facile synthesis of Mo_2N quantum dots embedded N-doped carbon nanosheets composite as advanced anode materials for lithium-ion batteries. *Mater. Lett.* **2020**, *276*, 128205. [CrossRef]
46. Zhang, C.; Li, Y.; Liu, X.; Zheng, X.; Zhao, Y.; Zhang, D. Facile synthesis of $Fe_{24}N_{10}$/porous carbon as a novel high-performance anode material for lithium-ion batteries. *Mater. Lett.* **2021**, *300*, 130196. [CrossRef]
47. Long, B.; Balogun, M.-S.; Luo, L.; Luo, Y.; Qiu, W.; Song, S.; Zhang, L.; Tong, Y. Encapsulated Vanadium-Based Hybrids in Amorphous N-Doped Carbon Matrix as Anode Materials for Lithium-Ion Batteries. *Small* **2017**, *13*, 1702081. [CrossRef]

48. Li, Y.; Wang, X.; Wang, L.; Jia, D.; Yang, Y.; Liu, X.; Sun, M.; Zhao, Z.; Qiu, J. Ni@Ni_3 N Embedded on Three-Dimensional Carbon Nanosheets for High-Performance Lithium/Sodium–Sulfur Batteries. *ACS Appl. Mater. Interfaces* **2021**, *13*, 48536–48545. [CrossRef]
49. Ma, C.; Jia, X.; Liu, X.; Wang, J.; Qiao, W.; Yu, J.; Ling, L. Ultrafine NbN nanoparticle decorated nitrogen-doped carbon nanosheets with efficient polysulfide catalytic conversion for superior Li–S batteries. *J. Power Sources* **2022**, *520*, 230764. [CrossRef]
50. Tarascon, J.-M.; Armand, M. Issues and challenges facing rechargeable lithium batteries. *Nature* **2001**, *414*, 359–367. [CrossRef] [PubMed]
51. Wang, X.; Ding, Y.L.; Deng, Y.P.; Chen, Z. *Ni-Rich/Co-Poor Layered Cathode for Automotive Li-Ion Batteries: Promises and Challenges*; Wiley-VCH Verlag: Weinheim, Germany, 2020. [CrossRef]
52. Xiao, J.; Huang, Y.; Ma, Y.; Li, C.; Fu, L.; Zeng, W.; Wang, X.; Li, X.; Wang, M.; Guo, B. Organic active materials in rechargeable batteries: Recent advances and prospects. *Energy Storage Mater.* **2023**, *63*, 103046. [CrossRef]
53. Jiang, B.; Su, Y.; Liu, R.; Sun, Z.; Wu, D. Calcium Based All-Organic Dual-Ion Batteries with Stable Low Temperature Operability. *Small* **2022**, *18*, 2200049. [CrossRef] [PubMed]
54. Wu, C.; Lu, X.; Peng, L.; Xu, K.; Peng, X.; Huang, J.; Yu, G.; Xie, Y. Two-dimensional vanadyl phosphate ultrathin nanosheets for high energy density and flexible pseudocapacitors. *Nat. Commun.* **2013**, *4*, 2431. [CrossRef]
55. Zhang, K.; Hu, Z.; Tao, Z.; Chen, J. Inorganic &organic materials for rechargeable Li batteries with multi-electron reaction. *Sci. China Mater.* **2014**, *57*, 42–58. [CrossRef]
56. Wu, H.; Cui, Y. Designing nanostructured Si anodes for high energy lithium ion batteries. *Nano Today* **2012**, *7*, 414–429. [CrossRef]
57. Xu, X.; Cao, R.; Jeong, S.; Cho, J. Spindle-like mesoporous α-Fe_2O_3 anode material prepared from MOF template for high-rate lithium batteries. *Nano Lett.* **2012**, *12*, 4988–4991. [CrossRef] [PubMed]
58. Sun, Y.; Hu, X.; Luo, W.; Xia, F.; Huang, Y. Reconstruction of conformal nanoscale MnO on graphene as a high-capacity and long-life anode material for lithium ion batteries. *Adv. Funct. Mater.* **2013**, *23*, 2436–2444. [CrossRef]
59. Gong, Z.; Yang, Y. Recent advances in the research of polyanion-type cathode materials for Li-ion batteries. *Energy Environ. Sci.* **2011**, *4*, 3223–3242. [CrossRef]
60. Mishra, A.; Mehta, A.; Basu, S.; Malode, S.J.; Shetti, N.P.; Shukla, S.S.; Nadagouda, M.N.; Aminabhavi, T.M. Electrode materials for lithium-ion batteries. *Mater. Sci. Energy Technol.* **2018**, *1*, 182–187. [CrossRef]
61. Sharma, Y.; Sharma, N.; Rao, G.V.S.; Chowdari, B.V.R. Nanophase $ZnCo_2O_4$ as a high performance anode material for Li-ion batteries. *Adv. Funct. Mater.* **2007**, *17*, 2855–2861. [CrossRef]
62. Hu, Y.S.; Kienle, L.; Guo, Y.G.; Maier, J. High lithium electroactivity of nanometer-sized rutile TiO_2. *Adv. Mater.* **2006**, *18*, 1421–1426. [CrossRef]
63. Kim, M.G.; Lee, S.; Cho, J. Highly Reversible Li-Ion Intercalating MoP_2 Nanoparticle Cluster Anode for Lithium Rechargeable Batteries. *J. Electrochem. Soc.* **2009**, *156*, A89. [CrossRef]
64. Tiwari, J.N.; Tiwari, R.N.; Kim, K.S. Zero-dimensional, one-dimensional, two-dimensional and three-dimensional nanostructured materials for advanced electrochemical energy devices. *Prog. Mater. Sci.* **2012**, *57*, 724–803. [CrossRef]
65. Ezema, F.I.; Lokhande, C.D.; Lokhande, A.C. *Chemically Deposited Metal Chalcogenide-Based Carbon Composites for Versatile Applications*; Springer International Publishing: Heidelberg, Germany, 2023; ISBN 978-3-03-123400-2. [CrossRef]
66. Yu, W.-J.; He, W.; Wang, C.; Liu, F.; Zhu, L.; Tian, Q.; Tong, H.; Guo, X. Confinement of TiO_2 quantum dots in graphene nanoribbons for high-performance lithium and sodium ion batteries. *J. Alloys Compd.* **2022**, *898*, 162856. [CrossRef]
67. Yin, X.; Zhi, C.; Sun, W.; Lv, L.-P.; Wang, Y. Multilayer NiO@Co_3O_4 @graphene quantum dots hollow spheres for high-performance lithium-ion batteries and supercapacitors. *J. Mater. Chem. A Mater.* **2019**, *7*, 7800–7814. [CrossRef]
68. Li, X.; Meng, X.; Liu, J.; Geng, D.; Zhang, Y.; Norouzi Banis, M.; Li, Y.; Yang, J.; Li, R.; Sun, X.; et al. Tin oxide with controlled morphology and crystallinity by atomic layer deposition onto graphene nanosheets for enhanced lithium storage. *Adv. Funct. Mater.* **2012**, *22*, 1647–1654. [CrossRef]
69. Wang, L.; Shi, Z.; Feng, X.; Zhang, Y.; Cao, S.; Zhang, J. Engineering the micro/nano structure of $Ca_3Co_4O_9$ anode material for lithium-ion batteries. *J. Mater. Sci. Mater. Electron.* **2024**, *35*, 17. [CrossRef]
70. Choi, Y.-H.; Park, H.; Lee, S.; Jeong, H.-D. Synthesis and Electrochemical Performance of π-Conjugated Molecule Bridged Silicon Quantum Dot Cluster as Anode Material for Lithium-Ion Batteries. *ACS Omega* **2020**, *5*, 8629–8637. [CrossRef] [PubMed]
71. Wang, D.; Li, X.; Wang, J.; Yang, J.; Geng, D.; Li, R.; Cai, M.; Sham, T.-K.; Sun, X. Defect-rich crystalline SnO_2 immobilized on graphene nanosheets with enhanced cycle performance for li ion batteries. *J. Phys. Chem. C* **2012**, *116*, 22149–22156. [CrossRef]
72. Zhang, G.; Hou, S.; Zhang, H.; Zeng, W.; Yan, F.; Li, C.C.; Duan, H. High-performance and ultra-stable lithium-ion batteries based on MOF-derived ZnO@ZnO quantum dots/C core-shell nanorod arrays on a carbon cloth anode. *Adv. Mater.* **2015**, *27*, 2400–2405. [CrossRef] [PubMed]
73. Kim, H.; Kim, S.-W.; Park, Y.-U.; Gwon, H.; Seo, D.-H.; Kim, Y.; Kang, K. SnO_2/graphene composite with high lithium storage capability for lithium rechargeable batteries. *Nano Res.* **2010**, *3*, 813–821. [CrossRef]

74. Rana, M.; Boaretto, N.; Mikhalchan, A.; Santos, M.V.; Marcilla, R.; Vilatela, J.J. Composite Fabrics of Conformal MoS2Grown on CNT Fibers: Tough Battery Anodes without Metals or Binders. *ACS Appl. Energy Mater.* **2021**, *4*, 5668–5676. [CrossRef]
75. Paek, S.M.; Yoo, E.J.; Honma, I. Enhanced cyclic performance and lithium storage capacity of SnO_2/graphene nanoporous electrodes with three-dimensionally delaminated flexible structure. *Nano Lett.* **2009**, *9*, 72–75. [CrossRef] [PubMed]
76. Zhou, G.; Wang, D.-W.; Li, F.; Zhang, L.; Li, N.; Wu, Z.-S.; Wen, L.; Lu, G.Q.; Cheng, H.-M. Graphene-wrapped Fe_3O_4 anode material with improved reversible capacity and cyclic stability for lithium ion batteries. *Chem. Mater.* **2010**, *22*, 5306–5313. [CrossRef]
77. Zhang, T.; Huang, T.-T.; Li, X.-J.; Wang, K.; Wang, L.-Y.; Liang, J.-F.; Song, Y.-X.; Li, P.-Y.; Zhang, Y.-G.; Zhang, Y.-H.; et al. Ultra-high rapid-charging performance of 1D germanium anode materials for lithium-ion batteries. *J. Alloys Compd.* **2024**, *976*, 173287. [CrossRef]
78. Pinilla, S.; Park, S.-H.; Fontanez, K.; Márquez, F.; Nicolosi, V.; Morant, C. 0D-1D Hybrid Silicon Nanocomposite as Lithium-Ion Batteries Anodes. *Nanomaterials* **2020**, *10*, 515. [CrossRef]
79. Zhang, L.; Zhang, Y.; Xu, S.; Zhang, C.; Hou, L.; Yuan, C. Scalable Synthesis of One-Dimensional Mesoporous $ZnMnO_3$ Nanorods with Ultra-Stable and High Rate Capability for Efficient Lithium Storage. *Chem. A Eur. J.* **2019**, *25*, 16683–16691. [CrossRef] [PubMed]
80. Liu, W.; Zong, K.; Ghani, U.; Saad, A.; Liu, D.; Deng, Y.; Raza, W.; Li, Y.; Hussain, A.; Ye, P.; et al. Ternary Lithium Nickel Boride with 1D Rapid-Ion-Diffusion Channels as an Anode for Use in Lithium-Ion Batteries. *Small* **2024**, *20*, 2309918. [CrossRef] [PubMed]
81. Cho, J.S.; Hong, Y.J.; Kang, Y.C. Design and Synthesis of Bubble-Nanorod-Structured Fe_2O_3 –Carbon Nanofibers as Advanced Anode Material for Li-Ion Batteries. *ACS Nano* **2015**, *9*, 4026–4035. [CrossRef] [PubMed]
82. Carbone, M. Zn defective $ZnCo_2O_4$ nanorods as high capacity anode for lithium ion batteries. *J. Electroanal. Chem.* **2018**, *815*, 151–157. [CrossRef]
83. Cui, Z.; Wang, S.; Zhang, Y.; Cao, M. High-performance lithium storage of Co_3O_4 achieved by constructing porous nanotube structure. *Electrochim. Acta* **2015**, *182*, 507–515. [CrossRef]
84. Cao, F.-F.; Deng, J.-W.; Xin, S.; Ji, H.-X.; Schmidt, O.G.; Wan, L.-J.; Guo, Y.-G. Cu-Si Nanocable Arrays as High-Rate Anode Materials for Lithium-Ion Batteries. *Adv. Mater.* **2011**, *23*, 4415–4420. [CrossRef]
85. Cheong, J.Y.; Kim, C.; Jung, J.-W.; Yoon, K.R.; Kim, I.-D. Porous SnO_2-CuO nanotubes for highly reversible lithium storage. *J. Power Sources* **2018**, *373*, 11–19. [CrossRef]
86. Xu, J.; Wu, H.; Wang, F.; Xia, Y.; Zheng, G. Zn_4 Sb_3 Nanotubes as Lithium Ion Battery Anodes with High Capacity and Cycling Stability. *Adv. Energy Mater.* **2013**, *3*, 286–289. [CrossRef]
87. Choi, W.; Oh, S.; Hwang, S.; Chae, S.; Park, H.; Lee, W.; Woo, C.; Dong, X.; Choi, K.H.; Ahn, J.; et al. One-dimensional van der Waals transition metal chalcogenide as an anode material for advanced lithium-ion batteries. *J. Mater. Chem. A Mater.* **2024**, *12*, 7122–7131. [CrossRef]
88. Liu, Y.; Zhao, Y.; Jiao, L.; Chen, J. A graphene-like MoS_2 /graphene nanocomposite as a highperformance anode for lithium ion batteries. *J. Mater. Chem. A* **2014**, *2*, 13109–13115. [CrossRef]
89. Wang, B.; Li, X.; Zhang, X.; Luo, B.; Jin, M.; Liang, M.; Dayeh, S.A.; Picraux, S.T.; Zhi, L. Adaptable Silicon–Carbon Nanocables Sandwiched between Reduced Graphene Oxide Sheets as Lithium Ion Battery Anodes. *ACS Nano* **2013**, *7*, 1437–1445. [CrossRef] [PubMed]
90. Mahmood, N.; Zhang, C.; Liu, F.; Zhu, J.; Hou, Y. Hybrid of Co_3 Sn_2 @Co Nanoparticles and Nitrogen-Doped Graphene as a Lithium Ion Battery Anode. *ACS Nano* **2013**, *7*, 10307–10318. [CrossRef]
91. Ni, J.; Zhao, Y.; Li, L.; Mai, L. Ultrathin MoO_2 nanosheets for superior lithium storage. *Nano Energy* **2015**, *11*, 129–135. [CrossRef]
92. Sun, F.; Huang, K.; Qi, X.; Gao, T.; Liu, Y.; Zou, X.; Wei, X.; Zhong, J. A rationally designed composite of alternating strata of Si nanoparticles and graphene: A high-performance lithium-ion battery anode. *Nanoscale* **2013**, *5*, 8586. [CrossRef] [PubMed]
93. Lee, D.G.; Cho, J.Y.; Kim, J.H.; Ryoo, G.; Yoon, J.; Jo, A.; Lee, M.H.; Park, J.H.; Yoo, J.-K.; Lee, D.Y.; et al. Dispersant-Free Colloidal and Interfacial Engineering of Si-Nanocarbon Hybrid Anode Materials for High-Performance Li-Ion Batteries. *Adv. Funct. Mater.* **2024**, *34*, 2311353. [CrossRef]
94. Zhou, M.; Li, X.; Wang, B.; Zhang, Y.; Ning, J.; Xiao, Z.; Zhang, X.; Chang, Y.; Zhi, L. High-Performance Silicon Battery Anodes Enabled by Engineering Graphene Assemblies. *Nano Lett.* **2015**, *15*, 6222–6228. [CrossRef] [PubMed]
95. Chen, Y.; Zhu, J.; Qu, B.; Lu, B.; Xu, Z. Graphene improving lithium-ion battery performance by construction of $NiCo_2O_4$/graphene hybrid nanosheet arrays. *Nano Energy* **2014**, *3*, 88–94. [CrossRef]
96. Fei, L.; Lin, Q.; Yuan, B.; Chen, G.; Xie, P.; Li, Y.; Xu, Y.; Deng, S.; Smirnov, S.; Luo, H. Reduced Graphene Oxide Wrapped FeS Nanocomposite for Lithium-Ion Battery Anode with Improved Performance. *ACS Appl. Mater. Interfaces* **2013**, *5*, 5330–5335. [CrossRef] [PubMed]
97. McNulty, D.; Carroll, E.; O'Dwyer, C. Rutile TiO_2 Inverse Opal Anodes for Li-Ion Batteries with Long Cycle Life, High-Rate Capability, and High Structural Stability. *Adv. Energy Mater.* **2017**, *7*, 1602291. [CrossRef]

98. Yang, S.; Liang, Q.; Wu, H.; Pi, J.; Wang, Z.; Luo, Y.; Liu, Y.; Long, Z.; Zhou, D.; Wen, Y.; et al. Lead-Free Double Perovskite $Cs_2NaErCl_6$: Li^+ as High-Stability Anodes for Li-Ion Batteries. *J. Phys. Chem. Lett.* **2022**, *13*, 4981–4987. [CrossRef] [PubMed]
99. Liu, S.; Zhang, K.; Tan, L.; Qi, S.; Liu, G.; Chen, J.; Lou, Y. All-inorganic halide perovskite $CsPbBr_3$@CNTs composite enabling superior lithium storage performance with pseudocapacitive contribution. *Electrochim. Acta* **2021**, *367*, 137352. [CrossRef]
100. Paul, T.; Maiti, S.; Chatterjee, B.K.; Bairi, P.; Das, B.K.; Thakur, S.; Chattopadhyay, K.K. Electrochemical Performance of 3D Network $CsPbBr_3$ Perovskite Anodes for Li-Ion Batteries: Experimental Venture with Theoretical Expedition. *J. Phys. Chem. C* **2021**, *125*, 16892–16902. [CrossRef]
101. Kaisar, N.; Paul, T.; Chi, P.-W.; Su, Y.-H.; Singh, A.; Chu, C.-W.; Wu, M.-K.; Wu, P.M. Electrochemical Performance of Orthorhombic $CsPbI_3$ Perovskite in Li-Ion Batteries. *Materials* **2021**, *14*, 5718. [CrossRef] [PubMed]
102. Zhang, J.; Li, C.M. Nanoporous metals: Fabrication strategies and advanced electrochemical applications in catalysis, sensing and energy systems. *Chem. Soc. Rev.* **2012**, *41*, 7016–7031. [CrossRef] [PubMed]
103. Zhao, Y.; Liu, B.; Pan, L.; Yu, G. 3D nanostructured conductive polymer hydrogels for high-performance electrochemical devices. *Energy Environ. Sci.* **2013**, *6*, 2856–2870. [CrossRef]
104. Huang, X.; Chen, J.; Lu, Z.; Yu, H.; Yan, Q.; Hng, H.H. Carbon inverse opal entrapped with electrode active nanoparticles as high-performance anode for lithium-ion batteries. *Sci. Rep.* **2013**, *3*, 2317. [CrossRef]
105. Wu, H.; Pi, J.; Liu, Q.; Liang, Q.; Qiu, J.; Guo, J.; Long, Z.; Zhou, D.; Wang, Q. All-Inorganic Lead Free Double Perovskite Li-Battery Anode Material Hosting High Li^+ Ion Concentrations. *J. Phys. Chem. Lett.* **2021**, *12*, 4125–4129. [CrossRef] [PubMed]
106. Pal, P.; Ghosh, A. Three-Dimensional Perovskite Anode for Quasi-Solid-State-Ion and Dual-Ion Batteries: Mechanism of Conversion Process in Perovskite. *Phys. Rev. Appl.* **2020**, *14*, 064010. [CrossRef]
107. Ma, Y.; Xie, X.; Yang, W.; Yu, Z.; Sun, X.; Zhang, Y.; Yang, X.; Kimura, H.; Hou, C.; Guo, Z.; et al. *Recent Advances in Transition Metal Oxides with Different Dimensions as Electrodes for High-Performance Supercapacitors*; Springer Science and Business Media B.V.: Dordrecht, The Netherlands, 2021. [CrossRef]
108. Chen, W.; Zhang, X.; Mo, L.-E.; Zhang, Y.; Chen, S.; Zhang, X.; Hu, L. $NiCo_2S_4$ quantum dots with high redox reactivity for hybrid supercapacitors. *Chem. Eng. J.* **2020**, *388*, 124109. [CrossRef]
109. Wang, Y.; Wang, Y.; Hosono, E.; Wang, K.; Zhou, H. The design of a $LiFePO_4$/carbon nanocomposite with a core-shell structure and its synthesis by an in situ polymerization restriction method. *Angew. Chem. Int. Ed.* **2008**, *47*, 7461–7465. [CrossRef]
110. Fang, T.; Duh, J.-G.; Sheen, S.-R. Improving the Electrochemical Performance of $LiCoO_2$ Cathode by Nanocrystalline ZnO Coating. *J. Electrochem. Soc.* **2005**, *152*, A1701. [CrossRef]
111. Fu, L.J.; Liu, H.; Li, C.; Wu, Y.P.; Rahm, E.; Holze, R.; Wu, H.Q. *Surface Modifications of Electrode Materials for Lithium Ion Batteries*; Elsevier Masson SAS: Amsterdam, The Netherlands, 2006. [CrossRef]
112. Wang, B.; Luo, B.; Li, X.; Zhi, L. Graphene-Inorganic Composites as Electrode Materials for Lithium-Ion Batteries. In *Chemical Synthesis and Applications of Graphene and Carbon Materials*; Wiley: Hoboken, NJ, USA, 2016; pp. 217–249. [CrossRef]
113. Yang, S.; Feng, X.; Ivanovici, S.; Müllen, K. Fabrication of Graphene-Encapsulated Oxide Nanoparticles: Towards High-Performance Anode Materials for Lithium Storage. *Angew. Chem.* **2010**, *122*, 8586–8589. [CrossRef]
114. Yang, S.; Cui, G.; Pang, S.; Cao, Q.; Kolb, U.; Feng, X.; Maier, J.; Mllen, K. Fabrication of cobalt and cobalt oxide/graphene composites: Towards high-performance anode materials for lithium ion batteries. *ChemSusChem* **2010**, *3*, 236–239. [CrossRef] [PubMed]
115. Kim, H.; Seo, D.H.; Kim, S.W.; Kim, J.; Kang, K. Highly reversible Co_3O_4/graphene hybrid anode for lithium rechargeable batteries. *Carbon* **2011**, *49*, 326–332. [CrossRef]
116. Xia, F.; Hu, X.; Sun, Y.; Luo, W.; Huang, Y. Layer-by-layer assembled MoO_2-graphene thin film as a high-capacity and binder-free anode for lithium-ion batteries. *Nanoscale* **2012**, *4*, 4707–4711. [CrossRef]
117. Mai, Y.J.; Tu, J.P.; Gu, C.D.; Wang, X.L. Graphene anchored with nickel nanoparticles as a high-performance anode material for lithium ion batteries. *J. Power Sources* **2012**, *209*, 1–6. [CrossRef]
118. Tao, H.C.; Fan, L.Z.; Mei, Y.; Qu, X. Self-supporting Si/Reduced Graphene Oxide nanocomposite films as anode for lithium ion batteries. *Electrochem. Commun.* **2011**, *13*, 1332–1335. [CrossRef]
119. Cheng, J.; Du, J. Facile synthesis of germanium-graphene nanocomposites and their application as anode materials for lithium ion batteries. *CrystEngComm* **2012**, *14*, 397–400. [CrossRef]
120. Zhu, X.; Zhu, Y.; Murali, S.; Stoller, M.D.; Ruoff, R.S. Nanostructured reduced graphene oxide/Fe_2O_3 composite as a high-performance anode material for lithium ion batteries. *ACS Nano* **2011**, *5*, 3333–3338. [CrossRef] [PubMed]
121. Liu, R.; Cho, S.I.; Lee, S.B. Poly(3,4-ethylenedioxythiophene) nanotubes as electrode materials for a high-powered supercapacitor. *Nanotechnology* **2008**, *19*, 215710. [CrossRef] [PubMed]
122. Nam, K.T.; Kim, D.-W.; Yoo, P.J.; Chiang, C.-Y.; Meethong, N.; Hammond, P.T.; Chiang, Y.-M.; Belcher, A.M. Virus-Enabled Synthesis and Assembly of Nanowires for Lithium Ion Battery Electrodes. *Science* **2006**, *312*, 885–888. [CrossRef] [PubMed]
123. Park, M.; Wang, G.; Kang, Y.; Wexler, D.; Dou, S.; Liu, H. Preparation and Electrochemical Properties of SnO_2 Nanowires for Application in Lithium-Ion Batteries. *Angew. Chem. Int. Ed.* **2007**, *46*, 750–753. [CrossRef]

124. Sides, C.R.; Croce, F.; Young, V.Y.; Martin, C.R.; Scrosati, B. A High-Rate, Nanocomposite $LiFePO_4$/Carbon Cathode. *Electrochem. Solid-State Lett.* **2005**, *8*, A484. [CrossRef]
125. Cho, S.I.; Lee, S.B. Fast Electrochemistry of Conductive Polymer Nanotubes: Synthesis, Mechanism, and Application. *ACC Chem. Res.* **2008**, *41*, 699–707. [CrossRef]
126. Kim, D.K.; Muralidharan, P.; Lee, H.-W.; Ruffo, R.; Yang, Y.; Chan, C.K.; Peng, H.; Huggins, R.A.; Cui, Y. Spinel $LiMn_2O_4$ Nanorods as Lithium Ion Battery Cathodes. *Nano Lett.* **2008**, *8*, 3948–3952. [CrossRef] [PubMed]
127. Yan, J.; Khoo, E.; Sumboja, A.; Lee, P.S. Facile Coating of Manganese Oxide on Tin Oxide Nanowires with High-Performance Capacitive Behavior. *ACS Nano* **2010**, *4*, 4247–4255. [CrossRef] [PubMed]
128. Li, J.; Cui, L.; Zhang, X. Preparation and electrochemistry of one-dimensional nanostructured MnO_2/PPy composite for electrochemical capacitor. *Appl. Surf. Sci.* **2010**, *256*, 4339–4343. [CrossRef]
129. Li, X.; Cheng, F.; Guo, B.; Chen, J. Template-Synthesized $LiCoO_2$, $LiMn_2O_4$, and $LiNi_{0.8}Co_{0.2}O_2$ Nanotubes as the Cathode Materials of Lithium Ion Batteries. *J. Phys. Chem. B* **2005**, *109*, 14017–14024. [CrossRef]
130. Wei, Q.; An, Q.; Chen, D.; Mai, L.; Chen, S.; Zhao, Y.; Hercule, K.M.; Xu, L.; Minhas-Khan, A.; Zhang, Q. One-Pot Synthesized Bicontinuous Hierarchical $Li_3V_2(PO_4)_3$ /C Mesoporous Nanowires for High-Rate and Ultralong-Life Lithium-ion Batteries. *Nano Lett.* **2014**, *14*, 1042–1048. [CrossRef]
131. Zhu, C.; Yu, Y.; Gu, L.; Weichert, K.; Maier, J. Electrospinning of Highly Electroactive Carbon-Coated Single-Crystalline $LiFePO_4$ Nanowires. *Angew. Chem. Int. Ed.* **2011**, *50*, 6278–6282. [CrossRef] [PubMed]
132. Wang, H.; Yang, Y.; Liang, Y.; Cui, L.-F.; Sanchez Casalongue, H.; Li, Y.; Hong, G.; Cui, Y.; Dai, H. $LiMn_1$-xFexPO$_4$ Nanorods Grown on Graphene Sheets for Ultra-High Rate Performance Lithium Ion Batteries. *arXiv* **2011**, arXiv:1107.0111. [CrossRef]
133. Trinh, D.V.; Nguyen, M.T.T.; Dang, H.T.M.; Dang, D.T.; Le, H.T.T.; Le, H.T.N.; Tran, H.V.; Huynh, C.D. Hydrothermally synthesized nanostructured $LiMn_xFe1-xPO_4(x = 0 - 0.3)$ cathode materials with enhanced properties for lithium-ion batteries. *Sci. Rep.* **2021**, *11*, 12280. [CrossRef] [PubMed]
134. Qiao, Y.Q.; Tu, J.P.; Mai, Y.J.; Cheng, L.J.; Wang, X.L.; Gu, C.D. Enhanced electrochemical performances of multi-walled carbon nanotubes modified $Li_3V_2(PO_4)_3$/C cathode material for lithium-ion batteries. *J. Alloys Compd.* **2011**, *509*, 7181–7185. [CrossRef]
135. Su, X.; Wu, Q.; Li, J.; Xiao, X.; Lott, A.; Lu, W.; Sheldon, B.W.; Wu, J. Silicon-Based Nanomaterials for Lithium-Ion Batteries: A Review. *Adv. Energy Mater.* **2014**, *4*, 1300882. [CrossRef]
136. Cui, L.-F.; Ruffo, R.; Chan, C.K.; Peng, H.; Cui, Y. Crystalline-Amorphous Core−Shell Silicon Nanowires for High Capacity and High Current Battery Electrodes. *Nano Lett.* **2009**, *9*, 491–495. [CrossRef] [PubMed]
137. Wu, H.; Chan, G.; Choi, J.W.; Ryu, I.; Yao, Y.; McDowell, M.T.; Lee, S.W.; Jackson, A.; Yang, Y.; Hu, L.; et al. Stable cycling of double-walled silicon nanotube battery anodes through solid–electrolyte interphase control. *Nat. Nanotechnol.* **2012**, *7*, 310–315. [CrossRef] [PubMed]
138. Naguib, M.; Mashtalir, O.; Carle, J.; Presser, V.; Lu, J.; Hultman, L.; Gogotsi, Y.; Barsoum, M.W. Two-Dimensional Transition Metal Carbides. *ACS Nano* **2012**, *6*, 1322–1331. [CrossRef] [PubMed]
139. Eames, C.; Islam, M.S. Ion Intercalation into Two-Dimensional Transition-Metal Carbides: Global Screening for New High-Capacity Battery Materials. *J. Am. Chem. Soc.* **2014**, *136*, 16270–16276. [CrossRef]
140. Naguib, M.; Mochalin, V.N.; Barsoum, M.W.; Gogotsi, Y. 25th Anniversary Article: MXenes: A New Family of Two-Dimensional Materials. *Adv. Mater.* **2014**, *26*, 992–1005. [CrossRef] [PubMed]
141. Naguib, M.; Come, J.; Dyatkin, B.; Presser, V.; Taberna, P.-L.; Simon, P.; Barsoum, M.W.; Gogotsi, Y. MXene: A promising transition metal carbide anode for lithium-ion batteries. *Electrochem. Commun.* **2012**, *16*, 61–64. [CrossRef]
142. Luo, J.; Tao, X.; Zhang, J.; Xia, Y.; Huang, H.; Zhang, L.; Gan, Y.; Liang, C.; Zhang, W. Sn^{4+} Ion Decorated Highly Conductive Ti_3C_2 MXene: Promising Lithium-Ion Anodes with Enhanced Volumetric Capacity and Cyclic Performance. *ACS Nano* **2016**, *10*, 2491–2499. [CrossRef] [PubMed]
143. Luo, J.; Tao, X.; Zhang, J.; Xia, Y.; Huang, H.; Zhang, L.; Gan, Y.; Liang, C.; Zhang, W. Ultra-High-Rate Pseudocapacitive Energy Storage in Two-Dimensional Transition Metal Carbides. In *MXenes*; Jenny Stanford Publishing: New York, NY, USA, 2023; pp. 723–743. [CrossRef]
144. Novoselov, K.S.; Geim, A.K.; Morozov, S.V.; Jiang, D.; Zhang, Y.; Dubonos, S.V.; Grigorieva, I.V.; Firsov, A.A. Electric Field Effect in Atomically Thin Carbon Films. *Science* **2004**, *306*, 666–669. [CrossRef]
145. Chabot, V.; Higgins, D.; Yu, A.; Xiao, X.; Chen, Z.; Zhang, J. A review of graphene and graphene oxide sponge: Material synthesis and applications to energy and the environment. *Energy Environ. Sci.* **2014**, *7*, 1565–1596. [CrossRef]
146. Wei, D.; Haque, S.; Andrew, P.; Kivioja, J.; Ryhänen, T.; Pesquera, A.; Centeno, A.; Alonso, B.; Chuvilin, A.; Zurutuza, A.; et al. Ultrathin rechargeable all-solid-state batteries based on monolayer graphene. *J. Mater. Chem. A Mater.* **2013**, *1*, 3177. [CrossRef]
147. Kim, H.; Park, K.-Y.; Hong, J.; Kang, K. All-graphene-battery: Bridging the gap between supercapacitors and lithium ion batteries. *Sci. Rep.* **2014**, *4*, 5278. [CrossRef] [PubMed]
148. Shu, K.; Wang, C.; Wang, M.; Zhao, C.; Wallace, G.G. Graphene cryogel papers with enhanced mechanical strength for high performance lithium battery anodes. *J. Mater. Chem. A* **2014**, *2*, 1325–1331. [CrossRef]

149. Cohn, A.P.; Oakes, L.; Carter, R.; Chatterjee, S.; Westover, A.S.; Share, K.; Pint, C.L. Assessing the improved performance of freestanding, flexible graphene and carbon nanotube hybrid foams for lithium ion battery anodes. *Nanoscale* **2014**, *6*, 4669–4675. [CrossRef] [PubMed]
150. Hassoun, J.; Bonaccorso, F.; Agostini, M.; Angelucci, M.; Betti, M.G.; Cingolani, R.; Gemmi, M.; Mariani, C.; Panero, S.; Pellegrini, V.; et al. An Advanced Lithium-Ion Battery Based on a Graphene Anode and a Lithium Iron Phosphate Cathode. *Nano Lett.* **2014**, *14*, 4901–4906. [CrossRef]
151. Liu, M.; Yan, C.; Zhang, Y. Fabrication of Nb_2O_5 Nanosheets for High-rate Lithium Ion Storage Applications. *Sci. Rep.* **2015**, *5*, 8326. [CrossRef]
152. Xiao, J.; Choi, D.; Cosimbescu, L.; Koech, P.; Liu, J.; Lemmon, J.P. Exfoliated MoS_2 Nanocomposite as an Anode Material for Lithium Ion Batteries. *Chem. Mater.* **2010**, *22*, 4522–4524. [CrossRef]
153. Qu, B.; Ma, C.; Ji, G.; Xu, C.; Xu, J.; Meng, Y.S.; Wang, T.; Lee, J.Y. Layered SnS_2-Reduced Graphene Oxide Composite—A High-Capacity, High-Rate, and Long-Cycle Life Sodium-Ion Battery Anode Material. *Adv. Mater.* **2014**, *26*, 3854–3859. [CrossRef] [PubMed]
154. Zhang, Y.; Zhu, P.; Huang, L.; Xie, J.; Zhang, S.; Cao, G.; Zhao, X. Few-Layered SnS_2 on Few-Layered Reduced Graphene Oxide as Na-Ion Battery Anode with Ultralong Cycle Life and Superior Rate Capability. *Adv. Funct. Mater.* **2015**, *25*, 481–489. [CrossRef]
155. Prikhodchenko, P.V.; Yu, D.Y.W.; Batabyal, S.K.; Uvarov, V.; Gun, J.; Sladkevich, S.; Mikhaylov, A.A.; Medvedev, A.G.; Lev, O. Nanocrystalline tin disulfide coating of reduced graphene oxide produced by the peroxostannate deposition route for sodium ion battery anodes. *J. Mater. Chem. A Mater.* **2014**, *2*, 8431. [CrossRef]
156. Zhu, F.-F.; Chen, W.-J.; Xu, Y.; Gao, C.-L.; Guan, D.-D.; Liu, C.-H.; Qian, D.; Zhang, S.-C.; Jia, J.-F. Epitaxial growth of two-dimensional stanene. *Nat. Mater.* **2015**, *14*, 1020–1025. [CrossRef] [PubMed]
157. Li, W.; Yang, Y.; Zhang, G.; Zhang, Y.-W. Ultrafast and Directional Diffusion of Lithium in Phosphorene for High-Performance Lithium-Ion Battery. *Nano Lett.* **2015**, *15*, 1691–1697. [CrossRef]
158. Mannix, A.J.; Zhou, X.-F.; Kiraly, B.; Wood, J.D.; Alducin, D.; Myers, B.D.; Liu, X.; Fisher, B.L.; Santiago, U.; Guest, J.R.; et al. Synthesis of borophenes: Anisotropic, two-dimensional boron polymorphs. *Science* **2015**, *350*, 1513–1516. [CrossRef]
159. Tritsaris, G.A.; Kaxiras, E.; Meng, S.; Wang, E. Adsorption and Diffusion of Lithium on Layered Silicon for Li-Ion Storage. *Nano Lett.* **2013**, *13*, 2258–2263. [CrossRef] [PubMed]
160. Jiang, H.R.; Lu, Z.; Wu, M.C.; Ciucci, F.; Zhao, T.S. Borophene: A promising anode material offering high specific capacity and high rate capability for lithium-ion batteries. *Nano Energy* **2016**, *23*, 97–104. [CrossRef]
161. Mahmoud, Z.H.; Ajaj, Y.; Ghadir, G.K.; Al-Tmimi, H.M.; Jasim, H.H.; Al-Salih, M.; Alubiady, M.H.S.; Al-Ani, A.M.; Jumaa, S.S.; Azat, S.; et al. Carbon-doped titanium dioxide (TiO_2) as Li-ion battery electrode: Synthesis, characterization, and performance. *Results Chem.* **2024**, *7*, 101422. [CrossRef]
162. Stein, A.; Wilson, B.E.; Rudisill, S.G. Design and functionality of colloidal–crystal–templated materials—Chemical applications of inverse opals. *Chem. Soc. Rev.* **2013**, *42*, 2763–2803. [CrossRef]
163. Xu, J.; Wang, X.; Wang, X.; Chen, D.; Chen, X.; Li, D.; Shen, G. Three-Dimensional Structural Engineering for Energy-Storage Devices: From Microscope to Macroscope. *ChemElectroChem* **2014**, *1*, 975–1002. [CrossRef]
164. Peng, C.; Chen, B.; Qin, Y.; Yang, S.; Li, C.; Zuo, Y.; Liu, S.; Yang, J. Facile ultrasonic synthesis of coo quantum dot/graphene nanosheet composites with high lithium storage capacity. *ACS Nano* **2012**, *6*, 1074–1081. [CrossRef] [PubMed]
165. Wang, X.; Wu, X.-L.; Guo, Y.-G.; Zhong, Y.; Cao, X.; Ma, Y.; Yao, J. Synthesis and lithium storage properties of Co_3O_4 nanosheet-assembled multishelled hollow spheres. *Adv. Funct. Mater.* **2010**, *20*, 1680–1686. [CrossRef]
166. Wang, J.; Yang, N.; Tang, H.; Dong, Z.; Jin, Q.; Yang, M.; Kisailus, D.; Zhao, H.; Tang, Z.; Wang, D. Accurate control of multishelled Co_3O_4 hollow microspheres as high-performance anode materials in lithium-ion batteries. *Angew. Chem. Int. Ed.* **2013**, *52*, 6417–6420. [CrossRef]
167. Liu, Z.; Mi, J.; Yang, Y.; Li, J.; Tan, X. Synthesis, characterization and electrochemical properties of three-dimensionally ordered macroporous α-Fe_2O_3. *Mater. Sci. Eng. B* **2012**, *177*, 1612–1617. [CrossRef]
168. Kasavajjula, U.; Wang, C.; Appleby, A.J. Nano- and bulk-silicon-based insertion anodes for lithium-ion secondary cells. *J. Power Sources* **2007**, *163*, 1003–1039. [CrossRef]
169. Park, M.-H.; Kim, M.G.; Joo, J.; Kim, K.; Kim, J.; Ahn, S.; Cui, Y.; Cho, J. Silicon nanotube battery anodes. *Nano Lett.* **2009**, *9*, 3844–3847. [CrossRef] [PubMed]
170. Zhang, T.; Fu, L.; Gao, J.; Yang, L.; Wu, Y.; Wu, H. Core-shell Si/C nanocomposite as anode material for lithium ion batteries. *Pure Appl. Chem.* **2006**, *78*, 1889–1896. [CrossRef]
171. Jin, Y.; Munakata, H.; Okada, N.; Kanamura, K. Design and evaluation of a three dimensionally ordered macroporous structure within a highly patterned cylindrical Sn-Ni electrode for advanced lithium ion batteries. *J. Nanomater.* **2013**, *2013*, 937019. [CrossRef]
172. Kim, D.; Suk, J.; Kim, D.W.; Kang, Y.; Im, S.H.; Yang, Y.; Park, O.O. Electrochemically grown three-dimensional porous Si@Ni inverse opal structure for higherperformance Li ion battery anode. *J. Mater. Chem. A* **2014**, *2*, 6396–6401. [CrossRef]

173. Wang, Z.L.; Xu, D.; Wang, H.G.; Wu, Z.; Zhang, X.B. In situ fabrication of porous graphene electrodes for high-performance energy storage. *ACS Nano* **2013**, *7*, 2422–2430. [CrossRef] [PubMed]
174. Lou, S.; Cheng, X.; Zhao, Y.; Lushington, A.; Gao, J.; Li, Q.; Zuo, P.; Wang, B.; Gao, Y.; Ma, Y.; et al. Superior performance of ordered macroporous $TiNb_2O_7$ anodes for lithium ion batteries: Understanding from the structural and pseudocapacitive insights on achieving high rate capability. *Nano Energy* **2017**, *34*, 15–25. [CrossRef]
175. Deng, Z.; Jiang, H.; Hu, Y.; Liu, Y.; Zhang, L.; Liu, H.; Li, C. 3D Ordered Macroporous MoS_2@C Nanostructure for Flexible Li-Ion Batteries. *Adv. Mater.* **2017**, *29*, 1603020. [CrossRef] [PubMed]
176. Liu, X.; Zhao, J.; Hao, J.; Su, B.L.; Li, Y. 3D ordered macroporous germanium fabricated by electrodeposition from an ionic liquid and its lithium storage properties. *J. Mater. Chem. A Mater.* **2013**, *1*, 15076–15081. [CrossRef]
177. Song, T.; Jeon, Y.; Samal, M.; Han, H.; Park, H.; Ha, J.; Yi, D.K.; Choi, J.-M.; Chang, H.; Choi, Y.-M.; et al. A Ge inverse opal with porous walls as an anode for lithium ion batteries. *Energy Environ. Sci.* **2012**, *5*, 9028–9033. [CrossRef]
178. Choi, H.; Takahashi, D.; Kono, K.; Kim, E. Evidence of supersolidity in rotating solid helium. *Science* **2010**, *330*, 1512–1515. [CrossRef] [PubMed]
179. Zhao, N.; Wang, G.; Huang, Y.; Wang, B.; Yao, B.; Wu, Y. Preparation of nanowire arrays of amorphous carbon nanotube-coated single crystal SnO_2. *Chem. Mater.* **2008**, *20*, 2612–2614. [CrossRef]
180. Zhao, N.H.; Yang, L.C.; Zhang, P.; Wang, G.J.; Wang, B.; Yao, B.D.; Wu, Y.P. Polycrystalline SnO_2 nanowires coated with amorphous carbon nanotube as anode material for lithium ion batteries. *Mater. Lett.* **2010**, *64*, 972–975. [CrossRef]
181. Liu, H.; Li, C.; Zhang, H.P.; Fu, L.J.; Wu, Y.P.; Wu, H.Q. Kinetic study on $LiFePO_4$/C nanocomposites synthesized by solid state technique. *J. Power Sources* **2006**, *159*, 717–720. [CrossRef]
182. Kong, F.; Longo, R.C.; Park, M.-S.; Yoon, J.; Yeon, D.-H.; Park, J.-H.; Wang, W.; KC, S.; Doo, S.-G.; Cho, K. Ab initio study of doping effects on $LiMnO_2$ and Li_2MnO_3 cathode materials for Li-ion batteries. *J. Mater. Chem. A* **2015**, *3*, 8489–8500. [CrossRef]
183. Shouji, E.; Buttry, D.A. New Organic-Inorganic Nanocomposite Materials for Energy Storage Applications. *Langmuir* **1999**, *15*, 669–673. [CrossRef]
184. Zhuang, G.V.; Yang, H.; Blizanac, B.; Ross, P.N., Jr. A Study of Electrochemical Reduction of Ethylene and Propylene Carbonate Electrolytes on Graphite Using ATR-FTIR Spectroscopy. *Electrochem. Solid-State Lett.* **2005**, *8*, A441. [CrossRef]
185. Wang, Q.; Wen, Z.; Li, J. A hybrid supercapacitor fabricated with a carbon nanotube cathode and a TiO_2-B nanowire anode. *Adv. Funct. Mater.* **2006**, *16*, 2141–2146. [CrossRef]
186. Ke, Q.; Wang, J. Graphene-based materials for supercapacitor electrodes—A review. *J. Mater.* **2016**, *2*, 37–54. [CrossRef]
187. Yadav, S.; Daniel, S. *Green Synthesis of Zero-Dimensional Carbon Nanostructures in Energy Storage Applications—A Review*; John Wiley and Sons Inc.: Hoboken, NJ, USA, 2024. [CrossRef]
188. Geim, A.K.; Grigorieva, I.V. Van der Waals heterostructures. *Nature* **2013**, *499*, 419–425. [CrossRef]
189. Wang, Z.L. Zinc oxide nanostructures: Growth, properties and applications. *J. Phys. Condens. Matter* **2004**, *16*, R829. [CrossRef]
190. Rolison, D.R.; Long, J.W.; Lytle, J.C.; Fischer, A.E.; Rhodes, C.P.; McEvoy, T.M.; Bour, M.E.; Lubers, A.M. Multifunctional 3D nanoarchitectures for energy storage and conversion. *Chem. Soc. Rev.* **2009**, *38*, 226–252. [CrossRef] [PubMed]

inorganics MDPI

Review

Recent Advances and Challenges in Hybrid Supercapacitors Based on Metal Oxides and Carbons

Lili Gao [1,*], Fuyuan Liu [1], Jiaxing Qi [2,3], Wenyue Gao [4] and Guobao Xu [2,3,*]

1 School of Materials Science and Engineering, Shenyang Jianzhu University, Shenyang 110168, China; fuyuanliu8895@163.com
2 State Key Laboratory of Electroanalytical Chemistry, Changchun Institute of Applied Chemistry, Chinese Academy of Sciences, Changchun 130022, China; jiaxingq@ciac.ac.cn
3 School of Applied Chemistry and Engineering, University of Science and Technology of China, Hefei 230026, China
4 Institute of Marine Science and Technology, Shandong University, Qingdao 266237, China; gaowy@email.sdu.edu.cn
* Correspondence: liligao@sjzu.edu.cn (L.G.); guobaoxu@ciac.ac.cn (G.X.)

Abstract: Hybrid supercapacitors (HSCs) are a novel type of supercapacitor composed of battery-type electrodes and capacitor-type electrodes, which have directly transformed the global energy landscape. On one hand, they can replace clean energy sources that are heavily dependent on climatic conditions in specific regions, thereby enhancing the effective utilization of intermittent energy sources. On the other hand, with their high energy density akin to secondary batteries and the long lifespan and high power density characteristic of supercapacitors, they perfectly bridge the gap between secondary batteries and supercapacitors. This article reviews the fundamental energy storage principles of HSCs and highlights the latest optimization strategies for HSCs based on transition metal oxides (TMOs) and carbon over the past two years. These strategies include heteroatom doping, heterostructured materials, nanocomposites, and metal–organic frameworks (MOF). Finally, prospects on future research directions of HSCs are discussed.

Keywords: hybrid supercapacitor; metal oxides; carbon-based electrodes

1. Introduction

The rapid development of society and the swift growth of the global population have led to a continuous increase in energy demand. Environmental pollution, ecological destruction, and energy shortages have become immense challenges facing human development today [1,2]. By 2050, the energy demand is projected to be more than double the current levels [3]. Currently, 80% of global energy supply is derived from non-renewable fossil fuels, including coal, oil, and natural gas. The use of these fossil fuels not only produces harmful and toxic gasses such as SO_2 and NO but also takes millions of years or even longer to regenerate [4,5]. Renewable energy sources like solar and wind power are intermittent and their power generation conditions vary, rendering them unsuitable for direct energy supply in power grids [6,7]. Therefore, converting other energy sources into electricity and supplying it to different energy-demanding applications has become an inevitable trend in energy development [8,9]. Searching for new, renewable energy sources that can replace traditional fossil fuels, and developing novel energy storage devices with high power and energy density has become one of the crucial challenges that humanity must address in the future for the storage and transportation of electrical energy [10,11].

Based on the current status of energy storage devices, secondary batteries possess a considerable energy density that can meet the pursuit of energy [12]. However, the power density of conventional batteries is relatively low, failing to meet the requirements of applications that necessitate high power density devices [13]. While traditional capacitors have a notable power density, their energy density is very low, making it challenging for them to become the energy storage devices needed for the future [14]. Encouragingly, the recently emerged hybrid ion capacitors represent a new type of supercapacitor that has directly changed the global energy landscape. On one hand, they can replace clean energy sources that are heavily dependent on weather conditions in specific regions, thereby enhancing the effective utilization of intermittent energy sources. On the other hand, with the high energy density of secondary batteries and the long lifespan and high power density of supercapacitors, they perfectly bridge the gap between the two [15–17]. Furthermore, to cater to the growing human living demands, HSCs hold immense industrial potential in applications such as electric vehicles, renewable energy sources (solar and wind power), wearable devices, and electronic products [18–22]. In applications such as new energy vehicles, large-scale medical devices, and heavy-duty machinery, energy storage devices with superior low-temperature performance and excellent instantaneous power characteristics are required during rapid startups. However, the impact of instantaneous high-current discharge, particularly in outdoor environments at -20 °C in northern regions, can significantly shorten the lifespan of storage batteries. During battery replacement in charging stations and commercial payment terminals, it is necessary to supplement the backup power supply to ensure data protection in the event of power failure, and the HSC batteries are well-suited to meet these demands. Therefore, designing and preparing high-performance electrode materials is key to improving the performance of hybrid ion capacitors, with common materials including conductive polymers [23], carbon materials [24], metal oxides [25], and metal sulfides [26]. Among these, transition metal oxides (TMOs), such as nickel oxide [27], iron oxide [28], cobalt oxide [29], zinc oxide [30], and manganese dioxide [31], are widely used as battery-type electrode materials due to their variable oxidation states, environmental friendliness, high specific capacity, and long lifespan. However, poor conductivity and slow charge transfer have limited the practical applications of TMOs. Furthermore, carbon materials generally exhibit behaviors related to pure non-Faradaic reactions, akin to Electric Double Layer Capacitors (EDLC) or pseudocapacitance, and are commonly used as electrodes in hybrid ion capacitors. However, the performance of carbon-based electrode materials is less than ideal, with their energy density, particularly the volumetric energy density, being low. This hinders the simultaneous achievement of high energy and power densities, severely limiting their practical applications in the field of energy storage.

In recent years, due to the unresolved economic aspects of HSC production and the enhancement of energy density/power density/specific capacity/safety, fastcharge–discharge rates, excellent cycling stability, and high rate capability have garnered significant attention from researchers to enhance capacitance and energy density. To address these issues and to assist a broad and interdisciplinary readership in deeper research within this field, this paper reviews the energy storage principles of hybrid supercapacitors, with a focus on HSCs based on TMOs and carbon. It highlights the latest optimization strategies, including heteroatom doping, heterostructures, nanocomposites, metal–organic frameworks, and electrolytes. Finally, the paper discusses and prospects future research directions and possibilities in this field.

2. Composition and Energy Storage Mechanism of HSC

HSC are primarily composed of current collectors, electrolytes, separators, and electrodes. The current collectors not only serve as a repository for electrons but also act as a substrate that allows for the uniform distribution of the electrode material on their surface. The electrolyte, separator, and electrodes are the core components that directly determine the performance of the device. The electrolyte, positioned between the anode and cathode, facilitates the transport and conduction of ions [32–34]. Typically, electrolytes should possess chemical and electrochemical stability, sufficient ion conductivity, and good thermodynamic stability. Therefore, electrolytes are pivotal in enhancing the electrochemical performance of HSC, directly influencing their energy density, power density, specific capacity, and safety. The impact of electrolytes on the performance of HSC under various conditions has been extensively studied and reviewed by researchers [35]. Electrolytes are generally categorized into four types: aqueous electrolytes, organic electrolytes, solid or quasi-solid electrolytes, and ionic liquid electrolytes [36]. Among these, aqueous electrolytes are widely used due to their low cost, simple preparation methods, high ionic conductivity, and superior safety. However, the electrochemical stability window of aqueous electrolytes is limited by the decomposition voltage of water [37]. Organic electrolytes offer a wider electrochemical stability window (>3.0 V), but their practical application is constrained by flammability, high cost, and low ionic conductivity [38]. Compared to organic electrolytes, ionic liquid electrolytes not only provide a comparable wide electrochemical stability window but are also safer and more environmentally friendly. Nevertheless, the high viscosity of ionic liquids often results in lower ionic conductivity. Additionally, both ionic liquids and organic electrolytes require an absolutely anhydrous environment and involve complex purification and drying procedures that increase costs. The large-scale application of these electrolytes still faces significant challenges. On the other hand, hydrogel electrolytes exhibit superior mechanical properties and perform well at lower temperatures. In the realm of solid electrolytes, gel polymer electrolytes, which exist in a liquid phase within a solid matrix, dominate the market due to their high ionic conductivity [39]. In these systems, a polymer framework serves as the support, with electrolyte salts and plasticizers uniformly distributed within the polymer network. Commonly used polymers include polyethylene oxide (PEO), polyvinyl alcohol (PVA), polyacrylic acid (PAA), and polyacrylamide (PAMM), among others. The separator is a film that allows only target ions to pass freely, and it must have stable physical and chemical properties to prevent contact between the anode and cathode. [40].

In fact, charge storage is achieved through two distinct mechanisms: the physical adsorption/desorption of ions on the surface of capacitor-type electrode materials and the battery-type Faradaic reactions in battery-type electrodes.

Carbon–based materials can be used as capacitor-type anodes or cathodes, with their energy storage mechanism primarily based on EDLC (Figure 1a). During the charge process, electrons reach the anode through the external circuit, under the influence of the applied electric field, free ions with opposite charges migrate to the surface of the carbon–based material. Then, via electrostatic interactions, an electric double layer spontaneously forms at the interface between the electrode and electrolyte, leading to electrochemical dynamic equilibrium [41]. During the discharge process, electrons and ions desorb through electrostatic interactions, and the cations and anions desorb from carbon materials [42]. In the case of zinc-ion hybrid capacitor, the energy storage mechanism of the carbon-based cathode is mainly based on EDLC. For cation adsorption/desorption, the interaction equation is as shown in Equation (1), while for anion (X^-) adsorption/desorption, the interaction equation is as shown in Equation (2) [43].

$$C + Zn^{2+} \rightleftharpoons C||Zn^{2+} \quad (1)$$

$$C + X^- \rightleftharpoons C||X^- \quad (2)$$

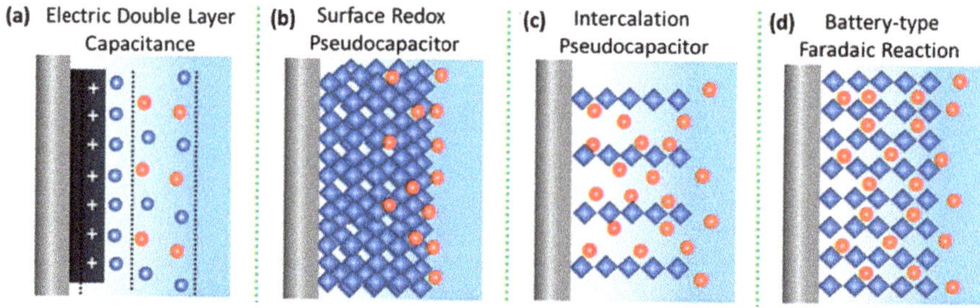

Figure 1. Energy storage mechanism of (**a**) EDLC, (**b**) surface redox pseudocapacitor, (**c**) intercalation pseudocapacitor, and (**d**) battery-type Faradaic reaction. Reproduced with permission from [41].

This is a static electrochemical energy storage process in which no electrode chemical reactions occur and no induced current is produced. During the charge–discharge process, an oriented double-layer distribution of electrons and dipoles occurs at the electrode/electrolyte interface, forming a capacitor-like structure similar to traditional plate capacitors to achieve energy storage. The charge–discharge process of EDLC is a pure non-Faradaic reaction involving simple adsorption/desorption of charges carries at the electrode/electrolyte interface, which can achieve fast charging or discharging. Therefore, it immediately responds to changes in charge storage on the electrode material surface, resulting in ideal symmetric triangular curves of the constant current charge–discharge (GCD) and rectangular cyclic voltammetry (CV) curves. Furthermore, carbon-based anodes or cathodes containing oxygen-containing functional groups or doped with heteroatoms exhibit pseudocapacitive mechanisms of surface redox reactions, which simultaneously display both Faradaic and EDLC mechanisms to achieve charge storage (Figure 1b). The GCD and CV curves are very similar to those of EDLC, being quasi-rectangular and quasi-triangular. The surface redox reaction is as shown in Equation (3) [44].

$$C–O + Zn^{2+} + 2e^- \rightleftharpoons C–O + Zn \quad (3)$$

For TMOs, in addition to the pseudocapacitive surface redox reactions mentioned earlier, they sometimes also exhibit intercalation pseudocapacitive mechanisms. A key feature of intercalation pseudocapacitance is that no phase transition occurs in the material during intercalation. Compared to EDLC, the charge storage process of intercalation pseudocapacitance involves reversible Faradaic electrochemical reactions of a large number of ions at the material–electrolyte interface. In a dual-ion HSC constructed with a Ni_3S_2/Ni_2O_3 heterostructure cathode using LiOH catholyte, two pseudocapacitive mechanisms—surface redox and intercalation—are present during discharge (Figure 1c) [45], while the charging process exhibits a different type of battery-like redox reaction (Figure 1d). Unlike the storage mechanism of surface redox pseudocapacitance, the redox process of battery-type anode/cathode materials involves phase transitions and stores more charge within the electrode material, thereby further enhancing the capacitance and energy density of the hybrid sensor [46]. Additionally, the battery-like redox process introduces buffer platforms in the GCD and forms a pair of distinct redox peaks in the CV curves.

Currently, the types of HSCs based on TMOs and carbon mainly include alkali HSCs, lithium-ion HSCs (Li-HSC), sodium-ion HSCs (Na-HSC), and zinc-ion HSCs (Zn-HSC). Alkaline HSCs are generally composed of carbon-based anodes and battery-type TMO cathodes. The alkaline electrolyte effectively inhibits the hydrogen evolution reaction (HER) at the electrode interface [47], and the TMOs undergo battery-like redox reactions with OH^- in the electrolyte to store charge. The energy storage mechanism of Li-HSC primarily includes the Faradaic reactions of insertion/extraction of lithium ions (Li^+ ions) into/from the negative electrodes, conversion and alloying reactions, and the physical adsorption/desorption of ions on the surface of the capacitor-type cathode material [48–51]. However, the intercalation-type graphite anode also faces some problems, such as sluggish Li^+ diffusion kinetics, and the development of highly crystalline, anisotropic lamellar structures under low voltage conditions, leading to the inability to achieve high capacities during fast charging [52–54]. In addition, Na-HSC is similar to Li-HSC in past reports. However, due to the larger radius of Na^+ (0.98 Å), the kinetics of anodic redox reactions are sluggish, severely affecting the electrochemical performance of Na-HSC. A typical Zn-HSC generally uses zinc metal as an anode that can accommodate Zn^{2+}, which displays highly reversible Zn^{2+} plating/stripping processes during the charge and discharge process. During the charge process, Zn^{2+} in the electrolyte migrates to the anode and deposits on the zinc anode, while metallic zinc releases electrons to form Zn^{2+} ions during the discharge process. The half-reaction for this reversible redox process is presented in the following Equation (4) [55]:

$$Zn^{2+} + 2e^- \rightleftharpoons Zn \quad (4)$$

Zn-HSCs typically have cathodes composed of carbon-based materials and non-carbon-based materials such as MXenes, TMOs, and conductive polymers. Zinc has a lower redox potential (−0.76 V, versus the standard hydrogen electrode), is much less reactive, and is safer to handle. Moreover, zinc has a theoretical gravimetric capacity of 820 mAh g^{-1} and a volumetric capacity of 5855 mAh cm^{-3}, which can serve as both a current collector and active material [56]. Zinc has merits of low cost, abundance in the earth, and environmental friendliness. Therefore, Zn-HSCs are not only a potential candidate for Li-HSCs and batteries but are also capable of achieving similar energy densities, higher charge–discharge rates, higher power densities, and a long cycling life. Nevertheless, Zn-HSCs also face several challenges that need to be addressed, such as the formation of zinc dendrites on the surface of the zinc anode after subsequent cycles. The eventual cracking of these dendrites can lead to dead zinc, rendering the capacitor unable to charge and discharge, thus, reducing cycle life. Overall, the aforementioned types of HSCs all have their shortcomings, and it is crucial to adopt effective optimization strategies to improve their electrical performance. The optimization strategies for the performance of these types of ion hybrid capacitors will be elaborated in detail in the subsequent sections of this paper.

3. Index of Characteristics

Several important metrics are commonly used to evaluate their electrochemical performance.

(1) Specific capacity

Specific capacity refers to the charge storage capability of the electroactive material or the entire capacitor per unit volume (or unit mass).

(2) Specific Energy

Specific energy refers to the energy that can be stored per unit volume or unit mass of the electroactive material or capacitor.

(3) Power density

Power density refers to the rated power of an electroactive material or capacitor per unit mass or volume, and it is a physical quantity used to represent the rate of discharge.

(4) Cycle Stability

In electrochemical testing, completing one cycle of the charge and discharge process constitutes a single cycle. The cycle life of an energy storage device is measured by the retention rate of its initial specific capacitance after multiple cycles of charge and discharge.

4. Performance Optimization Strategies and Applications of HSC Based on TMOs/Carbon Materials

4.1. Atomic Doping Optimization Strategy Based on TMOs/Carbon Materials

Atomic or ionic doping, as well as co-doping, have been regarded as a promising strategy to optimizing the physical and chemical properties of electrode materials. The specific location where heteroatoms are doped plays a crucial role in determining the electrochemical performance of these materials. However, excessive concentrations of heteroatom doping can disrupt the crystallinity of materials, thereby severely compromising their cycling stability. Consequently, the development of novel selective heteroatom doping strategies is imperative. Ding et al. [57] firstly demonstrated that the doping position can be adjusted with varying the heating rate. By controlling the heating rate of the urea-assisted annealing process, the doping site of nitrogen (N) atoms in the orthogonal Nb_2O_5 (t–Nb_2O_5) crystal structure (Figure 2a) was adjusted to enhance its rate capability. The study showed that rapid heating (10 °C min^{-1}) can effectively regulate the substitution of N within the t–Nb_2O_5 lattice (Figure 2b). This process not only results in a narrowing of the bandgap but also leads to an expansion of the transport layer spacing, facilitating the movement of Li^+. The N doping sites optimized (Figure 2c) at moderate heating rates (5 °C min^{-1}) can effectively enhance the migration of Li^+ and electrons. Additionally, the interstitial N formed (Figure 2d) during slower heating rates (1 °C min^{-1}) creates an electron-rich environment, reducing the diffusion barrier of Li^+ in the layer. However, an excessive amount of interstitial N can increase the steric hindrance for Li^+ diffusion, which is not favorable for the electron enrichment of the Li^+ transport layer. Furthermore, studies have demonstrated that a slower heating rate is conducive to increasing the percentage of nitrogen doping. Due to the synergistic effects of N doping at different positions, the N–doped t–Nb_2O_5 material exhibits excellent rate performance (104.6 mA h g^{-1} at 25 °C) and 70.5% capacity retention after 1000 cycles at 5 °C. Even at a high power density of 8.4 kW kg^{-1}, the constructed lithium–ion hybrid capacitor displayed a high energy density of 46.6 Wh kg^{-1}. This strategy provides a promising and feasible approach for the precise control of heteroatom doping positions.

Recently, TMOs of elements including Mn, Nb, V, Ni, Fe, Mo, Co, etc., have secured a significant portion in the fabrication of the cathode/anode in hybrid capacitors as battery–type materials. However, pure TMOs have some drawbacks when used as materials in hybrid capacitors, such as drastic volume expansion, low electronic conductivity, irreversible structural degradation, particle aggregation, and sluggish electrochemical reaction kinetics, leading to capacity decay and poor cycling stability [58]. These challenges can be mitigated by incorporating heteroatoms, ions, and conductive polymers into TMOs [59–65].

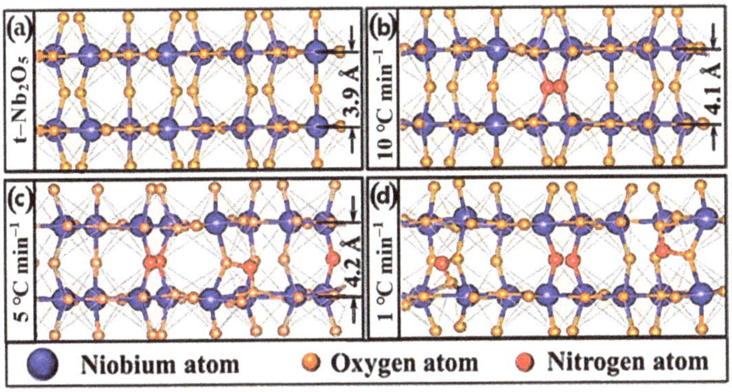

Figure 2. Schematic diagrams of N doping model of different samples, (**a**) t–Nb_2O_5, and different heating rates of (**b**) 10 °C min^{-1}, (**c**) 5 °C min^{-1}, (**d**) 1 °C min^{-1} control the doping site of N atoms in the orthogonal t-Nb_2O_5 crystal structure [57].

Manganese oxide (Mn_2O_3) is an excellent battery-type cathode material; it possesses low toxicity, environmental friendliness, cost effectiveness, superior chemical stability and electrochemical properties, a high theoretical capacity, and natural abundance. Moreover, manganese exists in various ionic states (Mn^{2+}, Mn^{3+}, Mn^{4+}, and Mn^{6+}). It can be used with several alkaline/acidic electrolytes [66–68]. Unfortunately, the high lattice energy of Mn_2O_3 nanoparticles hinders charge transport, resulting in lower charge storage capacity and conductivity. Additionally, due to phase transitions that cause continuous deformation and volume expansion at high charge–discharge rates, its electrochemical performance and durability are reduced [69,70]. Karuppaiah et al. [71] prepared neodymium (Nd)–doped α–Mn_2O_3 microspheres (MSs) with uniform morphology, high porosity, oxygen vacancies, and robust micro–nano structures using a simple hydrothermal method (Figure 3). The uniform morphology and porosity provide numerous electrochemically active sites for the insertion/deduction of OH^- ions, while Nd doping significantly increases the oxygen vacancy content in Mn_2O_3, effectively enhancing its energy density and conductivity. Furthermore, the robust micro–nano structure prevents the expansion and contraction of the active material during cycling, increasing the material's durability and lifespan. With the optimal doping condition (Nd5%–Mn_2O_3 3D–MSs), it exhibits a high specific capacitance of 862.14 F g^{-1} (431.07 C g^{-1}) at 0.5 A g^{-1} and an excellent cycle retention rate of 97.30% after 2000 cycles. Using Nd5%–Mn_2O_3 3D–MSs as a battery-type cathode and activated carbon (AC) as a capacitor-type anode, the constructed hybrid capacitor has a maximum energy density of 32.26 Wh kg^{-1} at a power density of 800 W kg^{-1}, with only a 4.56% capacity loss after 10,000 cycles, demonstrating good cycle retention performance.

The metal ion pre-intercalation doping strategy involves inserting metal cations into the layered structure of TMOs, which can effectively expand the interlayer spacing of TMOs and improve the storage performance and stability of the metal oxides [72,73]. Unfortunately, TMOs doped with single metal ions still have drawbacks such as slow electrochemical kinetics and poor conductivity. Recent studies have found that further inserting conductive polymers into TMOs based on the single metal ion doping strategy can enhance the electrochemical performance of the TMOs electrode materials. Xue et al. [74] prepared Zn$x$$MnO_2$/PPy (ZMOP) electrode materials using a hydrothermal method. The doping of Zn^{2+} and the addition of polypyrrole (PPy) enhanced the structural stability and conductivity of MnO_2. The prepared ZMOP cathode exhibited excellent specific capacity (156.4 mAh g^{-1} at 0.1 A g^{-1}) and outstanding cycle performance (82.6% retention after 5000 cycles at 0.2 A g^{-1}). Moreover, a flexible Zn-HSC was assembled using ZMOP nanowires as the cathode and PPy-derived porous carbon nanotubes (PCNTs) as the anode materials, which showed good cycle stability. The insertion–extraction processes on the ZMOP nanowire cathode and the adsorption–desorption processes on the PCNT anode both demonstrated excellent electrochemical performance, achieving a high capacitance of 109 F g^{-1} at 0.1 A g^{-1}. Meanwhile, the corresponding energy density reached 20 Wh kg^{-1} at a power density of 867 W kg^{-1}. After 5000 cycles at 0.2 A g^{-1}, the cycle performance of the Zn-HSC remained at 86.4%. The ZMOP nanowires prepared by this strategy are potential high-performance cathode materials for constructing aqueous zinc-ion capacitors. Similarly, Xu [75] combined Co^{2+} ions and polyaniline (PANI) with V_2O (CoVO–PANI) and synthesized CoVO–PANI cathode materials (CoVO–PANI@CC) in situ on carbon cloth via a simple one-step hydrothermal method (Figure 4). A flexible zinc-ion HSC was constructed using CoVO–PANI@CC as the cathode and AC@CC as the anode. The feathery morphology of CoVO–PANI@CC provided a large surface area, which increased the contact area with the electrolyte and was beneficial for the diffusion of Zn^{2+}. The synergistic interaction between Co^{2+} and polyaniline expanded the interlayer spacing of CoVO–PANI, reduced the binding energy and diffusion barrier for Zn^{2+}, and improved the electron transport conductivity of the material. Under optimal conditions, the CoVO–PANI@CC//2 M $ZnSO_4$//AC@CC exhibited a high areal capacitance of 847.3 mF cm^{-2} at 1 mA cm^{-2}. After 2000 cycles at 50 mA cm^{-2}, the retention rate of CoVO–PANI@CC//2M $ZnSO_4$//AC@CC was 81.37%, and the coulomb efficiency was 100%. The assembled flexible Zn-HSC device maintained excellent electrochemical performance, with high flexibility and stability, and an area capacity of 775.6 mF cm^{-2} at 1 mA cm^{-2}. This strategy provides a feasible method for the application of vanadium-based cathode materials in flexible zinc-ion supercapacitors.

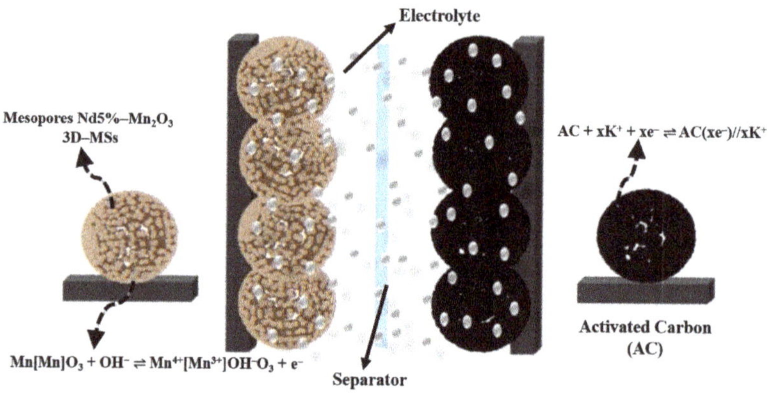

Figure 3. Schematic diagram shows the fabrication of a pouch–type HSC with mesoporous Nd5%–Mn_2O_3 3D–MSs as positive and AC as negative electrodes with filter paper as a separator in KOH electrolyte [71].

Carbon-based materials, including AC [76–78], porous carbon [79,80], graphene [81–86], carbon nanotubes (CNTs) [87–90], carbon nanofibers (CNFs) [91–93], and metal–organic framework-derived carbon [94], have been extensively studied as electrode materials for hybrid capacitors.

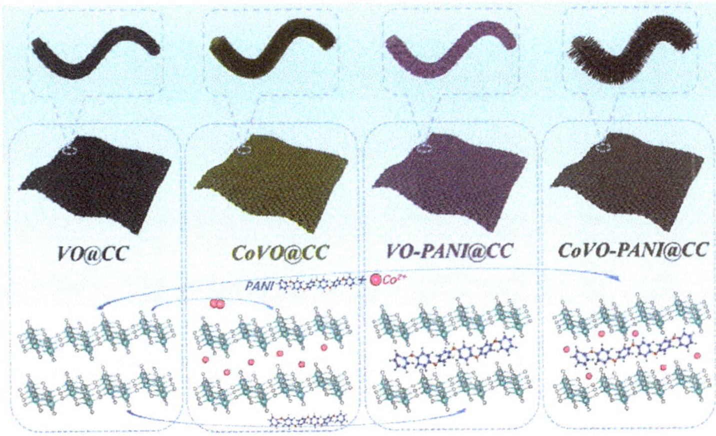

Figure 4. The preparation schematics of vo@cC, Covo@CC, VO–PANI@CC, and CoVO–PANI@CC [75].

Introducing heteroatoms (O, N, S, P, B, etc.) into the carbon matrix can improve the conductivity and surface hydrophilicity of carbon-based materials, thereby enhancing their capacitive performance. Li et al. [95] constructed a zinc-ion supercapacitor with a quadruple heteroatom self-doped hierarchical porous AC as the cathode. Through a carbonization process, four heteroatoms (O, N, S, P) were self-doped into paper mulberry (PM) to obtain the carbonized material (PMC), which was then treated with KOH as an activator to prepare PMACs rich in macropores, mesopores, and micropores (Figure 5). The micropores provided a large number of active sites for carrier adsorption, while the mesopores facilitated ion diffusion kinetics by providing abundant channels that could eliminate diffusion barriers. The macropores served as a reservoir for the electrolyte, thereby shortening the path for carrier transfer. Moreover, the porous structure produced by KOH activation had an ultra-high specific surface area (SSA) (1736.1 $m^2\ g^{-1}$), which not only effectively increased the number of reversible ion adsorption active sites but also enhanced hydrophilicity of the material due to the synergistic effect of quadruple heteroatom doping, promoting charge transfer rates and, thus, improving the material's storage. Functional groups containing O and N improved the surface hydrophilicity of the carbon-based material while providing additional sites for ion binding to enhance electronic conductivity. The S atom could increase the carbon spin density and induce structural defects in the carbon material. Phosphorus in the carbon matrix could disrupt the charge uniformity and increase the asymmetric spin density. Additionally, phosphorus could cause structural distortions, enhancing the storage capacity of Zn^{2+}. All these heteroatom dopants contributed significantly to the improvement of the electrode material's capacitance. Under optimal conditions, the prepared PMACs exhibited a considerable capacitance (240.7 F g^{-1}), an excellent energy density (108.3 Wh kg^{-1}), and a competitive power density (450 W kg^{-1}). At high mass loading, they demonstrated excellent capacitance, rate performance, and impressive cycle stability (97.1% retention after 10,000 cycles at 10 A/g under a mass loading of 14.46 mg cm^{-2}). The strategy employed in this study is of great significance for the development of low-cost, environmentally friendly, and electrochemically superior active materials.

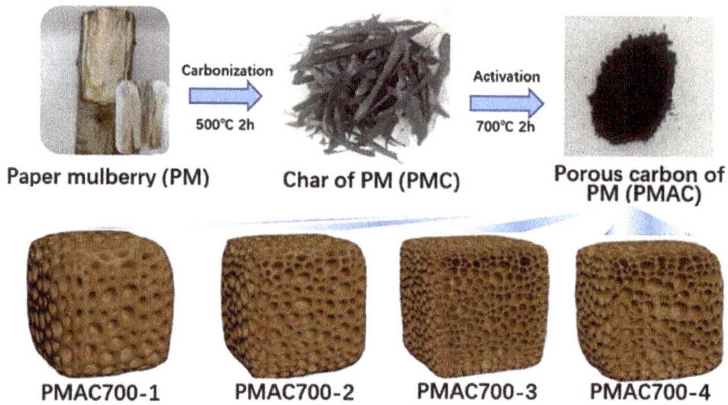

Figure 5. Schematic diagram of synthetic procedure of PMACs [95].

Graphene's enormous theoretical SSA, excellent conductivity, and chemical stability are considered promising for energy storage applications as an alternative material [96]. However, the severe aggregation problem caused by the strong van der Waals forces between graphene layers leads to significant barriers between the graphene layers, which severely reduces the actual SSA and conductivity of graphene [97]. Yao et al. [98] prepared a novel 3D phosphorus-doped carbon nanotube/reduced graphene oxide (P–CNT/rGO) composite aerogel using CNTs as intercalants and phosphoric acid as a phosphorus source through a simple one-step hydrothermal method and used it as the cathode material for Zn-HSC (Figure 6). The obtained P–CNT/rGO composite aerogel retained 45.5% of its capacitance at an ultra-high current density (Dc) of 100 A/g compared to 0.1 A/g. The group also compared the specific capacitance of the P–CNT/rGO composite aerogel (213.4 F/g) with that of rGO (105.9 F/g) and CNT/rGO (144.1 F/g) at 0.5 A/g. The excellent electrochemical performance of P–CNT/rGO could be attributed to the combined effects of carbon nanotube insertion and P doping modification, which led to improvements in SSA, expansion of interlayer spacing, and enhancement of pseudocapacitive reactions. The insertion and dispersion of carbon nanotubes between graphene sheets effectively prevented the aggregation of graphene. The synergistic effect of carbon nanotube insertion and P doping modification resulted in the P–CNT/rGO having the richest micropores and mesopores, increasing the SSA of graphene. Furthermore, since the atomic radius of the P atom is larger than that of the C atom and the electronegativity of the P atom is lower than that of the C atom, the incorporation of P atoms provided more structural defects for P–CNT/rGO and altered the charge density of graphene. The functionalization of the graphene surface (P, O atoms) could improve the wettability of the electrode with the electrolyte and the surface reactivity, providing more active sites for the chemisorption/desorption of Zn^{2+} and resulting in additional pseudocapacitance that significantly enhanced the capacitance performance of graphene. The zinc-ion supercapacitor assembled with P–CNT/rGO under optimal conditions exhibited a high energy density of 92.7 Wh/kg at a power density of 80 W/kg, a power density of 80 kW/kg at an energy density of 42.2 Wh/kg, and a capacitance retention rate of 94.2% after more than 10,000 cycles at 3 A/g, demonstrating very superior stability. This strategy provides valuable references for the in-depth optimization of energy density and power density of graphene-based cathode zinc-ion supercapacitors.

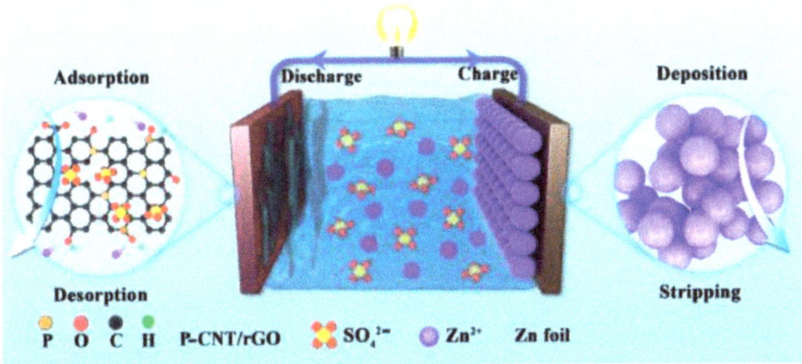

Figure 6. Schematic illustration of energy storage mechanism of the P–CNT/rGO based ZHC [98].

4.2. Heterostructure Optimization Strategy Based on TMOs

Single-phase TMOs are unable to fulfill the comprehensive performance requirements of a high capacity, high-rate performance, and long-term cycle stability. It is well known that crystalline samples exhibit strong resistance to electrolyte corrosion and better cycle stability [99], while amorphous samples contain more grain boundaries and ion diffusion channels, which can enhance electrochemical activity [100]. Therefore, if electroactive materials with different crystalline phases and chemical properties are combined rationally, it becomes feasible to achieve electrochemical performance superior to that of a single TMO [101]. Peng et al. [102] synthesized $Ni_{1.5}Co_{1.5}O_4$ through a hydrothermal reaction and then treated it with NaH_2PO_2 H_2O to acidify. Impressively, with the increase in the amount of NaH_2PO_2 H_2O, $Ni_{1.5}Co_{1.5}O_4$ underwent a redox reaction with H_3P gas to form $NiCoO_2$, as well as a dual substitution reaction where metal ions were phosphated to form NiCoP and P coordinated with O to form NiCo–PO_x. A heterostructure was obtained consisting of a large number of high-crystalline NiCoP nanoparticles dispersed in an amorphous NiCo–PO_x matrix, forming a NiCoP–$NiCoO_2$/NiCo–PO_x heterostructure (Figure 7). This crystalline/amorphous heterostructure simultaneously possesses high reactivity, performance, and long-term cycle stability. The voltage of the constructed rGO//6 M KOH//NiCoP–$NiCoO_2$/NiCo–PO_x hybrid supercapacitor could be extended to 1.56 V. It exhibited a specific capacitance of 56.1 mA h g^{-1} at 1 A g^{-1} and maintained a specific capacitance of 32.8 mA h g^{-1} at a high current density of 20 A g^{-1}. It still retained 78.7% of its capacitance after 10,000 cycles. At an energy density of 800 W kg^{-1}, it achieved a power density of 44.8 W h kg^{-1}, and at a power density of 1.6 kW kg^{-1}, it had an energy density of 26.2 Wh kg^{-1}. The strategy proposed in this heterostructure demonstrated excellent electrochemical performance, providing feasible insights for the optimization of advanced materials for energy storage applications.

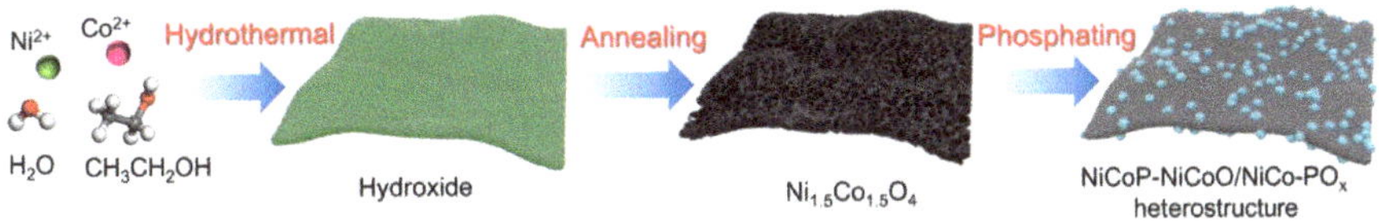

Figure 7. Schematic illustration of the synthesis procedure for NiCoP–$NiCoO_2$/NiCo–PO, heterostructure [102].

Compared to TMOs, transition metal sulfides (TMSs) have lower electronegativity and higher specific capacitance and conductivity, but they are not stable [103,104]. Transition bi-metallic oxides (TBMOs) [105] and transition bi-metallic sulfides (TBMSs) [106] contain multiple metal ions, which can provide a richer variety of active sites for redox reactions, thus, exhibiting excellent electrochemical performance. Min et al. [107] synthesized a core–shell mesoporous heterostructure nanosheet@ $NiZn_2O_4$@$NiZn_2S_4$ (NZO@NZS) on nickel foam (NF) as a cathode material for aqueous HSCs (Figure 8). Initially, NZO nanosheets were grown on NF as a framework to increase the stability of the HSC electrode material. Subsequently, the hydrothermal method used could uniformly and interconnectedly anchor non-crystalline NZS nanosheets onto the vertically aligned NZO nanosheet framework, forming a core–shell heterostructure with larger active sites and a richer variety of active sites. This expanded ion channels and improved the electron transfer rate of the composite material, further enhancing the electrochemical performance. It is worth mentioned that the proposed synthesis method of NZO@NZS, which saved any binder, significantly improved the conductivity of the electrode material. Under optimal conditions, at a Dc of 1 A g^{-1}, NZO@NZS exhibited a high specific capacitance of 1516.0 C g^{-1}. After 10,000 cycles (Dc = 15 A g^{-1}), the capacitance retention rate remained at 86.9%. Moreover, the HSC assembled with NZO as the cathode and AC as the anode achieved an energy density of 43.4 Wh kg^{-1} at a power density of 775.0 W kg^{-1}. This research strategy provided a reasonable new direction for the design and construction of core–shell structures of TMOs.

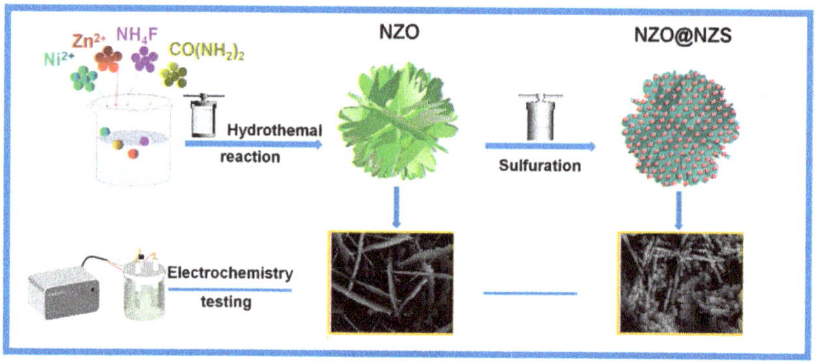

Figure 8. The fabrication process diagram of the NZO@NZS electrode [107].

4.3. Optimization Strategy of Nanocomposites Based on TMOs/Carbon Materials

Nanostructured TMOs combined with carbon-based nanomaterials (such as rGO) and non-carbon-based nanomaterials (such as MXene) to form binary or ternary nanocomposites are considered a new research direction for optimizing the electrochemical performance of HSCs.

Carbon-based materials not only improve the cycle life of TMO-based electrodes but also form a 3D network structure with TMO nanoparticles (NPs), providing an appropriate template for the growth of TMOs NPs. Ruman et al. [108] prepared SnO_2/CuO/FeO/PVP/rGO nanocomposites using a green synthesis method for use as battery-type electrode materials. The rGO sheets provided a larger conductive and porous platform for the growth of TMOs, and, thus, the obtained SnO_2/CuO/FeO/PVP/rGO exhibited significantly better electrochemical energy storage characteristics than SnO_2/CuO and SnO_2/CuO/PVP nanocomposites. Under optimal conditions, the SnO_2/CuO/FeO/PVP/rGO-modified nickel foam electrode achieved a specific capacitance of 249 C/g at 0.6 A/g. When used as a battery-type cathode with AC as the EDLC-type anode, the HSC constructed had a power density of 956 W/kg at 1.2 A/g and still retained 93.9% of its capacitance after 15,000 cycles. This group proposed a porous growth strategy, which involves embedding multi-metal redox substances into conductive porous materials and has injected new vitality into the optimization of energy storage performance for future HSCs.

MXenes typically have a chemical formula of $M_{n+1}X_nT_x$ (n = 1, 2, or 3), where M represents a transition metal, X represents carbon or N, and Tx represents surface terminations such as functional groups (e.g., = O, –OH, –F, and –Cl) [109]. They exhibit significant surface redox reaction pseudocapacitance in acidic electrolytes and different pseudocapacitance based on ion insertion mechanisms in neutral and alkaline electrolytes [110], making them widely applicable in HSCs. Ti_3C_2, as one of the lightest members of the MXene family, stands out among many MXenes. Pradeepa et al. [111] synthesized a heterostructured titanium carbide–zirconium oxide nanocomposite (Ti_3C_2–ZrO_2 NC) (Figure 9) and then precisely assembled a solid-state hybrid supercapacitor device (SHSC) using Ti_3C_2–ZrO_2 NC as the cathode, AC as the anode, Whatman filter paper as the separator, and PVA–KOH gel as the solid–state electrolyte. The synergistic interaction between the typical layered structure of Ti_3C_2 and ZrO_2 nanoparticles not only promoted the interlayer ion insertion/extraction rate but also provided a sufficient number of electrochemically active sites for surface redox reactions. Moreover, the mesoporous structure of Ti_3C_2–ZrO_2 NC facilitated the effective permeation of ions through the surface. This allowed Ti_3C_2–ZrO_2 NC to exhibit excellent charge transfer kinetics and high specific capacitance compared to single Ti_3C_2 and ZrO_2 NPs. The GCD curves of the Ti_3C_2–ZrO_2 NC electrode showed nonlinear behavior, revealing its battery charge storage process, with a high specific capacitance of 483.6 C/g at 1 A/g Dc, and it still retained a capacitance retention rate as high as 88.85% after 5000 charge–discharge cycles. The assembled Ti_3C_2–ZrO_2//AC solid-state hybrid supercapacitor (SHSC) device had an energy density and power density of 75.60 Wh/kg and 1445 W/kg, respectively. The strategy proposed in this two-dimensional layered MXene structure, which plays a key role in enhancing conductivity and reaction kinetics, may become a promising battery-type material for application in future energy storage devices.

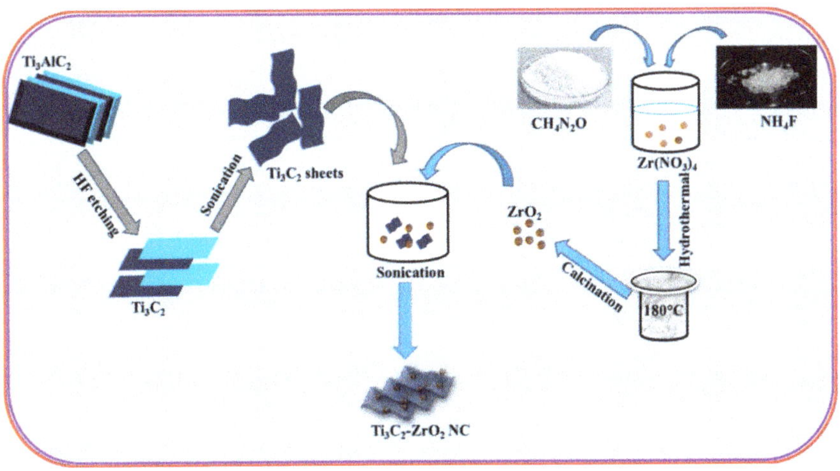

Figure 9. Schematic illustration for the synthesis procedure of $TisC_2$–ZrO_2 NC [111].

Unfortunately, the synthesis and processing of nanomaterials generally require specialized equipment and intricate fabrication techniques, which significantly increase the complexity and difficulty of manufacturing. Traditional metal oxide preparation often involves the mixing of hydroxides followed by high-temperature calcination/annealing/sintering processes, which are time-consuming. For instance, the synthesis of metal oxides from tin-based mixed hydroxides requires at least 2 h of reaction time. Therefore, the development of simple and efficient nanocomposite synthesis processes is also a key focus of research, aiming to transition technology towards industrialization and commercialization. To achieve a rapid reaction time of only 30 s, Sahoo et al. [112] employed a microwave method with a commercial microwave oven. During this short period, oxidized graphite was exfoliated and reduced to rGO, while the dehydration of $CoSn(OH)_6$ formed oxide nanoparticles that were deposited on the surface of the oxidized graphene. The microwave-prepared Co_3O_4/SnO_2/rGO (CSO–rGO) ternary nanocomposite constructed a hybrid electrode with a high specific capacitance (146.4 C/g) and exceptional cycle stability (103.2% after 15,000 cycles). Additionally, the group constructed an aqueous HSC using microwave-prepared rGO as the anode, due to the porous network structure of the CSO-rGO composite materials, as well as the gradual activation and electrochemical reconstruction of the composite electrodes during cycling, there is an increase charge storage, with the device achieving a maximum energy density of 27 Wh/kg and a cycle stability of 102.4% after 10,000 cycles.

4.4. Optimization Strategy of MOF-Derived TMOs/Carbon

Metal–organic frameworks (MOFs), also known as porous coordination polymers, are inorganic–organic hybrid materials formed by the self-assembly of metal ions or clusters with organic ligands through coordination bonds, possessing molecular-level porosity. In recent years, MOF-derived materials have received extensive research in the fields of electrochemical energy storage conversion. They can not only provide crucial supporting structures for a variety of electrode materials but also offer more active sites, larger SSA, and better promotion of electrochemical reactions due to their dispersed metal components, ultra-high surface area, and controllable porosity. Among them, MOF-derived electrode materials have received widespread attention in the field of energy storage.

Bhosale et al. [113] obtained a bimetallic MOF ($CoFe_2$–MOF) through simple thermal treatment. Impressively, they directly prepared a nano-porous carbon (NPC) anode material and a binary TMO ($CoFe_2O_4$) cathode material using a single bimetallic $CoFe_2$–MOF as the sole precursor through a method called "one-for-all". An HSC was constructed using 2 M potassium hydroxide as the alkaline aqueous electrolyte. Under optimal conditions, the constructed HSC exhibited an excellent specific capacitance at a Dc of 2 mA cm^{-2}, with an energy density of 56.2 Wh kg^{-1} and a power density of 1091.5 W kg^{-1}. Moreover, it showed an outstanding cycle stability of 97.91% over 5000 GCD cycles. The revolutionary "one-for-all" concept proposed by this strategy has created endless possibilities for developing the next generation of supercapacitors with high safety, low cost, wide voltage window, and long-term cycle stability.

Tin dioxide (SnO_2) and its derivatives have been widely used as anode materials for lithium-ion supercapacitors due to their high theoretical capacity, low working potential, abundant reserves, and environmental friendliness [114,115]. The storage mechanism of lithium in SnO_2 is typically divided into two stages. The first stage is an irreversible conversion reaction, where Li^+ ions react with SnO_2 to form metallic tin (Sn) and lithium oxide (Li_2O) [116]. The second stage involves an alloy reaction between Sn and Li^+ ions, providing higher capacity. Unfortunately, the volume expansion (approximately 260%) caused by the alloy formation severely affects the lifespan of the lithium-ion supercapacitor. Duan et al. [117] synthesized a Mn_2SnO_4@C anode composite material with a uniform carbon coating and rhombic structure through a two-step carbonization process and the addition of a tin source using a Mn-based metal–organic framework (Figure 10). During the charging process, the Mn_2SnO_4 undergoes an irreversible conversion reaction, and the resulting Mn and Sn are uniformly dispersed on the material surface at the atomic scale. Subsequently, The reaction between Sn and Li^+ ($Sn + xLi^+ + xe^- \rightarrow Li_xSn$ ($0 < x <$

4.4)) results in the formation of lithium alloys. Firstly, the alloying and conversion storage mechanisms contributed to the enhancement of electrode capacity. Secondly, the uniform carbon coating encapsulation design in Mn_2SnO_4@C not only favored the improvement of conductivity and prevented particle aggregation but also effectively buffered the volume during the Sn alloying process [118]. Moreover, the incorporation of Mn into SnO_2 significantly enhanced the intrinsic conductivity compared to pure SnO_2. During cycling, the "Mn" and "Sn" elements formed different metals (Mn, Sn), lithium alloys (Li_xSn), and metal oxides (MnO, SnO_2) at various redox potentials, creating a synergistic effect that acted as a buffering framework. At different current densities, the anode's capacity retention rate exceeded 100%, and after rate tests, the anode delivered 944 mAh g^{-1} discharge capacity. Under optimal conditions, a novel lithium-ion supercapacitor was assembled using a Mn_2SnO_4@C anode and coconut shell biomass carbon (CSBC) cathode, providing an ultra-high energy density of 217.9 Wh kg^{-1} at 210 W kg^{-1} and maintaining 25.1 Wh kg^{-1} even at a high power of 21 kW kg^{-1}. The research group's application of the Mn_2SnO_4@C anode material to lithium-ion integrated circuits offers a new approach for improving the performance of the next-generation lithium-ion integrated circuits.

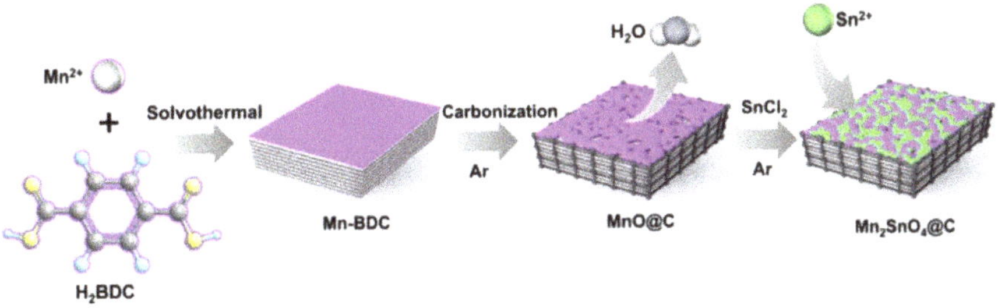

Figure 10. Schematic diagram of the synthesis route of Mn_2SnO_4@C [117].

Imidazole-based molecular sieve frameworks-8 (ZIF-8) possess a porous structure with uniformly distributed N atoms, making them an ideal precursor and template for the preparation of N-doped carbon materials [119,120]. In recent years, ZIF-8-derived N-doped porous carbons have been extensively studied and used as cathode materials for Zn-HSCs [121]. However, the limited N content in the ZIF-8 precursor has restricted the further improvement of the storage capacity of Zn-HSC. Jia et al. [122] selected a N-rich MOF (Zn-based metal-triazolate, Zn-MET) as a carbon precursor and N source, and prepared Zn-MET–x (x = pyrolysis temperature) with ultra-high N doping and hierarchical porous structures using a simple direct pyrolysis process, effectively accelerating the diffusion kinetics of zinc ions. Therefore, the optimal Zn-MET–800 used as a cathode demonstrated impressive performance. The constructed zinc-ion HSC had a high capacity of 164.2 mAh/g at 0.1 A/g and excellent rate capability of 64.4 mAh/g at 10 A/g, with a high cycle stability retention rate of 90.3% after 30,000 cycles at 10 A/g. This strategy provides a potential pathway for the preparation of advanced porous carbon materials with ultra-high N doping and offers new insights for the development of high-performance Zn-HSC.

4.5. Optimization Strategies Based on Electrolytes

The voltage window directly affects the performance and lifespan of supercapacitors. Previous studies have shown that novel electrolyte decoupling strategies can extend the voltage window, with Zhong et al. [123] using an electrolyte decoupling strategy to increase the open-circuit voltage of a Zn-MnO_2 battery from 1.50 V to 2.83 V, thereby enhancing energy density and lifespan. Additionally, electrolyte decoupling strategies can also utilize different electrolytes to replace traditional cathodes used for Zn^{2+} storage with other types of cathodes (such as those for Li^+ storage) [47]. Based on this, Wang et al. [45] proposed a new method that fixed a Ni_3S_2/Ni_2O_3 heterojunction on N-doped carbon nanosheets (Ni_3S_2/Ni_2O_3@N–CNA) and assembled a water-based dual-ion alkaline electrolyte zinc-ion supercapacitor (Figure 11). The zinc foil anode used a mixture of $Zn(AC)_2$ and NaOH as the anodic solution, while the Ni_3S_2/Ni_2O_3@N–CNA cathode used LiOH as the cathodic solution. An anion exchange membrane (AEM) was used as a separator to prevent the mixing of Zn^{2+} and Li^+ in the anodic and cathodic solutions, which could affect the reversibility of the electrode processes. During the charge storage process, compared to Ni_2O_3, the Ni_3S_2 in the Ni_3S_2/Ni_2O_3 heterostructure has a lower Ni atomic valence state, making it easier to undergo redox reactions with OH^-. During the discharge process, the reduction–oxidation reaction of Ni atoms is accompanied by the insertion and extraction of Li^+, with the Ni_3S_2/Ni_2O_3 heterojunction not only favoring the surface pseudocapacitance generated by the adsorption of hydrated Li^+ on the Ni_3S_2 side but also the dehydration behavior of hydrated Li^+ on the Ni_3S_2 and Ni_2O_3 sides, which is beneficial for the pseudocapacitance of Li^+ insertion. Under optimal conditions, the constructed water-based dual-ion alkaline electrolyte zinc-ion supercapacitor exhibited excellent specific capacitance (236 F g^{-1}) and energy density (64.2 Wh kg^{-1}). This strategy has opened up a new pathway for the development of high-performance Zn-HSCs.

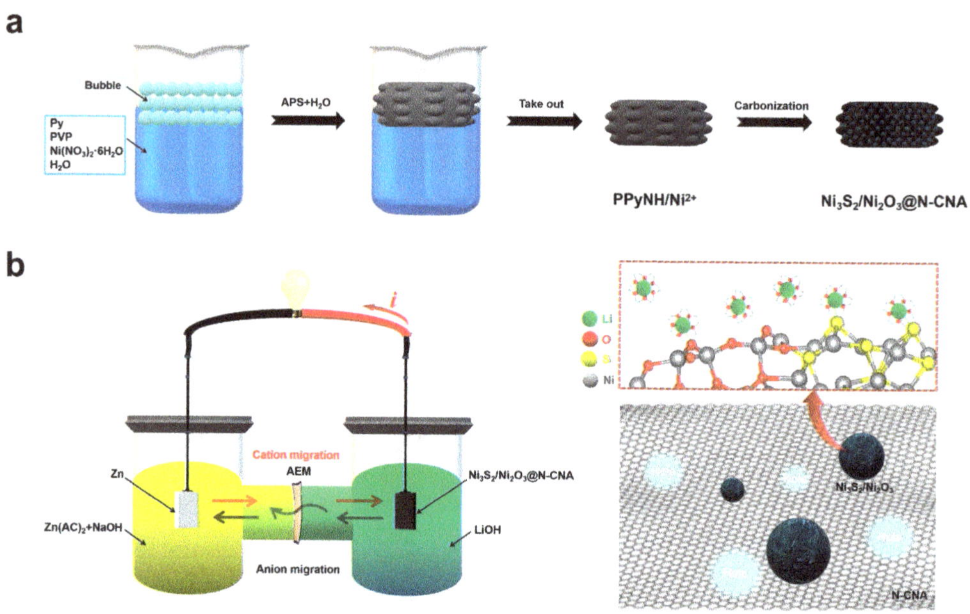

Figure 11. Schematic illustration for (**a**) the preparation process of Ni_3S_2/Ni_2O_3@N–CAN and (**b**) the charge storage mechanism of a Ni_2O_3@N–CAN //Zn Zn–HSC in the discharging process, including the device structure of the Zn–HSC (left) and the adsorption mechanism of the Ni_2O_3@N–CAN cathode for adsorbing the hydrated Li^+ ions (right) [45].

The addition of redox-active molecules to the electrolyte, which enables the storage and release of charge within the electrolyte, is another feasible strategy to enhance the specific capacitance of supercapacitors [124,125]. Due to the redox activity of thiocyanates in aqueous electrolytes and their solubility in organic solvents, Maćkowiak et al. [126] used thiocyanates as redox additives in the electrolyte. Under a current condition of C/10, they successfully achieved the insertion of thiocyanate cations into the anode using a one-step method to assemble composite electrodes and construct a metal-ion HSC. The addition of thiocyanates compensated for the charge on the cathode and efficiently inserted metal ions into the structure of the anode. The study showed that the one-step assembly system had lower resistance than the traditional two-step assembly system, reduced assembly time and cost, and improved specific energy. Impressively, among the successfully constructed lithium-ion, sodium-ion, and potassium-ion HSCs, the Li^+-HSC with a thiocyanate electrolyte produced an exceptionally high specific energy equivalent to that of a lithium-ion battery (115 Wh kg^{-1}), which was far superior to traditional hybrid capacitors (71 Wh kg^{-1}), and it maintained efficiency above 80% of the initial capacitance even after 1000 cycles. This strategy could potentially promote further research and application of metal-ion capacitors in electrolyte optimization.

Similarly, anthraquinone compounds have received widespread attention in the preparation of redox-enhanced electrolytes [127,128]. Gan et al. [129] selected sodium anthraquinone-2-sulfonate (AQ) with high solubility, excellent stability, and unique solvation behavior as an additive to a 2 M $ZnSO_4$ aqueous solution electrolyte and constructed a zinc-ion HSC. The redox behavior of the additive originated from the reversible oxidation and reduction of the AQ/AQH_2 couple, with the conjugated electrons in the AQ molecule promoting its own redox behavior [130], and the aromatic framework enhancing the stability of the Zn anode. Additionally, the added AQ had a higher coordination binding ability with Zn^{2+} than H_2O molecules [131], effectively reducing the Zn^{2+}–H_2O interactions and enhancing the diffusion coefficient of the $ZnSO_4$–AQ system. Notably, the passivation of Zn^{2+}–additive complexes produced a robust intermediate layer with a concentration gradient, and the AQ additive also inhibited the formation of $Zn_4SO_4(OH)_6$ and ZnO by–products. The study showed that the AQ additive significantly improved the lifespan of the zinc-ion battery at 2 mA cm^{-2} and 1 mAh cm^{-2}, increasing the cycle number from 50 to 1600. Importantly, the redox behavior of the electrolyte modulated the charge transfer behavior and electrochemical kinetics of the battery. The constructed zinc-ion supercapacitor achieved a capacity of 195.0 mAh g^{-1} with 5 mM AQ, while the capacity of the ZIC without the additive was 149.6 mAh g^{-1}.

5. Conclusions and Future Perspectives

Currently, research on HSCs (Table 1) primarily focuses on the optimization of surface/interface chemistry based on TMOs and carbon-based materials, while studies on electrolytes are relatively scarce. Achieving high electron transfer and ion diffusion rates has emerged as the current research priority. Although HSC devices have made significant progress, long-term research and exploration are still required. The Zn-HSC holds great promise as a pivotal area for the further development of HSC. In terms of optimization strategies, future developments may concentrate on leveraging strategies such as heteroatom doping, pre-intercalation, nanocomposites, and binder-free approaches to develop novel pseudocapacitive materials based on TMOs, aimed at enhancing the conductivity, ion diffusion rates, and electrochemical performance of pseudocapacitive materials. Regarding the enhancement of carbon-based material properties, the main research directions are to fabricate atomically precise, reproducible, doped carbon-based materials with simple preparation processes and to develop templated carbon with a controllable three-dimensional hierarchical porous structure that matches the electrolyte ion mobility. In the industrial sector, the primary challenge in the development of HSCs is to devise methods that reduce costs and enhance sustainability. Scalable manufacturing techniques can further lower the cost of hybrid capacitors by reducing the expenses associated with electrode and electrolyte materials. For instance, Mandal et al. [132] extracted activated carbon and transition metal oxides (TMOs) from industrial energy crops such as jute, kenaf, hemp, and bamboo, which were subsequently utilized as electrode materials for hydrogen storage capacitors. Hong et al. [133] developed a roll-to-roll dry transfer process that simultaneously controls the peeling speed and tension for large-scale production of graphene. Scaling up the production of nanocomposites to an industrial level may encounter both technical and economic challenges [134]; it is imperative to develop production processes for synthesizing efficient and selectively controlled nanocomposites to address the issues of uniformity and reproducibility encountered in batch production. Moreover, to enable flexible HSCs to operate over a wide temperature range and meet the needs of next-generation flexible, portable, and wearable electronic products, the development of flexible carbon/TMOs electrode materials and solid-state electrolytes with high ductility, self-healing ability, mechanical strength, stretchability, and compressibility is a critical issue that urgently needs to be addressed. The development of HSCs with a small volume and high energy density is also a future trend. Three-dimensional and four-dimensional 4D printing, as simple, compatible, and controllable manufacturing techniques, are essential for achieving precise and compact positioning of hybrid materials in integrated devices within limited spaces and for maximizing their performance. However, there is currently limited research on 3D and 4D printing. Bigham et al. [135] proposed 3D and 4D printing techniques for customized applications based on Mxene materials, highlighting the strong mechanical strength and biocompatibility of Mxene composites, which make them an ideal choice for scaffolding and implantable carrier applications. Incorporating both high-temperature and low-temperature performance considerations, in-depth research into the reaction kinetics and de-solvation processes of electrode materials and electrolytes, and the development of all-weather HSC with rapid charging capabilities can significantly expand application scenarios. This is a challenging yet practical research direction. Generally, at low temperatures, the reversible diffusion of lithium ions is a bottleneck affecting all-weather HSCs; while at high temperatures, the main challenges are the chemical decomposition of the electrolyte itself and the loss of surface chemical passivation mechanisms between the electrolyte and the anode/cathode interfaces.

Table 1. Performance comparison of advanced HSCs based on above electrode materials.

Sr#	Cathode Material	Anode Material	Electrolyte	Potential Window	Capacitance/ Capacity	Specific Energy [Wh kg^{-1}]	Specific Power [W kg^{-1}]	Cycling Stability [%]	Publishing Year	Refs.
1	Ni_3S_2/ Ni_2O_3@N–CNA	Zn–foil	1 M LiOH (Cathode), 0.3 M $Zn(AC)_2$ + 3 M NaOH (Anode)	0.4–1.8 V	236 F/g @1 A/g	64.2	700	75%, 10,000	2024	[45]
2	AC	T–Nb_2O_5	1 M $LiPF_6$	1.0–3.5 V	82.4 F/g @0.25 C	46.6	8400	100%, 1000	2023	[57]
3	Nd5%–Mn_2O_3 3D–MSs	AC	1 M KOH	0.0–0.55 V	862.14 F/g @0.5 A/g	32.26	800	97.3%, 2000	2020	[71]
4	Zn_xMnO_2/Ppy (ZMOP)	PCNTs	2 M $ZnSO_4$ + 0.1 M $MnSO_4$	0.0–1.9 V	109 F/g @0.1 A/g	20	867	86.4%, 5000	2024	[74]
5	CoVO–PANI@CC	AC@CC	2 M $ZnSO_4$	0.0–2.0 V	847.3 MF/cm^2 @1 mA/cm^2	1.69 mWh/cm^2	3.6 W/cm^2	81.37%, 20,000	2024	[75]
6	PMAC700–3	Zn–foil	2 M $ZnSO_4$	0.0–1.6 V	240.7 F/g @0.5 A/g	108.3	450	97.1%, 10,000	2024	[95]
7	P–CNT/rGO	Zn–foil	2 M $ZnSO_4$	0.0–1.8 V	213.4 F/g @0.5 A/g	92.7	80 KW/kg	94.2%, 10,000	2023	[98]
8	NiCoP–$NiCoO_2$/ NiCo–PO_x	RGO	6 M KOH	0.0–1.56 V	56.1 mAh/g @1 A/g	44.8	800	78.7%, 10,000	2023	[102]
9	AC	NZO@NZS	1 M KOH	0.0–0.8 V	1516 C/g @1 A/g	43.4	774.9	86.9%, 10,000	2024	[107]
10	SnO_2/CuO/FeO/ PVP/rGO	AC	2 M KOH	0.0–0.6 V	249 C/g @0.6 A/g	41.7	956	93.9%, 15,000	2024	[108]

Table 1. *Cont.*

Sr#	Cathode Material	Anode Material	Electrolyte	Potential Window	Capacitance/ Capacity	Specific Energy [Wh kg^{-1}]	Specific Power [W kg^{-1}]	Cycling Stability [%]	Publishing Year	Refs.
11	Ti_3C_2–ZrO_2 NC	AC	PVA–KOH gel	0.0–1.8 V	483.6 C/g @1 A/g	75.6	1445	88.85%, 5000	2024	[111]
12	rGO	Co_3O_4/ SnO_2/rGO (CSO–rGO)	3 M KOH	0.0–1.5 V	146.4 C/g @1 A/g	27	751.7	102.4%, 10,000	2024	[112]
13	NPC	$CoFe_2O_4$	PVA–KOH	0.0–1.4 V	112.1 F/g @1 mA/cm^2	56.2	1091.5	97.91%, 5000	2024	[113]
14	CSBC	Mn_2SnO_4@C	1 M $LiPF_6$	0.0–4.2 V	88.95 F/g @0.1 A/g	217.9	210	79%, 5000	2023	[117]
15	Zn–MET–800	Zn–foil	2 M $ZnSO_4$	0.1–1.7 V	164.2 mAh/g @0.1 A/g	128.5	4700	90.3%, 30,000	2024	[122]

Author Contributions: Conceptualization, L.G.; software, L.G. and F.L.; formal analysis, L.G.; investigation, L.G.; resources, L.G.; writing—original draft preparation, L.G. and F.L.; writing—review and editing, L.G., F.L., J.Q., W.G. and G.X.; visualization, L.G.; supervision, L.G.; project administration, L.G. and F.L.; funding acquisition, L.G., W.G. and G.X. All authors have read and agreed to the published version of the manuscript.

Funding: Liaoning Provincial Department of Education Key Scientific Research Project for Higher Education Institutions (National Project Cultivation Fund) (No. LJ212410153033), the Outstanding Youth Science Foundation of Shandong Province (Overseas) (No. 2022HWYQ-058), and National Natural Science Foundation of China (No. 22174136).

Data Availability Statement: No new data were created or analyzed in this study. Data sharing is not applicable to this article.

Conflicts of Interest: The authors declare no conflicts of interest.

References

1. Shao, Q.; Wang, P.; Huang, X. Opportunities and Challenges of Interface Engineering in Bimetallic Nanostructure for Enhanced Electrocatalysis. *Adv. Funct. Mater.* **2018**, *29*, 1806419. [CrossRef]
2. Rajagopal, A.; Yao, K.; Jen, A.K.Y. Toward Perovskite Solar Cell Commercialization: A Perspective and Research Roadmap Based on Interfacial Engineering. *Adv. Mater.* **2018**, *30*, 1800455. [CrossRef] [PubMed]
3. Wang, D.-G.; Liang, Z.; Gao, S.; Qu, C.; Zou, R. Metal-organic framework-based materials for hybrid supercapacitor application. *Coord. Chem. Rev.* **2020**, *404*, 213093. [CrossRef]
4. Kittner, N.; Lill, F.; Kammen, D.M. Energy storage deployment and innovation for the clean energy transition. *Nat. Energy* **2017**, *2*, 17125. [CrossRef]
5. Liu, J.; Wang, J.; Xu, C.; Jiang, H.; Li, C.; Zhang, L.; Lin, J.; Shen, Z.X. Advanced Energy Storage Devices: Basic Principles, Analytical Methods, and Rational Materials Design. *Adv. Sci.* **2017**, *5*, 1700322. [CrossRef]
6. Choudhary, N.; Li, C.; Moore, J.; Nagaiah, N.; Zhai, L.; Jung, Y.; Thomas, J. Asymmetric Supercapacitor Electrodes and Devices. *Adv. Mater.* **2017**, *29*, 1605336. [CrossRef] [PubMed]
7. Thackeray, M.M.; Wolverton, C.; Isaacs, E.D. Electrical energy storage for transportation—Approaching the limits of, and going beyond, lithium-ion batteries. *Energy Environ. Sci.* **2012**, *5*, 7854–7863. [CrossRef]
8. Jia, B.; Zhang, B.; Cai, Z.; Yang, X.; Li, L.; Guo, L. Construction of amorphous/crystalline heterointerfaces for enhanced electrochemical processes. *eScience* **2023**, *3*, 100112. [CrossRef]
9. Zhao, J.; Yang, L.; Li, R.; Zhou, Y. One−Step Synthesis of Fe−Based Metal–Organic Framework (MOF) Nanosheet Array as Efficient Cathode for Hybrid Supercapacitors. *Inorganics* **2023**, *11*, 169. [CrossRef]
10. Cho, S.Y.; Yoon, H.J.; Kim, N.R.; Yun, Y.S.; Jin, H.-J. Sodium-ion supercapacitors based on nanoporous pyroproteins containing redox-active heteroatoms. *J. Power Sources* **2016**, *329*, 536–545. [CrossRef]
11. Islam, M.S.; Mubarak, M.; Lee, H.-J. Hybrid Nanostructured Materials as Electrodes in Energy Storage Devices. *Inorganics* **2023**, *11*, 183. [CrossRef]
12. Ye, J.; Chen, Z.; Liu, Q.; Xu, C. Tin sulphide nanoflowers anchored on three-dimensional porous graphene networks as high-performance anode for sodium-ion batteries. *J. Colloid Interface Sci.* **2018**, *516*, 1–8. [CrossRef] [PubMed]

13. Fleischmann, S.; Leistenschneider, D.; Lemkova, V.; Krüner, B.; Zeiger, M.; Borchardt, L.; Presser, V. Tailored Mesoporous Carbon/Vanadium Pentoxide Hybrid Electrodes for High Power Pseudocapacitive Lithium and Sodium Intercalation. *Chem. Mater.* **2017**, *29*, 8653–8662. [CrossRef]
14. Zhang, K.; Yang, X.; Li, D. Engineering graphene for high-performance supercapacitors: Enabling role of colloidal chemistry. *J. Energy Chem.* **2018**, *27*, 1–5. [CrossRef]
15. Wasiński, K.; Półrolniczak, P.; Walkowiak, M. Proof-of-concept study of a new type sodium-ion hybrid electrochemical capacitor with organic electrolyte. *Electrochim. Acta* **2018**, *259*, 850–854. [CrossRef]
16. Scaravonati, S.; Magnani, G.; Gaboardi, M.; Allodi, G.; Riccò, M.; Pontiroli, D. Electrochemical intercalation of fullerene and hydrofullerene with sodium. *Carbon* **2018**, *130*, 11–18. [CrossRef]
17. Liu, X.; Wang, H.; Cui, Y.; Xu, X.; Zhang, H.; Lu, G.; Shi, J.; Liu, W.; Chen, S.; Wang, X. High-energy sodium-ion capacitor assembled by hierarchical porous carbon electrodes derived from Enteromorpha. *J. Mater. Sci.* **2018**, *53*, 6763–6773. [CrossRef]
18. Kiran, A.S.; Shankar, E.G.; Nagaraju, M.; Yu, J.S. Binder-free phosphor-coated multiphase bismuth cobalt-selenide electrode for solar-charged quasi-solid-state hybrid supercapacitors. *Appl. Surf. Sci.* **2024**, *663*, 160174. [CrossRef]
19. Mohamed, M.M.; Aziz, M.A.; Hussain, A.; Hardianto, Y.P.; Yamani, Z.H. Dendrite-free zinc-ion hybrid supercapacitor with jute-derived carbon and nanostructured zinc on steel mesh for EVs. *J. Energy Storage* **2024**, *100*, 113635. [CrossRef]
20. Junied Arbaz, S.; Chandra Sekhar, S.; Ramulu, B.; Nagaraju, M.; Su Yu, J. Electrodeposited cobalt sulfide nanolayer fenced nickel-copper carbonate hydroxide nanowires as an electrode for hybrid supercapacitors: A wind turbine-driven energy storage system for portable applications. *Appl. Surf. Sci.* **2022**, *602*, 154288. [CrossRef]
21. Vaught, L.; Sellers, R.; Shirani Bidabadi, B.; Polycarpou, A.A.; Amiri, A. Flexible zinc-ion hybrid micro-supercapacitors with polymeric current collector for integrated energy storage in wearable devices. *Chem. Eng. J.* **2024**, *492*, 152234. [CrossRef]
22. Javaid, A.; Noreen, S. Mechanically robust structural hybrid supercapacitors with high energy density for electric vehicle applications. *J. Energy Storage* **2022**, *55*, 105818. [CrossRef]
23. Jung, Y.; Pyun, K.R.; Min, J.; Yoon, H.; Lee, M.; Kim, B.-W.; Lee, J.; Ko, S.H. An Ag–Au-PANI core–shell nanowire network for visible-to-infrared data encryption and supercapacitor applications. *J. Mater. Chem. A* **2023**, *11*, 7264–7275. [CrossRef]
24. Anil Kumar, Y.; Koyyada, G.; Ramachandran, T.; Kim, J.H.; Sajid, S.; Moniruzzaman, M.; Alzahmi, S.; Obaidat, I.M. Carbon Materials as a Conductive Skeleton for Supercapacitor Electrode Applications: A Review. *Nanomaterials* **2023**, *13*, 1049. [CrossRef]
25. Zhang, Z.; Sun, S.; Xu, Z.; Yin, S. Multicomponent Hybridization Transition Metal Oxide Electrode Enriched with Oxygen Vacancy for Ultralong-Life Supercapacitor. *Small* **2023**, *19*, 2302479. [CrossRef] [PubMed]
26. Das, A.; Raj, B.; Mohapatra, M.; Andersen, S.M.; Basu, S. Performance and future directions of transition metal sulfide-based electrode materials towards supercapacitor/supercapattery. *WIREs Energy Environ.* **2021**, *11*, e414. [CrossRef]
27. Wu, C.; Deng, S.; Wang, H.; Sun, Y.; Liu, J.; Yan, H. Preparation of Novel Three-Dimensional NiO/Ultrathin Derived Graphene Hybrid for Supercapacitor Applications. *ACS Appl. Mater. Interfaces* **2014**, *6*, 1106–1112. [CrossRef] [PubMed]
28. Sheng, S.; Liu, W.; Zhu, K.; Cheng, K.; Ye, K.; Wang, G.; Cao, D.; Yan, J. Fe_3O_4 nanospheres in situ decorated graphene as high-performance anode for asymmetric supercapacitor with impressive energy density. *J. Colloid Interface Sci.* **2019**, *536*, 235–244. [CrossRef]
29. Nguyen, Q.; Nakate, U.T.; Chen, J.; Park, S.; Park, S. Ceria nanoflowers decorated Co_3O_4 nanosheets electrodes for highly efficient electrochemical supercapacitors. *Appl. Surf. Sci.* **2023**, *613*, 156034. [CrossRef]
30. Li, X.; Wang, Z.; Qiu, Y.; Pan, Q.; Hu, P. 3D graphene/ZnO nanorods composite networks as supercapacitor electrodes. *J. Alloys Compd.* **2015**, *620*, 31–37. [CrossRef]
31. Luo, L.; Meng, W.; Wang, G.; Qin, J.; He, H.; Huang, H. MnO_2 nanoflowers-decorated MXene nanosheets with enhanced supercapacitor performance. *J. Alloys Compd.* **2023**, *957*, 170411. [CrossRef]
32. Kiruthiga, R.; Nithya, C.; Karvembu, R.; Venkata, B. Reduced Graphene Oxide Embedded V_2O_5 Nanorods and Porous Honey Carbon as High Performance Electrodes for Hybrid Sodium-ion Supercapacitors. *Electrochim. Acta* **2017**, *256*, 221–231.
33. Nti, F.; Han, J.I. Layered $Na_{2/3}Ni_{1/3}Mn_{2/3}O_2$ as electrode material with two redox active transition metals for high performance supercapacitor. *J. Alloys Compd.* **2017**, *728*, 78–87. [CrossRef]
34. Niu, J.; Liang, J.; Shao, R.; Liu, M.; Dou, M.; Li, Z.; Huang, Y.; Wang, F. Tremella-like N,O-codoped hierarchically porous carbon nanosheets as high-performance anode materials for high energy and ultrafast Na-ion capacitors. *Nano Energy* **2017**, *41*, 285–292. [CrossRef]
35. Peng, M.; Wang, L.; Li, L.; Peng, Z.; Tang, X.; Hu, T.; Yuan, K.; Chen, Y. Molecular crowding agents engineered to make bioinspired electrolytes for high-voltage aqueous supercapacitors. *eScience* **2021**, *1*, 83–90. [CrossRef]
36. Bagheri, A.; Taghavi, S.; Bellani, S.; Salimi, P.; Beydaghi, H.; Panda, J.K.; Isabella Zappia, M.; Mastronardi, V.; Gamberini, A.; Balkrishna Thorat, S.; et al. Venice's macroalgae-derived active material for aqueous, organic, and solid-state supercapacitors. *Chem. Eng. J.* **2024**, *496*, 153529. [CrossRef]

37. Zhao, Y.; Hu, X.; Stucky, G.D.; Boettcher, S.W. Thermodynamic, Kinetic, and Transport Contributions to Hydrogen Evolution Activity and Electrolyte-Stability Windows for Water-in-Salt Electrolytes. *J. Am. Chem. Soc.* **2024**, *146*, 3438–3448. [CrossRef] [PubMed]
38. Zhang, P.; Zhang, W.; Wang, Z.; Wang, X.; Ren, Q.; Zhang, S.; Wang, Y.; He, L.; Liu, P.; Zhang, Q.; et al. High-voltage, low-temperature supercapacitors enabled by localized "water-in-pyrrolidinium chloride" electrolyte. *eScience* **2023**, *3*, 100184. [CrossRef]
39. Magdum, S.S.; Thangarasu, S.; Oh, T.H. Three-Dimensional Ternary rGO/VS_2/WS_2 Composite Hydrogel for Supercapacitor Applications. *Inorganics* **2022**, *10*, 229. [CrossRef]
40. Liu, S.; Luo, Z.; Tian, G.; Zhu, M.; Cai, Z.; Pan, A.; Liang, S. TiO_2 nanorods grown on carbon fiber cloth as binder-free electrode for sodium-ion batteries and flexible sodium-ion capacitors. *J. Power Sources* **2017**, *363*, 284–290. [CrossRef]
41. Loh, K.H.; Liew, J.; Liu, L.; Goh, Z.L.; Pershaanaa, M.; Kamarulazam, F.; Bashir, S.; Ramesh, K.; Ramesh, S. A comprehensive review on fundamentals and components of zinc-ion hybrid supercapacitors. *J. Energy Storage* **2024**, *81*, 110370. [CrossRef]
42. Wang, G.; Zhang, L.; Zhang, J. A review of electrode materials for electrochemical supercapacitors. *Chem. Soc. Rev.* **2012**, *41*, 797–828. [CrossRef] [PubMed]
43. Liu, Q.; Zhang, H.; Xie, J.; Liu, X.; Lu, X. Recent progress and challenges of carbon materials for Zn-ion hybrid supercapacitors. *Carbon Energy* **2020**, *2*, 521–539. [CrossRef]
44. Wang, H.; Ye, W.; Yang, Y.; Zhong, Y.; Hu, Y. Zn-ion hybrid supercapacitors: Achievements, challenges and future perspectives. *Nano Energy* **2021**, *85*, 105942. [CrossRef]
45. Wang, W.; Yang, C.; Han, D.; Yu, S.; Qi, W.; Ling, R.; Liu, G. Ni_3S_2/Ni_2O_3 heterojunction anchored on N-doped carbon nanosheet aerogels for dual-ion hybrid supercapacitors. *J. Colloid Interface Sci.* **2024**, *654*, 709–718. [CrossRef] [PubMed]
46. Zhang, D.; Li, L.; Deng, J.; Guo, S.; Pang, H.; Lu, J.; Xia, D.; Ji, X. A new type of zinc ion hybrid supercapacitor based on 2D materials. *Nanoscale* **2021**, *13*, 11004–11016. [CrossRef] [PubMed]
47. Chao, D.; Qiao, S.Z. Toward High-Voltage Aqueous Batteries: Super- or Low-Concentrated Electrolyte? *Joule* **2020**, *4*, 1846–1851. [CrossRef]
48. Jeong, J.H.; Lee, G.W.; Kim, Y.H.; Choi, Y.J.; Roh, K.C.; Kim, K.B. A holey graphene-based hybrid supercapacitor. *Chem. Eng. J.* **2019**, *378*, 122126. [CrossRef]
49. Li, B.; Dai, F.; Xiao, Q.; Yang, L.; Shen, J.; Zhang, C.; Cai, M. Nitrogen-doped activated carbon for a high energy hybrid supercapacitor. *Energy Environ. Sci.* **2016**, *9*, 102–106. [CrossRef]
50. Wang, Y.; Song, Y.; Xia, Y. Electrochemical capacitors: Mechanism, materials, systems, characterization and applications. *Chem. Soc. Rev.* **2016**, *45*, 5925–5950. [CrossRef]
51. Shao, Y.; El-Kady, M.F.; Sun, J.; Li, Y.; Zhang, Q.; Zhu, M.; Wang, H.; Dunn, B.; Kaner, R.B. Design and Mechanisms of Asymmetric Supercapacitors. *Chem. Rev.* **2018**, *118*, 9233–9280. [CrossRef] [PubMed]
52. Divya, M.L.; Praneetha, S.; Lee, Y.; Aravindan, V. Next-generation Li-ion capacitor with high energy and high power by limiting alloying-intercalation process using SnO_2@Graphite composite as battery type electrode. *Compos. Part B* **2022**, *230*, 109487. [CrossRef]
53. Dai, Y.; Chen, L.; Babayan, V.; Cheng, Q.; Saha, P.; Jiang, H.; Li, C. Ultrathin MnO_2 nanoflakes grown on N-doped carbon nanoboxes for high-energy asymmetric supercapacitors. *J. Mater. Chem. A* **2015**, *3*, 21337–21342. [CrossRef]
54. Khomenko, V.; Raymundo-Piñero, E.; Béguin, F. High-energy density graphite/AC capacitor in organic electrolyte. *J. Power Sources* **2008**, *177*, 643–651. [CrossRef]
55. Wang, H.; Wang, M.; Tang, Y. A novel zinc-ion hybrid supercapacitor for long-life and low-cost energy storage applications. *Energy Storage Mater.* **2018**, *13*, 1–7. [CrossRef]
56. Wu, S.; Chen, Y.; Jiao, T.; Zhou, J.; Cheng, J.; Liu, B.; Yang, S.; Zhang, K.; Zhang, W. An Aqueous Zn-Ion Hybrid Supercapacitor with High Energy Density and Ultrastability up to 80000 Cycles. *Adv. Energy Mater.* **2019**, *9*, 1902915. [CrossRef]
57. Ding, X.; Huang, H.; Huang, Q.; Hu, B.; Li, X.; Ma, X.; Xiong, X. Doping sites modulation of T-Nb_2O_5 to achieve ultrafast lithium storage. *J. Energy Chem.* **2023**, *77*, 280–289. [CrossRef]
58. Reddy, M.V.; Subba Rao, G.V.; Chowdari, B.V.R. Metal Oxides and Oxysalts as Anode Materials for Li Ion Batteries. *Chem. Rev.* **2013**, *113*, 5364–5457. [PubMed]
59. Liu, W.; An, Y.; Zhang, X.; Wang, L.; Li, C.; Xu, Y.; Zhang, X.; Li, S.; Yi, S.; Gong, Y.; et al. General Synthesis of Graphene/Metal Oxide Heterostructures for Enhanced Lithium Storage Performance. *Adv. Funct. Mater.* **2023**, *34*, 2313274. [CrossRef]
60. Liu, W.; An, Y.; Wang, L.; Hu, T.; Lia, C.; Xu, Y.; Wang, K.; Sun, X.; Zhang, H.; Zhang, X.; et al. Mechanically flexible reduced graphene oxide/carbon composite films for highperformance quasi-solid-state lithium-ion capacitors. *J. Energy Chem.* **2023**, *80*, 68–76. [CrossRef]
61. Ma, F.; Yuan, A.; Xu, J.; Hu, P. Porous α-MoO_3/MWCNT Nanocomposite Synthesized via a Surfactant-Assisted Solvothermal Route as a Lithium-Ion-Battery High-Capacity Anode Material with Excellent Rate Capability and Cyclability. *ACS Appl. Mater. Interfaces* **2015**, *7*, 15531–15541. [CrossRef] [PubMed]

62. Xia, Q.; Zhao, H.; Du, Z.; Wang, J.; Zhang, T.; Wang, J.; Lv, P. Synthesis and electrochemical properties of MoO_3/C composite as anode material for lithium-ion batteries. *J. Power Sources* **2013**, *226*, 107–111. [CrossRef]
63. Wu, Y.P.; Rahm, E.; Holze, R. Carbon anode materials for lithium ion batteries. *J. Power Sources* **2003**, *114*, 228–236. [CrossRef]
64. Alemour, B.; Yaacob, M.H.; Lim, H.N.; Hassan, M.R. Review of Electrical Properties of Graphene Conductive Composites. *Int. J. Nanoelectron. Mater.* **2018**, *11*, 371.
65. Kinloch, I.A.; Suhr, J.; Lou, J.; Young, R.J.; Ajayan, P.M. Composites with carbon nanotubes and graphene: An outlook. *Sci. Educ.* **2018**, *362*, 547–553. [CrossRef] [PubMed]
66. Chen, S.; Liu, F.; Xiang, Q.; Feng, X.; Qiu, G. Synthesis of Mn_2O_3 microstructures and their energy storage ability studies. *Electrochimica Acta* **2013**, *106*, 360–371. [CrossRef]
67. Chidembo, A.T.; Aboutalebi, S.H.; Konstantinov, K.; Jafta, C.J.; Liu, H.K.; Ozoemena, K.I. In situ engineering of urchin-like reduced graphene oxide–Mn_2O_3–Mn_3O_4 nanostructures for supercapacitors. *RSC Adv.* **2014**, *4*, 886–892. [CrossRef]
68. Wei, W.; Cui, X.; Chen, W.; Ivey, D.G. Manganese oxide-based materials as electrochemical supercapacitor electrodes. *Chem. Soc. Rev.* **2011**, *40*, 1697–1721. [CrossRef] [PubMed]
69. Griffith, K.J.; Wiaderek, K.M.; Cibin, G.; Marbella, L.E.; Grey, C.P. Niobium tungsten oxides for high-rate lithium-ion energy storage. *Nature* **2018**, *559*, 556–563. [CrossRef]
70. Han, Y.; Heng, S.; Wang, Y.; Qu, Q.; Zheng, H. Anchoring Interfacial Nickel Cations on Single-Crystal $LiNi_{0.8}Co_{0.1}Mn_{0.1}O_2$ Cathode Surface via Controllable Electron Transfer. *ACS Energy Lett.* **2020**, *5*, 2421–2433. [CrossRef]
71. Karuppaiah, M.; Sakthivel, P.; Asaithambi, S.; Bharat, L.K.; Nagaraju, G.; Ahamad, T.; Balamurugan, K.; Yuvakkumar, R.; Ravi, G. Elevated energy density and cycle stability of α-Mn_2O_3 3D-microspheres with addition of neodymium dopant for pouch-type hybrid supercapacitors. *Electrochim. Acta* **2020**, *362*, 137169. [CrossRef]
72. Liu, T.; Chen, L.; Wang, X.; Huang, Y.; Wang, M.; Zhang, Y. Polypyrrole-coated V_2O_5 nanobelts arrays on carbon cloth for high performance zinc energy storage. *Electrochim. Acta* **2023**, *441*, 141806. [CrossRef]
73. Wang, Y.; Jiang, G.; Zhang, Z.; Chen, H.; Li, Y.; Kong, D.; Qin, X.; Li, Y.; Zhang, X.; Wang, H. Cable-like V_2O_5 Decorated Carbon Cloth as a High-Capacity Cathode for Flexible Zinc Ion Batteries. *Energy Technol.* **2022**, *10*, 2101170. [CrossRef]
74. Xue, Y.; Huo, J.; Wang, X.; Zhao, Y. Zn_xMnO_2/PPy Nanowires Composite as Cathode Material for Aqueous Zinc-Ion Hybrid Supercapacitors. *Battery Energy* **2024**, *3*, 20240035. [CrossRef]
75. Xu, Y.; Yang, X.; Li, X.; Gao, Y.; Wang, L.; Lü, W. Flexible zinc-ion hybrid supercapacitor based on Co^{2+}-doped polyaniline/V_2O_5 electrode. *J. Power Sources* **2024**, *623*, 235399. [CrossRef]
76. Yang, G.; Huang, J.; Wan, X.; Zhu, Y.; Liu, B.; Wang, J. A low cost, wide temperature range, and high energy density flexible quasi-solid-state zinc-ion hybrid supercapacitors enabled by sustainable cathode and electrolyte design. *Nano Energy* **2021**, *90*, 106500. [CrossRef]
77. Sun, Y.; Ma, H.; Zhang, X.; Liu, B.; Liu, L.; Zhang, X.; Feng, J.; Zhang, Q.; Ding, Y.; Yang, B.; et al. Salty Ice Electrolyte with Superior Ionic Conductivity Towards Low-Temperature Aqueous Zinc Ion Hybrid Capacitors. *Adv. Funct. Mater.* **2021**, *31*, 2101277. [CrossRef]
78. Zhang, W.; Zhang, R.; Shi, M.; Ma, L.; Huang, Y. Hierarchical polygon Co_3O_4 flakes/N,O-dual doped porous carbon frameworks for flexible hybrid supercapacitors. *Electrochim. Acta* **2022**, *424*, 140631. [CrossRef]
79. Zhang, H.; Liu, Q.; Fang, Y.; Teng, C.; Liu, X.; Fang, P.; Tong, Y.; Lu, X. Boosting Zn-Ion Energy Storage Capability of Hierarchically Porous Carbon by Promoting Chemical Adsorption. *Adv. Mater.* **2019**, *31*, 1904948. [CrossRef]
80. Wang, L.; Peng, M.; Chen, J.; Tang, X.; Li, L.; Hu, T.; Yuan, K.; Chen, Y. High Energy and Power Zinc Ion Capacitors: A Dual-Ion Adsorption and Reversible Chemical Adsorption Coupling Mechanism. *ACS Nano* **2022**, *16*, 2877–2888. [CrossRef]
81. Zhang, L.L.; Zhao, X.; Ji, H.; Stoller, M.D.; Lai, L.; Murali, S.; McDonnell, S.; Cleveger, B.; Wallace, R.M.; Ruoff, R.S. Nitrogen doping of graphene and its effect on quantum capacitance, and a new insight on the enhanced capacitance of N-doped carbon. *Energy Environ. Sci.* **2012**, *5*, 9618. [CrossRef]
82. Shao, Y.; Sun, Z.; Tian, Z.; Li, S.; Wu, G.; Wang, M.; Tong, X.; Shen, F.; Xia, Z.; Tung, V.; et al. Regulating Oxygen Substituents with Optimized Redox Activity in Chemically Reduced Graphene Oxide for Aqueous Zn-Ion Hybrid Capacitor. *Adv. Funct. Mater.* **2020**, *31*, 2007843. [CrossRef]
83. Xu, H.; He, W.; Li, Z.; Chi, J.; Jiang, J.; Huang, K.; Li, S.; Sun, G.; Dou, H.; Zhang, X. Revisiting Charge Storage Mechanism of Reduced Graphene Oxide in Zinc Ion Hybrid Capacitor beyond the Contribution of Oxygen-Containing Groups. *Adv. Funct. Mater.* **2022**, *32*, 2111131. [CrossRef]
84. Luo, J.; Xu, L.; Liu, H.; Wang, Y.; Wang, Q.; Shao, Y.; Wang, M.; Yang, D.; Li, S.; Zhang, L.; et al. Harmonizing Graphene Laminate Spacing and Zinc-Ion Solvated Structure toward Efficient Compact Capacitive Charge Storage. *Adv. Funct. Mater.* **2022**, *32*, 2112151. [CrossRef]
85. Wu, Z.S.; Parvez, K.; Feng, X.; Müllen, K. Graphene-based in-plane micro-supercapacitors with high power and energy densities. *Nat. Commun.* **2013**, *4*, 2487. [CrossRef] [PubMed]

86. Nagar, B.; Dubal, D.P.; Pires, L.; Merkoçi, A.; Gómez-Romero, P. Design and Fabrication of Printed Paper-Based Hybrid Micro-Supercapacitor by using Graphene and Redox-Active Electrolyte. *ChemSusChem* **2018**, *11*, 1849–1856. [CrossRef]
87. Dubal, D.P.; Chodankar, N.R.; Caban-Huertas, Z.; Wolfart, F.; Vidotti, M.; Holze, R.; Lokhande, C.D.; Gomez-Romero, P. Synthetic approach from polypyrrole nanotubes to nitrogen doped pyrolyzed carbon nanotubes for asymmetric supercapacitors. *J. Power Sources* **2016**, *308*, 158–165. [CrossRef]
88. Fan, X.; Peng, Z.; Yang, Y.; Zhou, H.; Guo, X. Atomic H-induced cutting and unzipping of single-walledcarbon nanotube carpets with teepee structure and theirenhanced supercapacitor performance. *J. Mater. Chem. A* **2015**, *3*, 10077–10084. [CrossRef]
89. Choi, C.; Lee, J.A.; Choi, A.Y.; Kim, Y.T.; Lepró, X.; Lima, M.D.; Baughman, R.H.; Kim, S.J. Flexible Supercapacitor Made of Carbon Nanotube Yarn with Internal Pores. *Adv. Mater.* **2013**, *26*, 2059–2065. [CrossRef]
90. Sun, G.; Yang, H.; Zhang, G.; Gao, J.; Jin, X.; Zhao, Y.; Jiang, L.; Qu, L. A capacity recoverable zinc-ion micro-supercapacitor. *Energy Environ. Sci.* **2018**, *11*, 3367–3374. [CrossRef]
91. Li, W.; Zhang, F.; Dou, Y.; Wu, Z.; Liu, H.; Qian, X.; Liu, D.; Xia, Y.; Tu, B.; Zhao, D. A Self-Template Strategy for the Synthesis of Mesoporous Carbon Nanofibers as Advanced Supercapacitor Electrodes. *Adv. Energy Mater.* **2011**, *1*, 382–386. [CrossRef]
92. Jost, K.; Stenger, D.; Perez, C.R.; McDonough, J.K.; Lian, K.; Gogotsi, Y.; Dion, G. Knitted and screen printed carbon-fiber supercapacitors for applications in wearable electronics. *Energy Environ. Sci.* **2013**, *6*, 2698. [CrossRef]
93. Ren, J.; Bai, W.; Guan, G.; Zhang, Y.; Peng, H. Flexible and Weaveable Capacitor Wire Based on a Carbon Nanocomposite Fiber. *Adv. Mater.* **2013**, *25*, 5965–5970. [CrossRef] [PubMed]
94. Li, H.; Wu, J.; Wang, L.; Liao, Q.; Niu, X.; Zhang, D.; Wang, K. A zinc ion hybrid capacitor based on sharpened pencil-like hierarchically porous carbon derived from metal–organic framework. *Chem. Eng. J.* **2022**, *428*, 131071. [CrossRef]
95. Li, T.; Wang, D.; Zhang, B.; Ding, P.; Li, H.; Wang, T.; Pei, Q.; Liu, X.; Guo, R. Tetra-heteroatom self-doped and hierarchical porous carbon toward high-performance zinc-ion hybrid supercapacitors. *Chem. Eng. Sci.* **2024**, *298*, 120352. [CrossRef]
96. Li, B.; Yu, M.; Li, Z.; Yu, C.; Wang, H.; Li, Q. Constructing Flexible All-Solid-State Supercapacitors from 3D Nanosheets Active Bricks via 3D Manufacturing Technology: A Perspective Review. *Adv. Funct. Mater.* **2022**, *32*, 2201166. [CrossRef]
97. Sun, Z.; Fang, S.; Hu, Y.H. 3D Graphene Materials: From Understanding to Design and Synthesis Control. *Chem. Rev.* **2020**, *120*, 10336–10453. [CrossRef] [PubMed]
98. Yao, J.; Li, F.; Zhou, R.; Guo, C.; Liu, X.; Zhu, Y. Phosphorous-doped carbon nanotube/reduced graphene oxide aerogel cathode enabled by pseudocapacitance for high energy and power zinc-ion hybrid capacitors. *Chin. Chem. Lett.* **2024**, *35*, 108354. [CrossRef]
99. Jiang, S.; Yun, S.; Cao, H.; Zhang, Z.; Feng, H.; Chen, H. Porous carbon matrix-encapsulated MnO in situ derived from metal-organic frameworks as advanced anode materials for Li-ion capacitors. *Sci. China Mater.* **2021**, *65*, 59–68. [CrossRef]
100. Huang, C.; Song, X.; Qin, Y.; Xu, B.; Chen, H.C. Cation exchange reaction derived amorphous bimetal hydroxides as advanced battery materials for hybrid supercapacitors. *J. Mater. Chem. A* **2018**, *6*, 21047–21055. [CrossRef]
101. Chen, H.C.; Qin, Y.; Cao, H.; Song, X.; Huang, C.; Feng, H.; Zhao, X.S. Synthesis of amorphous nickel–cobalt–manganese hydroxides for supercapacitor-battery hybrid energy storage system. *Energy Storage Mater.* **2019**, *17*, 194–203. [CrossRef]
102. Peng, Z.; Diao, Y.; Huang, J.; Zhang, C.; Chen, H.-C. Construction of NiCoP-NiCoO/NiCo–PO heterostructure by controllable phosphating effect for high-performance hybrid supercapacitors and alkaline zinc batteries. *J. Alloys Compd.* **2023**, *965*, 171502. [CrossRef]
103. Guo, Z.C.; Zhao, Y.X.; Mu, J.P.; Zhang, Z.X.; Mu, J.B.; Che, H.W.; Wang, Y.M.; Zhang, X.L.; Wang, G.S.; Zhang, M.Y. Fabrication of $NiPS_3$ coupled with hollow porous nitrogen-doped carbon capsules for highperformance asymmetric supercapacitor. *J. Energy Storage* **2023**, *61*, 106805. [CrossRef]
104. Wang, T.; Chen, H.C.; Yu, F.; Zhao, X.S.; Wang, H. Boosting the cycling stability of transition metal compounds-based supercapacitors. *Energy Storage Mater.* **2019**, *16*, 545–573. [CrossRef]
105. Zhai, X.; Pan, H.; Wang, F.; Gao, X.; Xiong, Z.; Li, L.; Chang, Q.; Cheng, S.; Zuo, Z.; Li, Y. Controlled Growth of 3D Interpenetrated Networks by $NiCo_2O_4$ and Graphdiyne for High-Performance Supercapacitor. *ACS Appl. Mater. Interfaces* **2022**, *14*, 18283–18292. [CrossRef]
106. Zhao, Y.; Shi, Z.; Li, H.; Wang, C.-A. Designing pinecone-like and hierarchical manganese cobalt sulfides for advanced supercapacitor electrodes. *J. Mater. Chem. A* **2018**, *6*, 12782–12793. [CrossRef]
107. Min, X.; Liu, W.-n.-y.; Wen, B.; Feng, J.; Lin, X. Nanosheet@nanosheet $NiZn_2O_4$@$NiZn_2S_4$ core-shell mesoporous heterostructure as an electrode material for aqueous hybrid supercapacitors. *J. Energy Storage* **2024**, *81*, 110410. [CrossRef]
108. Ruman, U.E.; Khan, A.; Fahad, H.M.; Asif, M.; Shaheen, F.; Aziz, M.H.; Ahmad, R.; Alam, M.; Sharif, S.; Afzal, S. Biogenic-ecofriendly synthesized SnO_2/CuO/FeO/PVP/RGO nanocomposite for enhancing energy density performance of hybrid supercapacitors. *J. Energy Storage* **2024**, *89*, 111643. [CrossRef]
109. Li, X.; Huang, Z.; Shuck, C.E.; Liang, G.; Gogotsi, Y.; Zhi, C. MXene chemistry, electrochemistry and energy storage applications. *Nat. Rev. Chem.* **2022**, *6*, 389–404. [CrossRef] [PubMed]

110. Huang, J.; Yuan, K.; Chen, Y. Wide Voltage Aqueous Asymmetric Supercapacitors: Advances, Strategies, and Challenges. *Adv. Funct. Mater.* **2021**, *32*, 2108107. [CrossRef]
111. Pradeepa, S.S.; Sutharthani, K.; Suba, D.R.; Yen, P.F.; Sivakumar, M. Two-dimensional layered Ti_3C_2 MXene nanosheets decked with ZrO_2 nanospheres for the high-performance solid-state hybrid supercapacitors. *J. Energy Storage* **2024**, *98*, 112821.
112. Sahoo, S.; Milton, A.; Sood, A.; Kumar, R.; Choi, S.; Maity, C.K.; Han, S.S. Microwave-assisted synthesis of perovskite hydroxide-derived Co_3O_4/SnO_2/reduced graphene oxide nanocomposites for advanced hybrid supercapacitor devices. *J. Energy Storage* **2024**, *99*, 113321. [CrossRef]
113. Bhosale, R.; Bhosale, S.; Chavan, V.; Jambhale, C.; Kim, D.-k.; Kolekar, S. Hybrid Supercapacitors Based on Nanoporous Carbon and $CoFe_2O_4$ Derived from a Bimetallic Organic Framework. *ACS Appl. Nano Mater.* **2024**, *7*, 2244–2257.
114. Gao, C.; Jiang, Z.; Wang, P.; Jensen, L.R.; Zhang, Y.; Yue, Y. Optimized assembling of MOF/SnO_2/Graphene leads to superior anode for lithium ion batteries. *Nano Energy* **2020**, *74*, 104868. [CrossRef]
115. Kim, H.; Kim, H.; Muhammad, S.; Um, J.H.; Sher Shah, M.S.A.; Yoo, P.J.; Yoon, W.S. Catalytic effect of reduced graphene oxide on facilitating reversible conversion reaction in SnO_2 for next-generation Li rechargeable batteries. *J. Power Sources* **2020**, *446*, 227321. [CrossRef]
116. Kebede, M.A. Tin oxide–based anodes for both lithium-ion and sodium-ion batteries. *Curr. Opin. Electrochem.* **2020**, *21*, 182–187. [CrossRef]
117. Duan, Y.K.; Li, Z.W.; Zhang, S.C.; Su, T.; Ma, Y.F.; Jiao, A.J.; Fu, Z.H. Metal-organic frameworks (MOFs)-derived Mn_2SnO_4@C anode based on dual lithium storage mechanism for high-performance lithium-ion capacitors. *Chem. Eng. J.* **2023**, *477*, 146914. [CrossRef]
118. Chen, J.S.; Lou, X.W. SnO_2-Based Nanomaterials: Synthesis and Application in Lithium-Ion Batteries. *Small* **2013**, *9*, 1877–1893. [CrossRef]
119. Wang, Q.; Sarkar, A.; Li, Z.; Lu, Y.; Velasco, L.; Bhattacharya, S.S.; Brezesinski, T.; Hahn, H.; Breitung, B. High entropy oxides as anode material for Li-ion battery applications: A practical approach. *Electrochem. Commun.* **2019**, *100*, 121–125. [CrossRef]
120. Chen, W.; Pei, J.; He, C.T.; Wan, J.; Ren, H.; Wang, Y.; Dong, J.; Wu, K.; Cheong, W.C.; Mao, J.; et al. Single Tungsten Atoms Supported on MOF-Derived N-Doped Carbon for Robust Electrochemical Hydrogen Evolution. *Adv. Mater.* **2018**, *30*, 1800396. [CrossRef] [PubMed]
121. Wei, M.; Jiang, Z.; Yang, C.; Jiang, T.; Zhang, L.; Zhao, G.; Zhu, G.; Yu, L.; Zhu, Y. A Hollow-Shaped ZIF-8-N-Doped Porous Carbon Fiber for High-Performance Zn-Ion Hybrid Supercapacitors. *Batteries* **2023**, *9*, 405. [CrossRef]
122. Jia, D.; Shen, Z.; Zhou, W.; Li, Y.; He, J.; Jiang, L.; Wei, Y.; He, X. Ultrahigh N-doped carbon with hierarchical porous structure derived from metal-organic framework for high-performance zinc ion hybrid capacitors. *Chem. Eng. J.* **2024**, *485*, 149820. [CrossRef]
123. Zhong, C.; Liu, B.; Ding, J.; Liu, X.; Zhong, Y.; Li, Y.; Sun, C.; Han, X.; Deng, Y.; Zhao, N.; et al. Decoupling electrolytes towards stable and high-energy rechargeable aqueous zinc–manganese dioxide batteries. *Nat. Energy* **2020**, *5*, 440–449. [CrossRef]
124. Byeon, J.; Ko, J.; Lee, S.; Kim, D.H.; Kim, S.W.; Kim, D.; Oh, W.; Hong, S.; Yoo, S.J. Solubility-Enhancing Hydrotrope Electrolyte with Tailor-Made Organic Redox-Active Species for Redox-Enhanced Electrochemical Capacitors. *ACS Energy Lett.* **2023**, *8*, 2345–2355. [CrossRef]
125. Evanko, B.; Boettcher, S.W.; Yoo, S.J.; Stucky, G.D. Redox-Enhanced Electrochemical Capacitors: Status, Opportunity, and Best Practices for Performance Evaluation. *ACS Energy Lett.* **2017**, *2*, 2581–2590. [CrossRef]
126. Maćkowiak, A.; Jeżowski, P.; Matsui, Y.; Ishikawa, M.; Fic, K. Redox-active electrolytes as a viable approach for the one-step assembly of metal-ion capacitors. *Energy Storage Mater.* **2024**, *65*, 103163. [CrossRef]
127. Jing, Y.; Zhao, E.W.; Goulet, M.-A.; Bahari, M.; Fell, E.M.; Jin, S.; Davoodi, A.; Jónsson, E.; Wu, M.; Grey, C.P.; et al. In situ electrochemical recomposition of decomposed redox-active species in aqueous organic flow batteries. *Nat. Chem.* **2022**, *14*, 1103–1109. [CrossRef] [PubMed]
128. Chen, Q.; Lv, Y.; Yuan, Z.; Li, X.; Yu, G.; Yang, Z.; Xu, T. Organic Electrolytes for pH-Neutral Aqueous Organic Redox Flow Batteries. *Adv. Funct. Mater.* **2021**, *32*, 2108777. [CrossRef]
129. Gan, X.; Zhang, C.; Ye, X.; Qie, L.; Shi, K. Unveiling the potential of redox electrolyte additives in enhancing interfacial stability for Zn-ion hybrid capacitors. *Energy Storage Mater.* **2024**, *65*, 103175. [CrossRef]
130. Zhao, K.; Fan, G.; Liu, J.; Liu, F.; Li, J.; Zhou, X.; Ni, Y.; Yu, M.; Zhang, Y.-M.; Su, H.; et al. Boosting the Kinetics and Stability of Zn Anodes in Aqueous Electrolytes with Supramolecular Cyclodextrin Additives. *J. Am. Chem. Soc.* **2022**, *144*, 11129–11137. [CrossRef]
131. Yu, X.; Li, Z.; Wu, X.; Zhang, H.; Zhao, Q.; Liang, H.; Wang, H.; Chao, D.; Wang, F.; Qiao, Y.; et al. Ten concerns of Zn metal anode for rechargeable aqueous zinc batteries. *Joule* **2023**, *7*, 1145–1175. [CrossRef]
132. Mandal, S.; Hu, J.; Shi, S.Q. A comprehensive review of hybrid supercapacitor from transition metal and industrial crop based activated carbon for energy storage applications. *Mater. Today Commun.* **2023**, *34*, 105207. [CrossRef]

133. Hong, N.; Kireev, D.; Zhao, Q.; Chen, D.; Akinwande, D.; Li, W. Roll-to-Roll Dry Transfer of Large-Scale Graphene. *Adv. Mater.* **2021**, *34*, 2106615. [CrossRef] [PubMed]
134. Choi, C.; Yun, T.G.; Hwang, B. Dispersion Stability of Carbon Nanotubes and Their Impact on Energy Storage Devices. *Inorganics* **2023**, *11*, 383. [CrossRef]
135. Bigham, A.; Zarepour, A.; Khosravi, A.; Iravani, S.; Zarrabi, A. 3D and 4D printing of MXene-based composites: From fundamentals to emerging applications. *Mater. Horiz.* **2024**, *11*, 6257–6288. [CrossRef]

MDPI AG
Grosspeteranlage 5
4052 Basel
Switzerland
Tel.: +41 61 683 77 34

Inorganics Editorial Office
E-mail: inorganics@mdpi.com
www.mdpi.com/journal/inorganics

www.ingramcontent.com/pod-product-compliance
Lightning Source LLC
Chambersburg PA
CBHW061629210926
51850CB00009B/2487